R. EDSON

Polar Oceanography

Polar Oceanography

Part B
Chemistry, Biology, and Geology

Edited by

Walker O. Smith, Jr.

*Botany Department and
Graduate Program in Ecology
University of Tennessee
Knoxville, Tennessee*

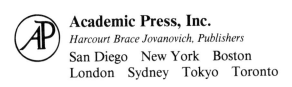

Academic Press, Inc.
Harcourt Brace Jovanovich, Publishers
San Diego   New York   Boston
London   Sydney   Tokyo   Toronto

This book is printed on acid-free paper. ∞

Copyright © 1990 by Academic Press, Inc.
All Rights Reserved.
No part of this publication may be reproduced or transmitted in any form or by any means, electronic or mechanical, including photocopy, recording, or any information storage and retrieval system, without permission in writing from the publisher.

Academic Press, Inc.
San Diego, California 92101

United Kingdom Edition published by
Academic Press Limited
24–28 Oval Road, London NW1 7DX

Library of Congress Cataloging-in-Publication Data

(Revised for vol. 2 pt. B)

Polar oceanography.

   Includes bibliographical references.
   Contents: pt. A. Physical science -- pt. B. Chemistry, biology, and geology.
   1. Oceanography--Polar regions.   I. Smith, Walker O.
GC401.P633   1990   551.45'8   89-18391
ISBN   0-12-653031-9   (pt. A : alk. paper)
ISBN   0-12-653032-7   (pt. B : alk. paper)

Printed in the United States of America
90 91 92 93   9 8 7 6 5 4 3 2 1

Contents

Contents of Part A	*vii*
Contributors	*ix*
Preface	*xi*

8. Chemical Oceanography — 407
E. P. Jones, David M. Nelson, and Paul Treguer

    I. Introduction — 407
    II. The Arctic Ocean — 408
    III. The Southern Ocean — 432
    References — 469

9. Polar Phytoplankton — 477
Walker O. Smith, Jr. and Egil Sakshaug

    I. Introduction — 477
    II. Pelagic Production — 480
    III. Processes Affecting Phytoplankton Distribution — 511
    IV. Ice-Algal Production — 512
    V. Coupling of Ice Community with Water Column — 515
    VI. Conclusions — 516
    References — 517

10. Polar Zooplankton — 527
Sharon L. Smith and Sigrid B. Schnack-Schiel

    I. Introduction — 527
    II. The Arctic — 528
    III. The Antarctic — 554
    IV. Concluding Remarks — 579
    References — 581

11. The Upper Trophic Levels in Polar Marine Ecosystems — 599
David G. Ainley and Douglas P. DeMaster

    I. Introduction — 599
    II. The Arctic and Antarctic Environments — 600

III.	Polar Communities and Food Webs	602
IV.	Factors That Concentrate Upper-Level Trophic Interactions in Polar Waters	609
V.	Competitive Interactions	620
VI.	Summary and Future Directions	623
	References	625

12. Polar Benthos 631
Paul K. Dayton

I.	Introduction	632
II.	On Early Exploration and Collections	634
III.	Origin, Evolution, and Historical Background of Arctic Habitats and Species	635
IV.	Arctic Patterns and Processes	641
V.	Arctic Macroalgae	651
VI.	Origin, Evolution, and Historical Background of Antarctic Habitats and Species	653
VII.	Antarctic Patterns and Processes	658
VIII.	Antarctic Benthic Macroalgae	669
IX.	Discussion	672
	References	676

13. Particle Fluxes and Modern Sedimentation in the Polar Oceans 687
Susumu Honjo

I.	Introduction	688
II.	Ocean Particles and Mechanisms of Settling	691
III.	Methods of Particle Flux Studies in the Polar Oceans: Field Experiments	694
IV.	Arctic Oceans and Their Marginal Seas: Pelagic Particle Fluxes	697
V.	Antarctic Oceans: Pelagic Particle Fluxes	707
VI.	Comparison of Pelagic Fluxes in the Arctic and Antarctic Oceans	715
VII.	Processes of Neritic Sedimentation in the Polar Oceans	722
VIII.	Summary and Conclusions	732
	References	734

Index   741

Contents of Part A
Physical Science

1. **Meteorology**
   Robert A. Brown

2. **Sea Ice in the Polar Regions**
   Anthony J. Gow and Walter B. Tucker III

3. **Remote Sensing of the Polar Oceans**
   Robert A. Shuchman and Robert G. Onstott

4. **Large-Scale Physical Oceanography of Polar Oceans**
   Eddy C. Carmack

5. **Mesoscale Phenomena in the Polar Oceans**
   Robin D. Muench

6. **Small-Scale Processes**
   Miles G. McPhee

7. **Models and Their Applications to Polar Oceanography**
   Sirpa Häkkinen

Contributors

Numbers in parentheses indicate the pages on which the authors' contributions begin.

David G. Ainley (599), Pt. Reyes Bird Observatory, Stinson Beach, California 94970

Paul K. Dayton (631), Scripps Institution of Oceanography A-001, La Jolla, California 92093

Douglas P. DeMaster (599), National Marine Fisheries Service, La Jolla, California 92038

Susumu Honjo (687), Woods Hole Oceanographic Institution, Woods Hole, Massachusetts 02543

E. P. Jones (407), Department of Fisheries and Oceans, Bedford Institute of Oceanography, Dartmouth, Nova Scotia, Canada B3A 2R1

David M. Nelson (407), College of Oceanography, Oregon State University, Corvallis, Oregon 97331

Egil Sakshaug (477), Trondhjem Biological Station, The Museum, University of Trondheim, N-7018, Trondheim, Norway

Sigrid B. Schnack-Schiel (527), Alfred-Wegener-Institut für Polar-und Meeresforschung, 2850 Bremerhaven, Federal Republic of Germany

Sharon L. Smith (527), Oceanographic Sciences Division, Department of Applied Science, Brookhaven National Laboratory, Upton, New York 11973

Walker O. Smith, Jr. (477), Botany Department and Graduate Program in Ecology, University of Tennessee, Knoxville, Tennessee 37996

Paul Treguer (407), Institut d'Etudes Marines, Universite de Bretagne Occidentale, 29287 Brest Cedex, France

Preface

The study of the world's oceans has rapidly expanded in the past decade through the use of new and exciting technologies (including the use of remote sensing, moored samplers, acoustic current meters, sonic arrays, and many others), in conjunction with more traditional data collection methods from ships. The merger of these techniques has led to a greater understanding of oceanographic processes at all scales. Yet, in many respects, the study of polar oceanography has lagged behind the study of temperate and tropical regions. The reason for this is simple: the severe logistic constraints imposed by the harsh environments of the Arctic and Antarctic simply preclude a simple extension of temperate oceanography into polar oceans. Many oceanographers who have never studied within the ice do not fully appreciate this fact, and we who do have often wondered how simple our science would be if only we did not have to drill ice holes through 1.5 m of ice to sample, to worry about water freezing as it is drained from Niskin bottles, or to expect the breakdown of equipment at the surface because of the change of physical properties of plastics at low temperatures.

Despite the numerous logistic and environmental difficulties inherent in the study of polar waters, these regions continue to be the focus of study for many oceanographers. In large part this is due to the realization that polar regions play a critical role in many global phenomena. For example, they are implicated as critical regions in controlling the global carbon dioxide cycle, being large sources of CO_2 in winter and spring and potential sinks in summer. They also play a dominant role in the global hydrologic cycle by accounting for approximately 98% of the world's fresh water. Polar regions are major sites of deposition within a number of biogeochemical cycles; they also are the sites for the formation of the ocean's deep water and hence are critical heat sinks to the atmosphere. In short, polar oceans have a profound effect on many large-scale oceanographic processes, and these effects are evident throughout the world's oceans.

Polar oceans are also being critically examined for exploitation of their mineral and biological resources, and there is extensive evidence that global environmental problems will disproportionately impact polar regions. In order to understand the influences of human activities in the Arctic and Antarctic, an appreciation of the role of polar oceans is required, as is the response of polar systems to human-induced perturbations.

Substantial differences exist between the Arctic and Antarctic, a fact which makes a synthesis of the existing knowledge extremely difficult. One major difference between the two areas is that the Arctic is an oceanic basin with its own coherent and restricted circulation, whereas the Antarctic is a continent with the surrounding seas dominated by a circumpolar current which exchanges large amounts of mass and heat with the Pacific, Atlantic, and Indian oceans (Fig. A shows the bathymetry of the two regions). As a corollary to the physiography, the seasonal climate and irradiance variations are more extreme in the Arctic Ocean than in the waters of Antarctica. However, the seasonal difference in the pack ice area is much greater in the Antarctic than in the Arctic (Fig. B shows the average minimum and maximum ice extent and concentration for the Arctic and Antarctic for the years 1978–1986). Hence the oceanographic impacts of ice-related processes tend to be different in each. The unifying and distinguishing characteristics of the Arctic and Antarctic are a major feature of each of the following chapters.

Prior to the initiation of these volumes *(Polar Oceanography, Part A* and *Part B)*, a number of researchers commented that our knowledge of polar oceanography is expanding at such a rapid pace that such a synthesis might be premature. It is clear that polar research is now a major portion of oceanographic research; evidence for this can be seen in the budgets of funding organizations, the numbers of polar-related papers presented at oceanographic meetings and published in peer-reviewed journals, and the attention given to international polar expeditions within the oceanographic community. Recent and present large programs in polar waters include MIZEX (Marginal Ice Zone Experiment), AIWEX (Arctic International Wave Experiment), the Greenland Sea Project, Pro Mare, AMERIEZ (Antarctic Marine Ecosystem Research at the Ice-edge Zone), WEPOLEX (Weddell Sea Polynya Experiment), CEAREX (Co-ordinated Eastern Arctic Experiment), ISHTAR (Inner Shelf Transfer and Recycling), and numerous others. These experiments are largely international in nature; in addition, many are interdisciplinary and merge the disciplines of physical, chemical, geological, and biological oceanography in order to understand the processes inherent in each region and their interaction. In this respect polar oceanography is perhaps even more interactive than temperate and tropical oceanography, although similar interdisciplinary projects are underway in nonpolar regions. Yet because of the cost of research on ice-breakers (and their relative rarity), as well as the number of nations which have national interests in polar waters, international projects are the most expedient manner in which to study polar regions. The inclusion of many non-U.S. contributors emphasizes the active role of many nations in polar research.

The two volumes have been organized into thirteen chapters. The first chapter in Part A deals with polar meteorology and air–sea–ice interactions,

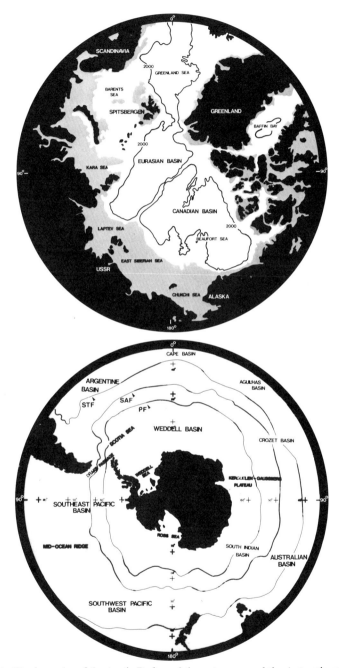

Figure A Physiography of the Arctic Basin and the waters around the Antarctic continent.

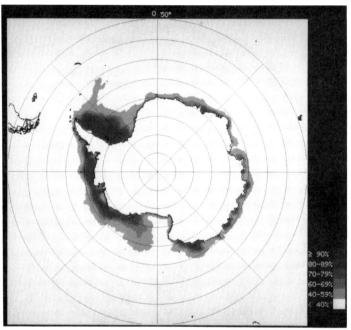

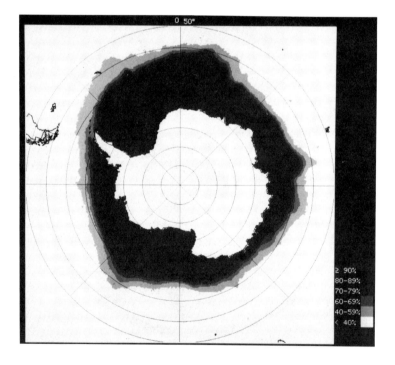

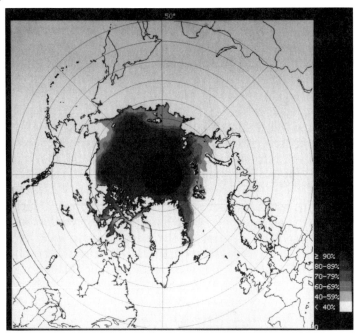

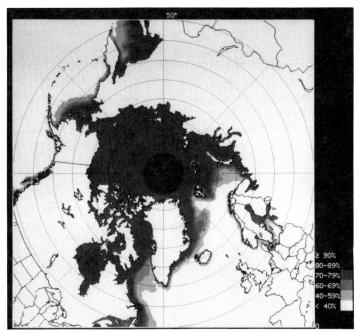

Figure B The average ice distribution and concentration at the minimum and maximum extent for the Antarctic and Arctic. Values represent the means for the years 1978–1986. Figures kindly provided by J. Comiso (NASA Goddard Space Flight Center).

whereas the second covers the properties of sea ice. The third chapter reviews the uses of remote sensing in polar regions, and the following three chapters cover physical oceanographic processes on three different scales: large, meso-, and small. This does not imply that these scales do not interact but provides a convenient manner in which to synthesize the existing information. The last chapter of Part A deals with models in polar systems. Models have played (and will continue to play) an extremely important role in polar oceanography because of the difficulty and expense of conducting long-term studies within and near ice-covered regions; furthermore, they play an extremely important role in hypothesis generation and testing in polar oceanography.

Part B begins with a chapter on chemical oceanography which follows those chapters in Part A on physical oceanography but precedes those in Part B dealing with biological oceanography, because the chapter serves as a transition from a description of physical processes to the effects of biological processes on chemical properties. The following four chapters on biological oceanography are divided by trophic structure (i.e., a chapter on autotrophic processes is first, followed by one on zooplankton, then one which synthesizes the information currently available on higher trophic levels, and finally one on the benthos). These chapters are followed by one on particle fluxes in polar regions, since flux is a function of the physical, chemical, and biological regimes of the water column.

The two volumes are not meant to be all inclusive, and some topics (e.g., bacterial processes, acoustics, optics) have not been exhaustively addressed. However, we hope the material presented provides an updated synthesis of most of the important research areas in polar oceanography.

This work has been encouraged by many oceanographers, both within and outside the polar community, including those within funding agencies of various countries. The editor and authors appreciate and acknowledge their support, both direct and indirect, and hope that this synthesis plays some role in adequately describing past work and providing a basis for discussion of future studies. Dr. Josefino Comiso kindly provided Fig. B. All of the chapters have been reviewed both formally and informally by a great number of colleagues, and without their help the quality of the chapters would have been seriously compromised.

Walker O. Smith, Jr.

8 Chemical Oceanography

E. P. Jones
Department of Fisheries and Oceans
Bedford Institute of Oceanography
Dartmouth, Nova Scotia, Canada

David M. Nelson
College of Oceanography
Oregon State University
Corvallis, Oregon

Paul Treguer
Institut d'Etudes Marines
Universite de Bretagne Occidentale
29287 Brest Cedex, France

I. Introduction   407
II. The Arctic Ocean   408
   A. The Continental Shelves   410
   B. Water Masses in the Central Arctic Ocean: Chemical Characteristics   412
   C. Water Mass Dating   421
   D. Budgets   426
   E. Trace Metals   427
   F. Artificial Radionuclides   429
   G. Sea Ice Chemistry   431
III. The Southern Ocean   432
   A. Chemical Distributions as Indicators of Circulation   433
   B. Elemental Cycles   449
   References   469

I. Introduction

Low temperatures are a dominating characteristic of the Arctic and Antarctic oceans. Both regions experience considerable cooling, and ice plays a major role in determining many of their features. The differences between the two oceans are, however, perhaps more striking than the similarities. The Arctic Ocean (in this chapter meaning the seas north of Bering and Fram

Straits) is an enclosed ocean, whereas the southern ocean's boundaries are open. The topographically constrained and limited exchange of water between the Arctic Ocean and the rest of the world's oceans is in marked contrast to the Antarctic, where exchange is not strongly restricted and the waters formed in the Antarctic are found in many of the deep regions of the major oceans. The Arctic Ocean is mostly covered by ice much of the year, whereas the Southern Ocean undergoes considerable seasonal variation in ice cover. Finally, the Arctic Ocean is bordered by vast continental shelves with a large input of fresh water from rivers, whereas there is little freshwater input into the seas surrounding Antarctica. The distribution of many chemicals is determined not only directly by the unique characteristics of the polar oceans but also indirectly by how these characteristics control biological processes that in turn influence chemical distributions.

II. The Arctic Ocean

Much of what is known about the distributions of chemicals in the Arctic Ocean, including the central basins as well as the continental shelves, has been learned from the relatively few measurements of chemical constitutents made from ice camps in the central Arctic Ocean (Fig. 8.1). The first reasonably extensive data set was obtained from Ice Island T-3 in the Canada basin (Kinney *et al.,* 1970). Other chemical data sets were collected during the Lomonosov Ridge Experiment (LOREX) near the North Pole (e.g., Moore, 1981; Moore *et al.,* 1983; Livingston *et al.,* 1984), during the Canadian Experiment to Study the Alpha Ridge (CESAR) over the Alpha Ridge separating the Canada basin from the Makarov basin (e.g., Wallace and Moore, 1985; Jones and Anderson, 1986; Moore and Smith, 1986; Wallace *et al.,* 1987), and during the Arctic Internal Wave Experiment (AIWEX) in the Canada basin (e.g., Ostlund *et al.,* 1987b). In 1980, the Swedish icebreaker *Ymer* surveyed chemical constituents in waters of the northern part of Fram Strait and southern edge of the Nansen basin just north of the Yermak Plateau (Anderson and Dyrssen, 1980, 1981). In 1983 and 1984, the German research icebreaker F.S. *Polarstern* occupied several stations just north of Fram Strait. During the summer of 1987, F.S. *Polarstern* collected data across the Nansen basin, the first oceanographic section taken across a deep ocean basin within the Arctic Ocean. The sampling program was much more extensive than those from the ice camps. In addition to more standard chemical measurements, a carefully selected suite of natural and anthropogenic tracers designed to describe the fate and mixing history of the water masses with respect to the full-depth circulation of the Nansen basin was taken (e.g., Anderson *et al.,* 1989). Russian nutrient measurements carried

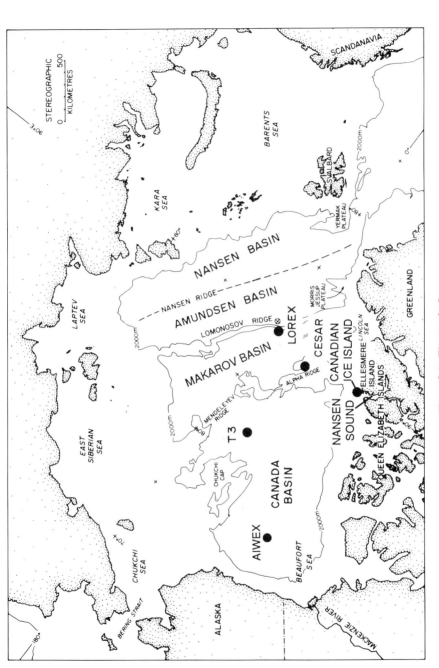

Figure 8.1 The Arctic Ocean.

out from ice stations for the past several decades are represented in an atlas (Gorshkov, 1983).

A. The Continental Shelves

The distributions of many chemicals not only are determined by the unique features of the Arctic Ocean but also in many instances reveal these features. The physical and chemical processes that take place on the continental shelves determine several physical and chemical features of the central regions of the Arctic Ocean and have been discovered in part through studies of these features.

The Arctic Ocean continental shelves are large. Although their extent depends on how boundaries are defined, the continental shelves constitute about one-third of the total area of the Arctic Ocean. Not only are the shelves large, they also experience quite different conditions from those found in the central regions of the Arctic Ocean. For example, much of the shelf area is ice free during part of the year. The shelf water is mixed by wind and by brine produced during the formation of ice. There is a large freshwater source from rivers, and regions of the shelves are relatively active biologically. All of these processes affect the chemical properties of the shelf water and subsequently other water masses as the shelf water is advected into central regions.

Little is known of the distributions of chemicals in the water masses of the vast continental shelves off northern Asia and Europe from measurements made on the shelves themselves. A large number of measurements of nutrients and oxygen were made in the 1960s (Codispoti and Richards, 1968, 1971). Some measurements have been reported for the Beaufort Sea (e.g., Macdonald *et al.,* 1987), but because the Beaufort Sea continental shelf is small compared to the Asian continental shelf, it is not thought to be significant in determining the chemical distributions in other regions. Nevertheless, the processes that occur on the Beaufort Sea shelf ought to be similar to those on the Asian shelves, since the Beaufort Sea shelf also has a considerable supply of fresh water from rivers and tends to be fairly ice free during the summer. Despite the general lack of measurements of chemical constituents on the continental shelves, observations in the central regions of the Arctic Ocean allow several inferences to be made regarding shelf processes.

In addition to the influence of source waters (from the Pacific and North Atlantic), the chemical characteristics of shelf water that have been revealed by studies in central regions are determined mostly by three processes: river runoff, biological production, and interactions between the sediments and overlying water. River runoff introduces several materials into the ocean, including some trace metals, artificial and naturally occurring radionuclides, and calcium carbonate, which has been of special interest recently in con-

8 Chemical Oceanography

nection with global carbon budgets. Biological processes over the continental shelves influence the distributions of carbon and nutrients that are fixed in the water column and released with the consumption of oxygen at the sediment–water interface during decay. Trace metals also are likely involved in this process. Sediment–water interactions influence chemical distributions through interactions of some trace metals and radionuclides with particulate matter as well as through dissolution of various materials in the sediments.

Because biological processes are especially active over some shelf regions of the Arctic Ocean and because they involve many of the chemical constituents measured in the water, it is worthwhile to summarize some of the chemical reactions that take place. A stoichiometric model for the soft parts of biogenic matter based on the Redfield–Ketchum–Richards model (Redfield et al., 1963) but modified to include explicitly metals and lipids (e.g., Dyrssen, 1977; Jones et al., 1984) is

$$(CH_2O)_s(CH_2)_t(NHCH_2CO)_u(CHPO_4M) \tag{8.1}$$

where M represents metals, typically magnesium as well as some others including trace metals. The subscripts s, t, and u are closely related to the "Redfield ratios." They represent the concentrations of the components relative to phosphate. The oxidation and dissolution of the components of this model are given by the equations

$$CH_2O + O_2 \longrightarrow CO_2 + H_2O \tag{8.2}$$

$$CH_2 + 1.5O_2 \longrightarrow CO_2 + H_2O \tag{8.3}$$

$$NHCH_2CO + 3.5O_2 \longrightarrow NO_3^- + H^+ + 2CO_2 + H_2O \tag{8.4}$$

$$CHPO_4M + O_2 \longrightarrow CO_2 + MHPO_4 \tag{8.5}$$

$$MHPO_4 + H_2O \longrightarrow M^{2+} + H^+ + PO_4^{3-} + OH^- \tag{8.6}$$

The hard parts of biogenic matter, silica and calcium carbonate, dissolve, producing silicate, Ca^{2+}, and CO_2. The carbon dioxide becomes part of the total carbonate (C_t), mostly as HCO_3^-. Calcium carbonate from geological sources also contributes to C_t. Carbon dioxide required during the dissolution can come directly from the atmosphere or indirectly through decaying organic matter according to the equations

$$CaCO_3(s) + CO_2 + H_2O \longrightarrow Ca^{2+} + 2HCO_3^- \tag{8.7}$$

$$CaCO_3(s) + CH_2O + O_2 \longrightarrow Ca^{2+} + 2HCO_3^- \tag{8.8}$$

The N:P:O:C ratios according to the Redfield–Ketchum–Richards model are 16:1:138:106. These will vary slightly according to the composition of the decaying matter and have been found to be somewhat different in

Baffin Bay (Jones *et al.*, 1984) and the deep North Atlantic (Takahashi *et al.*, 1985).

Ice formation over the continental shelves may be the process by which the continental shelves most influence the chemical characteristics of water masses in the central Arctic Ocean. The cold, relatively saline halocline in the central regions of the Arctic Ocean almost certainly results from shelf processes. Earlier studies concluded that the halocline could not be formed from the surface water mixing with the underlying Atlantic layer. The source of the halocline water was postulated to be Atlantic water, which has sufficient salt content, that had upwelled via canyons of the continental shelves, was cooled and freshened, and was then advected into central regions (Coachman and Barnes, 1962). More recent work focuses on an alternative process. The brine produced during the seasonal formation of sea ice mixes with the water on the continental shelves, which then advects into central regions. Much of this water goes to form the halocline, but some could penetrate to deeper regions, even reaching the bottom of the main ocean basins (Aagaard *et al.*, 1981; Melling and Lewis, 1982; Swift *et al.*, 1983; Midttun, 1985; Anderson and Jones, 1986).

Two studies in shelf regions have examined nutrient regeneration under conditions that should be typical of the continental shelves. An incubation study using sediments collected from the continental shelf north of Ellesmere Island showed that regeneration of nutrients from buried organic matter was too slow to account for the rate of supply of nutrients to the upper halocline water (Anderson and Jones, 1990). On the other hand, measurements of several chemical constituents in the water column in Storfjorden, a large fjord in Svalbard, indicated that brine production during sea ice formation and biological activity during summer occur. This would seem to confirm that the processes described above can take place over much of the continental shelf region (Anderson *et al.*, 1988).

B. Water Masses in the Central Arctic Ocean: Chemical Characteristics

As discussed in Chapter 4, exchange of water in the Arctic Ocean with other oceans is mostly through Fram Strait, where water from the North Atlantic and from the Norwegian–Greenland seas flows in and water from the Arctic Ocean flows out. Water whose source is the Pacific enters through the shallow Bering Strait, and near-surface Arctic Ocean water flows out through the Canadian Archipelago. Although some of the chemical characteristics of the inflowing waters are traceable in the Arctic Ocean, to a large extent they are modified on the continental shelves, including the Barents Sea, or within the main Arctic Ocean basins.

The Arctic Ocean is a comparatively low-energy regime with relatively well-defined layers throughout most of its volume. The extensive year-round ice cover over much of the surface of the Arctic Ocean inhibits heat exchange with the atmosphere and mixing by winds. This, together with considerable freshwater runoff, results in a shallow mixed layer above a pronounced halocline. Underneath the halocline is the warner Atlantic layer, water of Atlantic origin that is thought to have been cooled by the surface layer. Under the Atlantic layer is cold water whose exact origin is not yet well determined, but which is generally thought to have a major component of Norwegian Sea Deep Water. This deep water is often referred to as Arctic Ocean Deep Water, and the boundary between it and the Atlantic layer is somewhat arbitrarily defined traditionally by the 0° isotherm or more recently by the sigma-1 = 32.785 surface (Aagaard *et al.*, 1985).

It has become apparent that there are differences in the chemical properties of the water masses in the different basins of the Arctic Ocean and that physical and chemical processes, particularly those operating on the large shelf regions, affect the chemical properties of each basin in different ways (e.g., Kinney *et al.*, 1970; Moore *et al.*, 1983; Jones and Anderson, 1986; Ostlund *et al.*, 1987b). A major, though perhaps not unexpected, finding from the Nansen basin section was that there is not only variability in chemical characteristics from basin to basin within the Arctic Ocean but also considerable variability in many properties within a single basin itself (Fig. 8.2) (Anderson *et al.*, 1989). In spite of these inter- and intrabasin differences, the general character of the different water masses seems to be fairly well preserved throughout much of the Arctic Ocean away from the continental shelves. It is therefore reasonable to discuss commonly measured chemical constituents (Table 8.1), primarily nutrients and oxygen but including other appropriate constituents, for water masses in the central region of the Arctic Ocean defined by salinity and temperature (Fig. 8.3).

1. The surface layer

The surface layer contains water flowing in from the Bering Sea as well as from the Norwegian–Greenland seas. To these source waters are added contributions from river runoff and sea ice meltwater. The chemical characteristics of these components are modified to some extent by processes on the continental shelves.

The most obvious change in the surface layer within the Arctic Ocean is salinity S, which ranges from a fresh $S = 30$ near Bering Strait to above $S = 34$ north of Fram Strait. This reflects the differences between regions influenced mainly by Pacific water to which river runoff has been added and those influenced mainly by North Atlantic water. Oxygen isotope ratios can determine the relative amounts of sea ice meltwater and meteoric water (mostly river runoff) and have been used to designate regions with different

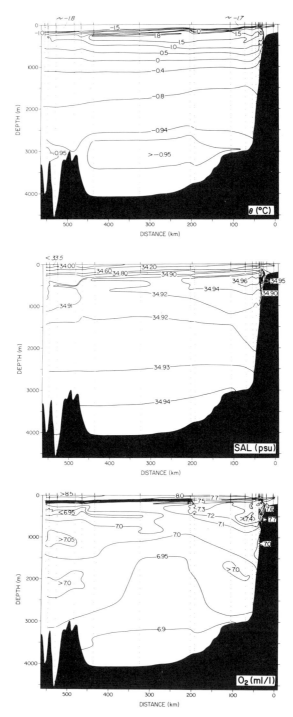

Figure 8.2 Oceanographic sections across the Nansen basin of the Arctic Ocean from the 1987 Arktis IV/3 cruise of F.S. *Polarstern*. (After Anderson *et al.*, 1989a.) *(Figure continues.)*

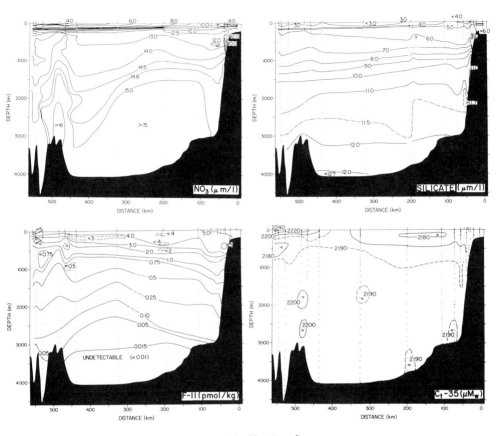

Figure 8.2 *(Continued)*

amounts of these two components (Tan *et al.*, 1983; Ostlund and Hut, 1984; Anderson and Dyrssen, 1988). The nutrient values also reflect the different sources of water in the surface layer and have been used to clarify the origins of surface water in the southern part of Nansen basin (Anderson and Jones, 1986).

The chemical characteristics of the surface layer are differentiated from those of the other Arctic Ocean water masses perhaps most by the freshwater component from river runoff. This is indicated by the excess total alkalinity and total carbonate values that have been ascribed to dissolved calcium carbonate and atmospheric carbon dioxide that enter the Arctic Ocean through river input, mostly from the large drainage basins of Asia (Anderson and Dyrssen, 1981). In recent work, total alkalinity and total carbonate measurements traced river runoff water through the northern half of the section across Nansen basin (Fig. 8.2), showing that it followed quite closely

Table 8.1 Some Characteristics of Arctic Ocean Water Masses Below 2000 m

		Nansen basin*				Canada basin*	
Variables	Fram Strait NSDW	Fram 3[3]	Ymer stations 104 and 105[4]	Polarstern 1987	Amundsen basin[5]	Makarov basin[5]	Ice Island T-3[6]	AIWEX[7,8]
θ (°C)	−0.93 to −1.06[1]	−0.99 to −1.02 (a) −0.95 to −0.96 (b)	−0.90 to 0.96	−0.85 to −0.96 (a) −0.95 (b)	−0.8 to −0.9	−0.5	−0.53	−0.521 to −0.515 (a) −0.515 (b)
S	34.908 to 34.911[1]	34.918 to 34.927 (a) 34.944 to 34.953 (b)	34.919 to 34.937	34.925 to 34.936 (a)	34.932 to 34.951	34.953 to 34.954		34.942 to 34.953 (a) 34.954 (b)
O_2 (μmol/kg)	305[1] 308 to 316[2]	301 (a)	304 to 316	303 (a) 301 to 298 (b)			278 to 289	300 to 293 (a) 293 (b)
Si (μmol/kg)	11.2 to 12.9[1] 10.9 to 11.9[2]	11.2 to 11.5 (a) 10.6 to 10.9 (b)	10.4 to 11.2	11.3 to 11.8 (a) 12.0 (b)	9.7 to 10	12.4 to 12.8	13 to 14	12.6 to 14.7 (a) 14.8 (b)
NO_3^- (μmol/kg)		14.9 to 15.2 (a) 14.3 to 14.7 (b)		14.9			14.3 to 15.0	14.7 to 15.0 (a) 15.0 (b)
PO_4^{3-} (μmol/kg)	0.94 to 1.00[2]	1.02 to 1.07 (a) 0.98 to 1.04 (b)	0.84 to 0.90	1.00				0.94 to 1.00

* Where indicated, (a) represents data from 2000 to 3000 m and (b) represents data below 3000 m.
[1] Swift et al., 1983; [2] Ymer Station 122 (Anderson and Dyrssen, 1980); [3] Anderson and Dyrssen, 1980; [4] Anderson and Jones, 1986; [5] Moore et al., 1983; [6] Kinney et al., 1970; [7] Ostlund et al., 1987; [8] J.H. Swift, personal communication.

8 Chemical Oceanography

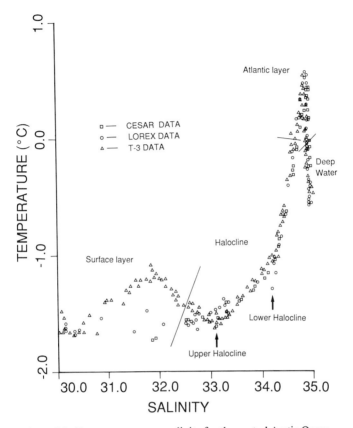

Figure 8.3 Temperature versus salinity for the central Arctic Ocean.

a branch of the Transpolar Drift as indicated by ice movement (Anderson *et al.*, 1989). This finding indicates that these two chemical constituents should be useful in tracing river runoff water in the Arctic Ocean. Nutrient concentrations in the surface layer are relatively high and oxygen concentrations are in near equilibrium with the atmosphere, consistent with the general concept of very low primary productivity in central regions.

2. The halocline

The halocline in the central Arctic Ocean is a distinctive feature that separates the cold, relatively fresh surface layer from the underlying warmer, more saline Atlantic layer. Associated with the upper part of this halocline is a prominent nutrient maximum. The nutrient maximum was observed in the Canada basin as early as 1948 from the NP-2 Ice Station (Nikiforev *et al.*, 1966) and later from the NP-22 Ice Station (Mel'nikov and Pavlov, 1978). A

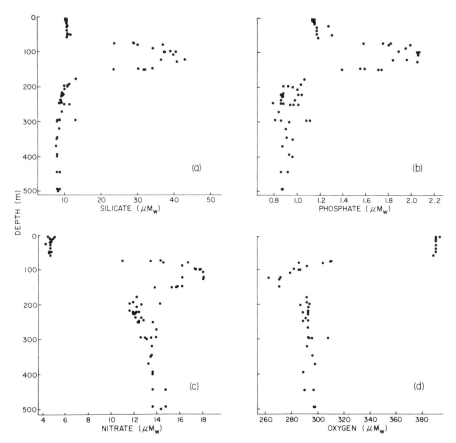

Figure 8.4 Profiles typical of the central Arctic Ocean obtained at the CESAR ice camp, 1983. *(Figure continues.)*

set of profiles of nutrient and oxygen concentrations was obtained from the T-3 Ice Island (Kinney *et al.*, 1970), and the same general features, but more sharply defined profiles, were observed during the LOREX (Moore *et al.*, 1983) and CESAR ice camps (Fig. 8.4) (Jones and Anderson, 1986). The nutrient maximum has also been observed in the Beaufort Sea (Macdonald *et al.*, 1987).

Central to understanding the distribution of chemicals in the halocline is the question of how the halocline itself is maintained. As described above, cold, saline shelf water is advected throughout the basins of the central Arctic Ocean, forming the halocline and, to a large degree, determining the chemical and physical properties of the halocline. Convincing evidence that a significant portion of halocline water has indeed originated on the continen-

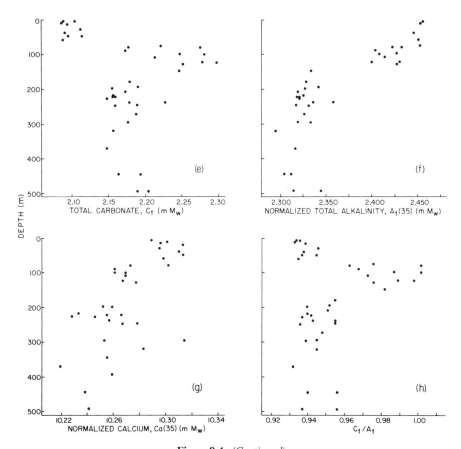

Figure 8.4 *(Continued)*

tal shelves is provided by measurements of radionuclides (Moore and Smith, 1986). Profiles of lead-210 and polonium-210 show distinct minima throughout the halocline to a depth of more than 200 m, extending below the nutrient maximum layer. The radionuclide minima were attributed to the influence of shelf sediments where these isotopes interact with particles.

The shelf-produced water carries with it the chemical signatures of the original shelf water. In the Chukchi–East Siberian seas region, this means that the water will carry the high nutrient signal of water of Pacific origin. In the Barents–Kara seas region, the water will have a lower nutrient content characteristic of water of Atlantic origin. Jones and Anderson (1986) argue that there are at least two distinct types of halocline water, with source waters originating in different regions of the continental shelves. Because of its high nutrient concentrations, the upper halocline water reflects its source in the Chukchi–East Siberian seas region. The lower halocline water has lower nutrient concentrations suggestive of an Atlantic source, but it also has a more subtle yet very distinctive signature, NO. [NO = $9NO_3^- + O_2$ is a conservative tracer (Broecker, 1974) based on the Redfield–Ketchum–Richards stoichiometric model for biogenic material in the ocean. The factor 9 corresponds very closely to the original Redfield–Ketchum–Richards model. A factor of 10.8 corresponds to more recently published values (Takahashi et al., 1985). Both factors seem equally suitable for distinguishing water masses in the Arctic Ocean.] A profile of NO shows a distinct minimum corresponding to the lower halocline (Jones and Anderson, 1986), with values lower than any others, including those in the Atlantic layer or those in North Atlantic water entering the Arctic Ocean through Fram Strait. The lower NO values provide additional evidence for a shelf origin of the lower halocline water. As nitrate is fixed during photosynthesis, oxygen is released, and if the water is well ventilated the oxygen in excess of saturation values can be ventilated to the atmosphere, resulting in lower NO values than would otherwise be the case. The low NO values of the lower halocline are therefore consistent with the origin of this water being in the Barents–Kara seas shelf region, which receives water from the North Atlantic and which in summer is mostly ice free and has considerable biological activity. The identification of NO as a tracer for water from the Barents–Kara seas region is supported by recent observations in the Nansen basin (Anderson et al., 1989). NO is an important, almost unique tracer of lower halocline water and Barents–Kara seas water circulation in the Arctic Ocean and flow out of Fram Strait.

3. The Atlantic layer

Although the salinity and temperature of the Atlantic layer are distinctive, its chemical characteristics are fairly similar to those of both the lower halocline water lying above and the top of the Arctic Ocean Deep Water lying below. Values of nutrients and NO are comparable to those found in the North Atlantic in the region where this water enters through Fram Strait. The temperature of the Atlantic layer is, however, much lower than that of the North Atlantic Water. Cooling is thought to occur mostly by mixing with the overlying cold halocline water, but the total carbonate values in the Atlantic layer give a strong indication that some shelf water penetrates directly to the Atlantic layer and may contribute to its heat loss (Anderson et al., 1990).

4. The deep water

The Arctic Ocean Deep Water extends from the bottom of the Atlantic layer near 1000 m to the bottom of each of the Arctic Ocean's four major basins. The Canada and Makarov basins are separated by the Alpha Ridge at a depth of about 2000 m. The Lomonosov Ridge between the Makarov and Amundsen basins has a relatively narrow gap at the same depth. As presently charted, the Arctic midocean Ridge between the Amundsen and Nansen basins would not appear to provide a major barrier to exchange of water between these two basins.

Above the ridges separating these basins and probably because there is no major barrier to inhibit exchange of water between the basins, the Arctic Ocean Deep Water seems to be reasonably homogeneous on a large scale (e.g., Kinney *et al.,* 1970; Moore *et al.,* 1983; Aagaard *et al.,* 1985; Jones and Anderson, 1986). On either side of the Lomonosov Ridge, there are distinct differences in temperature, salinity, and nutrients in the deep water. In the deepest parts of the basins, there have been very few high-quality measurements of any properties, including salinity. Although measurements were made to the bottom at the CESAR and LOREX Ice Camps, they did not reach the deepest regions of the Arctic Ocean, only 2000 m over the Alpha Ridge, 2500 m in the Makarov basin, and 3000 m in the Amundsen basin. Only at AIWEX in the Canada basin and recently during the F.S. *Polarstern* expedition in the Nansen basin were samples obtained at depths of nearly 3700 and 4000 m, respectively, corresponding to the deepest parts of these two basins. The salinity and nutrient concentrations in the deepest parts of the Canada basin are higher than those found on the other side of the Lomonosov Ridge. Although the higher nutrient concentrations could be explained by the greater age of the deep water in the Canada basin (see following section), it is harder to account for the higher salinity. One possibility is that the higher salinity results from brine produced on the continental shelves. Arguments based on oxygen isotope ratios that address this possibility, however, leave the question not completely resolved (Ostlund *et al.,* 1987b). These measurements indicate that no more than about 15% of the deep water originates from the shelves, which is not enough to account for the high salinity values compared to those on the other side of the Lomonosov Ridge, where most of the deep water in the Canada basin must have originated.

C. Water Mass Dating

Estimates of the ages or residence times of water masses in the Arctic Ocean have been based mostly on models of the distributions of transient tracers (i.e., tracers whose source function has varied with time, such as chlorofluoromethanes; tracers that undergo radioactive decay in the water, such as carbon-14; or tracers that do both, such as tritium, cesium-137, and carbon-14). All of these have anthropogenic sources; carbon-14 has a natural source as well. The determination of an "age" of a water mass depends very much on what is being considered — transit times, ventilation, mixing, etc. — and

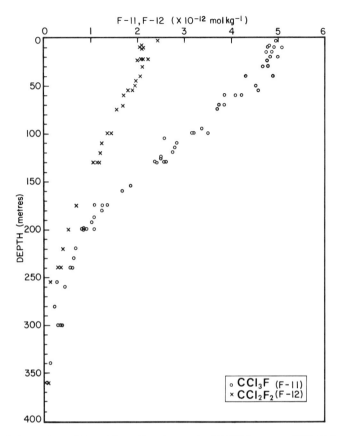

Figure 8.5 Freon 11 and Freon 12 profiles taken at the CESAR ice camp, 1983. (After Wallace and Moore, 1985.) *(Figure continues.)*

almost always on what type of model is used. Because measurements are generally sparse in the Arctic Ocean and models are usually extremely simplified, the sometimes subtle differences in concepts of an age may not be of great significance, but these differences should be kept in mind in any application of ages to other questions.

An early estimate of the residence time of fresh water in the surface layer that did not involve transient tracers was based on mass balances of salt and fresh water (Aagaard and Coachman, 1975). Assuming that water flowing from Bering Strait diluted underlying higher-salinity water, Aagaard and Coachman arrived at an age of 10 years. This must be considered a rough estimate, but it has proved to be consistent with later estimates based on transient tracers. An estimate of the residence time of the Atlantic layer of

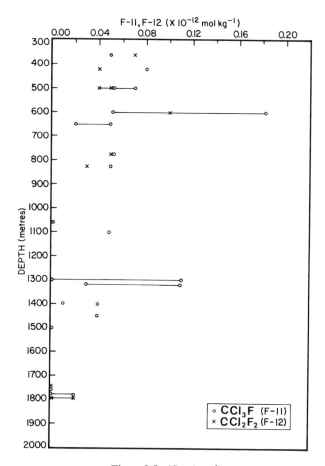

Figure 8.5 *(Continued)*

about 20–25 years is based on a similar approach (Aagaard and Greisman, 1975). This approach should be more widely applied as more is learned about circulation and rates of transport of water masses in studies such as that reported by Rudels (1986).

Chlorofluoromethanes (Freon 11 and Freon 12) have been measured over the Alpha Ridge in the central Arctic Ocean (Wallace and Moore, 1985), at the periphery of the Arctic Ocean just north of Fram Strait (Smethie *et al.*, 1988), and in the Nansen basin (Anderson *et al.*, 1989b). It is hoped that an assessment of the residence time of water masses can be made in order to determine rates of transport of the water masses and their chemical constituents as well as rates of reactions. In principle, this information can be obtained from chlorofluoromethane measurements through the use of models, though the results can be quite model dependent (e.g., Wallace and Moore, 1985).

At the CESAR site, the chlorofluoromethanes were seen to penetrate only to a depth of 900 m (Fig. 8.5). Because these compounds were introduced into the atmosphere in significant quantities only since the 1940s, water below 900 m must have an age of at least about 50 years. The ages of water masses above 900 m given by two "extreme case" models were much the same for shallower depths but began to diverge significantly below the halocline, near 200 m. Above 50 m the ventilation age was less than 5 years. Through the halocline, it varied from nearly 5 years at the top to about 15 years near the bottom. The ventilation age of the Atlantic layer was given as 30 years.

Chlorofluoromethane measurements at the southern boundary of the Amundsen and Nansen basins were used by Smethie et al. (1988) to estimate the circulation time of the deep water in the Eurasian basin. The deep water had chlorofluoromethane levels near 0.1 pmol kg^{-1}. Using a box model, Smethie et al. (1988) determined an age near 25 years. More recently, chlorofluoromethane measurements in the Nansen basin showed undetectable levels of Freon 11 and Freon 12 (<0.01 pmol kg^{-1}) below about 3000 m in the central part of the basin (Anderson et al., 1989a), implying a greater age for this water. Measurements of carbon tetrachloride made at the same time showed small but detectable concentrations in the deepest regions (Krysell and Wallace, 1988). Carbon tetrachloride was introduced into the atmosphere about 40 years earlier than the Freons, and if there is no natural source of carbon tetrachloride, these measurable quantities suggest an age of about 80 years for the deep water. Carbon-14 measurements suggest that the deep water may be even older (Anderson et al., 1989).

Tritium measurements have been used to estimate the residence time of the freshwater component from precipitation and runoff into the Arctic Ocean (Ostlund, 1982). Except for tritium, artificial radionuclides have not been used much for age determinations for water masses. Residence time estimates were based on a fairly large data set (by Arctic Ocean standards) gathered from ice camps and submarine patrols and many observations of tritium in precipitation. The model used determined the transfer time for the freshwater component to the upper waters, typically above 150 m, in Nansen basin to be close to 11 years. Slightly longer transfer times were found in more central regions and in the Atlantic layer. It should be noted that these times refer to the transfer of fresh water only. Since the freshwater component could have undergone a long transit before being mixed into deeper water, the transfer times do not necessarily correspond to the ages of the water masses determined by other methods.

The distributions of tritium and oxygen-18 isotopes have been used to determine the total net production and transport rates of ice and fresh water of meteoric origin (precipitation and runoff) in the Arctic Ocean (Ostlund and Hut, 1984). This work relies on mass balance equations for salt, water, and oxygen-18, the latter distinguishing between meteoric water and sea ice meltwater as sources of fresh water. Time scales are provided by tritium distributions. A residence time for halocline water of 10 years and freshwater transport from both sources together with total water transport were determined using only concentrations of these few constituents and without resorting to any hydrographic determinations of currents.

Carbon-14 from natural sources has been used to date water whose residence times exceed those that can be given by chlorofluoromethanes and tritium. An approach similar to that referred to above but using carbon-14 has been used to determine the ventilation rate of the deep water of the Canada basin (Ostlund et al., 1987b). Because anthropogenic carbon-14 was released during bomb tests along with tritium, the absence of tritium in a water sample is taken as an indication that any carbon-14 is of natural origin. In deep-water samples from the Eurasian basin collected north of Fram Strait (Ostlund et al., 1982) tritium was present at low levels, so the carbon-14 values were not solely of natural origin. Ages between 10 and 100 years were suggested for water below the halocline, with the older water corresponding to the deeper water below 2500 m. This is consistent with the chlorofluoromethane data from the Nansen basin, although the complete absence of chlorofluoromethane below 3000 m might suggest even greater ages. The absence of tritium in the deep waters of the Canada basin suggests long renewal times. The carbon-14 data give residence times of 500–800 years. These long residence times, together with oxygen-18 data, put an upper limit to the amount of shelf water that could have reached the deep regions of the Canada basin (no more than 10–15%). Both the long residence times and the small amount of shelf water are significant in trying to understand the salinity of the deep water, which is greater than that in the deep Eurasian basin, and in considering ventilation of the deep oceans in the context of climate questions.

Ages of water masses can lead to the determination of rates of transport through ocean basins. Examples applied to the Arctic Ocean are discussed in following sections, including the transport of carbon from the atmosphere to the ocean in the context of fossil fuel CO_2–climate questions, or rate of productivity on the continental shelves. Another interesting result from modeling chlorofluoromethanes is the determination of the vertical diffusion in the halocline (Wallace et al., 1987). However, a discussion of this is considered outside the scope of this chapter.

D. Budgets

Budgets for the chemical constituents of the Arctic Ocean have been based on water mass budgets or on models of rates of transport. The former suffer from imprecision of water mass balances as well as from lack of information about the chemical characteristics of the water masses in general. Nevertheless, credible budgets were constructed for salt, fresh water, alkalinity, and silica based on the chemical constituents of river runoff and assessments of water flowing into and out of the Arctic Ocean (Anderson *et al.*, 1983). The second approach involves rates of transport and ages. One example of this applied to salt, fresh water, and ice (Ostlund and Hut, 1984) was discussed above. This approach has been applied to two questions of wide interest: the carbon budget and biological productivity, the former of special concern in the context of the global CO_2 question.

1. Carbon dioxide

Of very strong current concern in the context of climate studies is the question of the buildup of fossil fuel carbon dioxide in the atmosphere and its transport to the deep oceans of the world. Several studies have been carried out to determine the rate of assimilation of carbon dioxide by the world's oceans. These studies have omitted the Arctic Ocean because it was thought not to provide a pathway for surface water to the deep oceans of the world. However, it has been recognized that the dense water formed on the continental shelves could provide a mechanism for the ventilation of the deep basins of the Arctic Ocean (Aagaard *et al.*, 1981; Melling and Lewis, 1982). The subsequent flow of deep Arctic Ocean water out through Fram Strait (Aagaard *et al.*, 1985) would result in the transfer of atmospheric carbon dioxide to the deep oceans via the Arctic Ocean.

An assessment of how effective this mechanism might be was made from total carbonate measurements (Anderson *et al.*, 1990). Carbon dioxide can enter the waters of the continental shelves of the Arctic Ocean in two ways: as part of the process of calcium carbonate dissolution according to Eqs. (8.7) and (8.8), or through fixation by biological processes [Eqs. (8.2)–(8.6)] and release during subsequent decay. In both instances, the carbon is advected from the shelves to the different water layers as described in Section II,A. These contributions to the total carbonate in the water, together with the ages of the water masses, lead to a determination of the rate of removal of carbon dioxide from the atmosphere to the ocean.

It was determined that a considerable amount of carbon is transferred from the atmosphere to the surface water and the halocline of the Arctic Ocean, with some penetrating to the Atlantic layer. Overall, approximately 60% of this carbon is associated with the decay of organic matter and 40%

with the dissolution of calcium carbonate that enters through river runoff. Clearly, biological processes play an important role in the transport of carbon from the atmosphere to the ocean in this region. The carbon exits the Arctic Ocean with these water masses through Fram Strait in the East Greenland Current. Unfortunately from the climate perspective, the rate of transfer of carbon to the deep ocean is very low, so the deep flow from the Arctic Ocean carries with it very little fossil fuel carbon dioxide from the atmosphere.

2. Biological processes

Determinations of biological productivity in the Arctic Ocean are very few. An assessment of the annual rate of new production, or the fraction of the total biological productivity which depends on resupply of nutrients to the continental shelves, can be made from chemical measurements in the central Arctic Ocean (Anderson *et al.,* 1989b). The new production can be estimated from the total carbonate in the upper halocline derived from the decay of organic matter on the continental shelves under the assumption that most of the decay products of organic matter fixed on the continental shelves enter these waters. This new production is an average value, integrated over regions contributing to the nutrient maximum in the upper halocline, i.e., the Chukchi, East Siberian, and Laptev seas. If the productivity over this region is assumed to be uniform, this contribution corresponds to about 45 g C m^{-2} yr^{-1}. This number can be compared to values for total production ranging between 12 and 98 g C m^{-2} yr^{-1} for Arctic waters (see Chapter 9). A second estimate of the new production was derived from the apparent oxygen utilization rates, which were determined from modeling profiles of chlorofluoromethanes, salinity, temperature, and oxygen in the water column (Wallace *et al.,* 1987). The value obtained for shelf production, subject to several qualifications, was between 8 and 21 g C m^{-2} yr^{-1}. Another estimate of new production based on chemical measurements was made for the shelf waters of the Beaufort Sea (Macdonald *et al.,* 1987). Values of 16 and 23 g C m^{-2} yr^{-1} were estimated for the years 1974 and 1975. The difference was ascribed to different ice conditions and may reflect the degree of interannual variability that can occur over the Arctic shelf.

E. Trace Metals

Trace metals measured in the Arctic Ocean include cadmium, zinc, copper, and aluminum at the LOREX Ice Camp (Moore, 1981); manganese, nickel, zinc, and cadmium at the Fram 3 and CESAR ice camps (Yeats, 1988); and cadmium, copper, nickel, zinc, and lead north of Svalbard during the Ymer-80 expedition (Danielsson and Westerlund, 1983; Mart *et al.,* 1984)

Table 8.2 Trace Metal Concentrations in Arctic Ocean Water

Metal	Layer	Concentration (nmol/liter)				
		Fram 3	CESAR	Ymer-80[a]	Ymer-80[b]	LOREX[c]
Cd	Surface layer	0.16 ±0.06 (6)[d]	0.48 ± 0.06 (3)	0.09	0.15	0.31
	Nutrient maximum		0.62 ± 0.14 (3)			0.58
	Atlantic layer	0.21 ±0.01 (4)	0.26 ± 0.06 (6)	0.14	0.18	0.19
	Deep water	0.27 ±0.06 (5)	0.23 ± 0.05 (5)	0.14	0.22	0.18
Ni	Surface layer	3.8 ± 0.3 (6)	5.2 ± 0.1 (2)	2.3	3.4	
	Nutrient maximum		6.3 ± 1.2 (3)			
	Atlantic layer	4.0 ± 0.4 (4)	3.0 ± 0.4 (6)	2.1	3.5	
	Deep layer	4.1 ±0.4 (4)	2.8 ± 0.3 (5)	3.2	3.6	
Zn	Surface layer	1.0 ±0.3 (3)	2.7 (1)		1	2.9
	Nutrient maximum		5.7 ± 1.2 (3)			5.5
	Atlantic layer	3.8 ±0.7 (3)	4.7 ± 0.7 (6)		1.5	3.1
	Deep layer	4.1 ±0.9 (4)	3.5 ± 1.5 (5)		3.5	2.6
Mn	Surface layer	4.7 ±1.1 (6)	2.0 ± 0.4 (3)			
	Nutrient maximum		1.0 ±0.3 (3)			
	Atlantic layer	2.7 ± 0.7 (4)	0.42 ± 0.18 (6)			
	Deep layer	4.3 ± 1.7 (4)	0.63 ± 0.18 (5)			

[a] Mart et al. (1984).
[b] Danielsson and Westerlund (1983).
[c] Moore (1981).
[d] Arithmetic mean ± standard deviation (number of samples).

(Table 8.2). These trace metals do not all behave in the same way chemically. Thus, although trace metal distributions in general have not been used as tracers of water masses, they can be used to help clarify sources of water masses and processes that might be occurring in them.

Prominent in the distributions is the correlation of cadmium, nickel, and perhaps zinc with the nutrient maximum in the upper halocline. The cadmium-to-phosphate ratio in the upper halocline is very similar to that of Pacific waters. This points to a Pacific source for this water, but it also argues against release of metals from the sediments, since if this were the case copper and manganese concentrations should also be elevated in the upper halocline water. The lack of elevated concentrations of these metals appears to contradict the role of shelf processes in maintaining the halocline and in producing the nutrient maximum and lead-210 minimum. A possible resolution of this difficulty lies in the fact that the nutrient and lead-210 distributions could arise from reactions at the sediment–water interface rather than from release of these materials from the sediments. The somewhat elevated levels of some trace metals in the lower halocline are consistent with mixing indicated by chlorofluoromethane distributions (Wallace and Moore, 1985; Wallace et al., 1987).

There is a clear distinction between the trace metal distributions at the CESAR and LOREX sites and those north of Svalbard, consistent with differences in most other properties also. The CESAR and LOREX data show a cadmium-to-phosphate ratio similar to that found in the Pacific Ocean, whereas the Fram 3 and Ymer-80 data show a cadmium-to-phosphate ratio different from those typical of either the Pacific or Atlantic Ocean (Yeats, 1988). This is a bit surprising for the near-surface water north of Svalbard, which has the Atlantic Ocean as its major source. It may be that shelf processes in the Barents–kara seas modify trace metal distributions north of Svalbard. How significant this ratio might be in terms of determining the sources of deeper water masses below the halocline is problematic, because no evidence exists to suggest that Pacific water is a major component of the deep water in any basin.

F. Artificial Radionuclides

Data on artificial radionuclides have been reported for four locations in the Arctic Ocean: the LOREX, CESAR, and AIWEX ice camps and just north of Fram Strait (e.g., Livingston, 1988) (Fig. 8.6). Although some other artificial radionuclides have been measures, cesium-137 and strontium-90 are the two which are most useful in addressing oceanographic questions. There are two sources of these radionuclides: fallout from nuclear weapons tests and

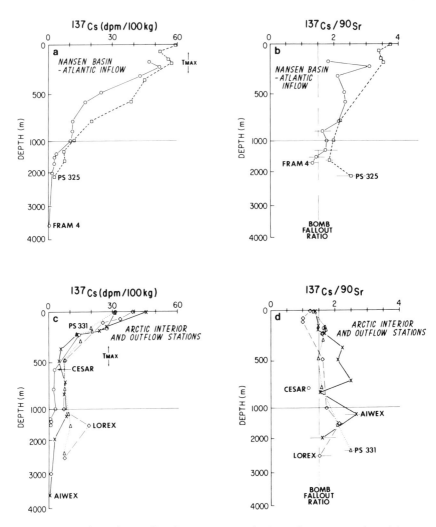

Figure 8.6 Radionuclide profiles from several Arctic Ocean ice camps. (After Livingston, 1988.)

waste from the Sellafield nuclear fuel reprocessing plant in Ireland. Fallout from the Chernobyl accident has not yet been reported for the Arctic Ocean.

Cesium-137 and strontium-90 in the ocean both behave in a conservative fashion; that is, they behave chemically much as salinity, although they both decay in time. Their half-lives are nearly the same (30 and 28 years, respectively), so their relative concentrations do not change much in time. Although their chemical behavior in the ocean is identical, on land strontium is less strongly removed by interactions with particles than is cesium. Thus, whereas the ratio of cesium-137 to strontium-90 in fallout is nearly 1.5, in river runoff the ratio is reduced. The ratio of these two radionuclides in waste from nuclear fuel reprocessing plants is much greater than in fallout, typically near 10. Because nuclear wastes enter the ocean in coastal regions with lower salinity, a ratio of cesium-137 to strontium-90 that is either greater or lower than 1.5 is an indication of freshwater input. Therefore, not only can measured concentrations in the ocean give a time scale associated with the source functions, but also the ratios of these radionuclides can help delineate sources and thereby trace the circulation patterns of water masses.

In the region just north of Fram Strait, the cesium-137 and strontium-90 distributions reflect fallout from nuclear weapons tests followed by a contribution from the Sellafield reprocessing plant that has been advected from the coastal waters of the North Sea to the Norwegian–Greenland Sea region. The Sellafield discharge becomes incorporated with water from the North Atlantic along the transit from the Irish Sea to Fram Strait, reducing the ratio of cesium-137 to strontium-90 to about 3. A maximum in the concentrations of these radionuclides corresponds to the temperature maximum of the Atlantic layer only in the region north of Fram Strait. At the AIWEX site below the halocline in the Atlantic layer, the ratio of cesium-137 to strontium-90 is greater than 1.5, indicating the presence of relatively recently introduced Atlantic water. Measurements at the Fram 3 Ice Camp in 1981 (Smith *et al.*, 1990) indicate that some of the Sellafield discharge that was incorporated in deeper water in the Norwegian–Greenland seas is advected through Fram Strait into the Arctic Ocean by a sharply defined boundary current along the continental slope between about 1000 and 2000 m. A maximum in cesium-137 and strontium-90 at the North Pole at 1500 m observed during LOREX in 1979 has been explained as a shelf signal reflecting Sellafield input. This maximum may result from the penetration of dense waters formed in the Barents Sea (Livingston, 1988). There must be very rapid ventilation of the deeper water if the source of the signal is from the Norwegian–Greenland seas, since the Sellafield signal does not appear to have reached the Norwegian–Greenland Sea region prior to 1978 (Livingston *et al.*, 1985).

G. Sea Ice Chemistry

Sea ice plays a major role in determining the distributions of chemicals in the Arctic Ocean. What has not been mentioned is the possible chemical effects of the changing of phase between ice and water. These effects would be present in both Arctic and Antarctic environments, or in any region where sea ice forms or melts. In general, there will be separations of dissolved constituents as well as isotopic separations when such changes in phase occur. Although any isotopic separation that might occur in the Arctic Ocean is probably small compared to scatter (e.g., Tan and Strain, 1980), several laboratory experiments show that there is preferential precipitation of some salts from seawater as it is frozen and ice is formed (e.g., Anderson and Jones, 1985). Distributions of several chemical constituents in seawater in regions that contain sea ice meltwater or in which sea ice was formed would be altered by this process. Because calcium carbonate is the first to precipitate from brine just below the freezing point of seawater, most obviously affected would be total alkalinity and total carbonate, both of which are involved in discussions of distributions of river runoff and carbon budgets in the Arctic Ocean. Despite fairly conclusive laboratory studies showing preferential precipitation of some salts in brine trapped in sea ice, evidence that this is an important process in nature is lacking. Some enrichment of some salts was measured in natural samples, but to date there is no evidence that this preferential precipitation is significant in determining distributions of chemicals in the Arctic Ocean (Reeburgh and Springer-Young, 1983; Anderson and Jones, 1985), although it could be important in the Sea of Okhotsk (Lyakhin, 1970) and possibly in Baffin Bay (Jones et al., 1983). Because ice under natural conditions can experience a complicated history of temperature and salinity variations, the significance of sea ice chemistry in determining chemical distributions cannot be ignored, but any quantitative geochemical interpretations based on the compositions of sea ice produced in the laboratory must be made with caution.

III. The Southern Ocean

Far from the main anthropogenic sources of the northern hemisphere and separated from the surface circulation of the other major oceans by the Subtropical Convergence, the Southern Ocean may appear to be relatively isolated. But this is merely an appearance: the Southern Ocean, which is the ultimate point reached by deep waters generated in the northern regions, is also a major area of bottom-water formation of the world ocean and is broadly opened to the Atlantic, Pacific, and Indian oceans at all depths greater than 200 m. Very large chemical fluxes are involved in these exchanges; for example, about 60 trillion moles of silicate are imported annually from the Atlantic Ocean (i.e., over a hundred times as great as the Amazon River input to the ocean). Because of the great geographic extent of cold surface waters (the temperature of about 20 million km^2 varies seasonally between -1.85 and $+4°C$), this area is a huge sink for dissolved gases and may play a significant role in regulating the composition of the atmosphere, especially with regard to the reequilibrating CO_2 content. This explains why the study of this "geoecosystem" is of basic interest in chemical oceanography. In this part of the chapter we focus our attention on the circulation of the chemical constituents, considered first as tracers of the physical processes occurring in the different water masses and then as indicators of biological activity. Finally, we consider some unusual aspects of major biogeochemical cycles in the Southern Ocean.

A. Chemical Distributions as Indicators of Circulation

1. The surface waters

For the purpose of this discussion, we define the surface layer as that bounded by the ocean surface and the subsurface temperature minimum (see Chapter 5). Three chemical processes in the surface waters of the Southern Ocean are of particular interest: delivery of large amounts of nutrients (especially nitrate, phosphate, and silicate) by upwelling; gas exchange with the atmosphere, as indicated by transient anthropogenic tracers; and absorption of carbon dioxide from the atmosphere, as regulated by physical and biological processes.

a. The nutrients: richness of surface waters as a result of deep-water upwelling The high nutrient concentrations of Antarctic Surface Water (ASW) are a striking feature of the Southern Ocean. In fact, the waters south of the Polar Front comprise by far the greatest geographic extent of nutrient-rich surface water on earth. The enrichment of the surface layer results from an oceanic divergence that brings the deep waters, originating mainly from the North Atlantic, to the surface. At the Antarctic Divergence (AD), nitrate, phosphate, and silicate concentrations, respectively, are over 25, 2, and 60 μM (Fig. 8.7), among the highest in any surface waters of the world ocean. Like temperature and salinity, the nutrients reflect the general circulation pattern in this area and show typical annular distributions (Fig. 8.7), although specific patterns occur in the cyclonic gyres of the Weddell Sea or near the Ross Sea ice edge and in the mesoscale eddies generated by the Kerguélen Plateau north of the Amery Sea.

The frontal zones of the Southern Ocean (Fig. 8.7) correspond not only to well-marked hydrographic structures but also to major boundaries in the nutrient distributions. The Subtropical Convervence (STC), which can be considered the northern limit of the Southern Ocean, coincides with a dramatic decrease in nitrate and phosphate, especially in the Indian sector. South of the Polar Front, the silicate concentration begins to increase progressively. The discrepancy between these different distributions is explained in Sections III,B,2,b and c below (see also Fig. 8.13).

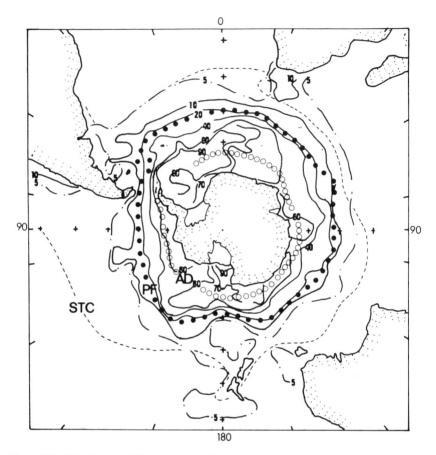

Figure 8.7 Distribution of silicate in Antarctic Surface Water (100 m depth). The concentrations of this nutrient (and also of nitrate and phosphate) are among the highest in the ocean. Their distributions are dependent on the Circumpolar Current and on the upwelling that takes place at the AD. The Polar Front (PF) coincides approximately with the isopleths of $10-20$ μM. After Gordon and Molinelli (1982).

A very significant biological consequence of the massive upwelling of deep waters south of the Polar Front is that inorganic nutrients, which are a major limiting resource for phytoplankton growth and productivity in most of the world's oceans, are virtually never depleted to potentially limiting concentrations in the Antarctic and subantarctic (see Section III,B,1).

b. Gas transfer at the sea surface and the processes forming the Antarctic surface waters Seasonal variations are observed in the characteristics of the surface waters in the Southern Ocean, accompanying the seasonal change in the hydrographic structure, which goes from vertically homogeneous during winter to stratified in summer. By using isotopic measurements, Weiss *et al.* (1979) gave the first general explanation of the origin and the dominant processes leading to the formation of both Winter Water (WW) and Summer Antarctic Surface Water (SSW). Diagrams of deuterium and oxygen-18 concentrations versus salinity (Fig. 8.8) show that the characteristic of WW are determined mainly by freezing rather than evaporation (both phenomena might explain a salinity increase compared with SW) and that the formation of SW results from the combined effects of evaporation–precipitation and freezing–melting. Western shelf water (a high-salinity surface component involved in Antarctic Bottom Water (AABW) formation in the Weddell and Ross seas) is exceptionally depleted in both deuterium and oxygen-18 (Weiss *et al.*, 1979). If freezing is involved in the formation of this high-salinity water mass, its isotopic composition also requires a significant admixture of meltwater from the base of the ice shelf.

Carbon-14 and tritium are of special interest in Antarctic regions, as they show striking contrasts between the atmosphere and the sea. The concentrations of both isotopes in the atmosphere have been greatly increased by atmospheric testing of thermonuclear weapons since 1954. However, because of the important of deep (and thus old on time scales measured by bomb isotopes) water in the formation of all Antarctic water masses, lower (prebomb) values must be expected in the surface waters of the oceanic and coastal domains of the Southern Ocean. In fact, Weiss *et al.* (1979) measured in WW and SW of the Weddell Sea concentrations of ^{14}C and tritium that were among the lowest of the world ocean [about 120% for ^{14}C and less than 1 tritium unit (TU) for tritium]. Low tritium values were later confirmed using the GEOSECS data in the Indian sector by Ostlund *et al.* (1980) and by Michel (1984) for the Weddell Sea. Thus, the main origin of surface waters of the Weddell Gyre is the upwelling of Warm Deep Water (WDW) during both summer and winter, and their characteristics are determined mostly by this process.

Recognition of a large input of WDW to the Antarctic surface layers may help explain its oxygen undersaturation. Until recently, chemical oceanographic studies in the Southern Ocean were performed mainly in summer, when the surface oxygen content is nearly saturated, whereas the underlying remnant WW is undersaturated by about 15% (i.e., about 50 μmol kg^{-1} below the full saturation value; Fig. 8.9a and b). If we consider the whole area covered by sea ice during winter, this corresponds to an approximate deficit of 50 trillion moles of oxygen per year in WW, which is comparable to the potential input of 100 trillion moles per year to SW due to photosynthesis (assuming a mean primary production of 50 g C m^{-2} yr^{-1}). In 1982 Levitus published the annual and seasonal mean oxygen levels at the sea surface,

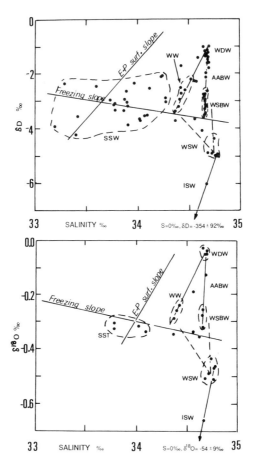

Figure 8.8 Diagrams of distributions of the stable isotopes (deuterium and oxygen-18) versus salinity for surface, deep, and bottom waters. *Surface waters:* also plotted in each part are lines representing the approximate slopes (dD/dS and dO/dS) due to evaporation–precipitation and freezing–melting in high-latitude surface waters; these lines are arbitrarily drawn to pass through the Summer Surface Water (SSW) envelopes and only the slopes are significant. Both deuterium and oxygen-18 distributions support the idea that SSW results from a combined action of freezing–melting and evaporation–precipitation (large scatter), unlike Winter Water (WW), which cannot be formed without significant freezing. *Deep and bottom waters:* WSBW = Weddell Sea Bottom Water; AABW = Antarctic Bottom Water. The isotopic composition of the high-salinity Western Shelf Water (WSW), which plays a main role in the formation of WSBW, is basically dependent on that of the potentially supercooled Ice Shelf Water (ISW) found at the base of the Filchner ice shelf; the latter, originating from the Antarctic continent, presents a severe isotopic depletion. For a detailed explanation of the formation of AABW, see Fig. 8.10. After Weiss *et al.* (1979).

8 Chemical Oceanography

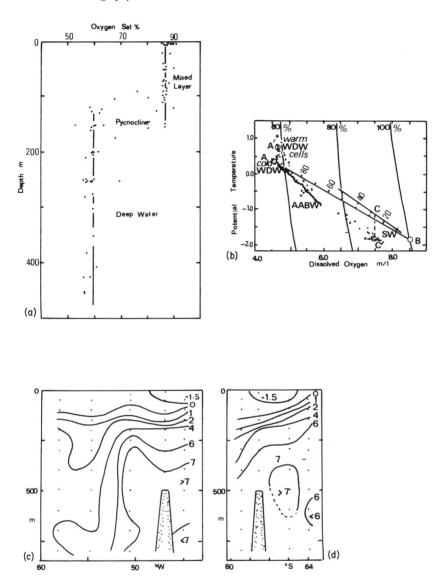

Figure 8.9 (a and b) Oxygen distributions. At the end of winter, the surface mixed layer (SW) of the Weddell Sea is undersaturated with oxygen (line CC'), which is explained by the blend of about one-fourth WDW (A : 0.5 °C, 4.50 ml l^{-1}) into three-fourths beginning of winter surface water (B : −1.87 °C, 8.54 ml l^{-1}) during the winter-ice coverage period (about 5 months). Redrawn from Gordon *et al.* (1984). (c and d) Helium-3 distributions in the Weddell Gyre reported as deviations from the isotopic helium ratio of air (^{3}He/^{4}He). Both sections showed that d^3He is maximum for the WDW core layer, the remnant WW being 0.5–3% higher than the solubility equilibrium (i.e., without any WDW entrainment). After Schlosser *et al.* (1987). Both studies estimated the corresponding yearly WDW upwelling to be about 30 m in the Weddell Sea.

which showed that the Southern Ocean annual mean level is a few percent undersaturated. The age and circulation rates of the deep and bottom waters of the world ocean, which contain a significant component of Antarctic surface waters (Foster and Carmack, 1976), are often calculated using "apparent oxygen utilization" (AOU) and basing this calculation on the assumption that complete saturation occurs at the sea surface. Thus, the degree of undersaturation of Southern Ocean surface water affects these calculations on a global scale.

Direct observations of ASW characteristics beneath the sea ice during winter confirm that oxygen is undersaturated (at levels comparable to those measured in the summer remnant WW). In the Weddell Gyre an input of one-third of upwelled, low-oxygen WDW is needed (Fig. 8.9a and b) to explain the final oxygen concentration in ASW during winter (Gordon et al., 1984). This leads to a mean entrainment rate of WDW into WW equal to 27 m yr^{-1}. For the Amery Sea, Treguer et al. (unpublished results) have calculated a rate of 45 m yr^{-1} to explain the undersaturation of the remnant WW. A balance calculated by Schlosser et al. (1987) for helium-3 (the lighter and much less abundant stable isotope of the noble gas helium; Fig. 8.9c, d) in the upper layers of the Weddell Sea and Bransfield Strait agrees well with the calculation of Gordon et al.; an entrainment rate between 15 and 35 m yr^{-1} is needed to explain a mean 2% excess of ^{3}He above the solubility equilibrium. Unlike oxygen, which is undersaturated in surface waters because of the seawater-ice coverage, helium-4 is supersaturated (110%) in the ice shelf water, as demonstrated by Schlosser (1986) for the southern Weddell Sea, as the result of high pressures on entrapped bubbles during ice formation.

The first data on Freon 11 and Freon 12 (the most common chlorofluoromethanes introduced into surface water from the atmosphere) were obtained during summer by Weiss and Bullister (1984) in the Scotia and Weddell seas. Because of the low but significant levels they measured in surface waters and the ratios of Freon 11 to Freon 12 in the deep water masses, these authors suggested that (1) there is admixture of WDW into WW beneath the sea ice cover, and (2) the ventilation time of the Weddell Gyre has to be on the order of a decade or less.

c. Excess CO_2 in Antarctic surface waters: an indicator of the interocean exchange of deep waters The steady increase of CO_2 in the atmosphere due to anthropogenic inputs since the beginning of the industrial age is now clearly demonstrated; the atmospheric CO_2 concentration is estimated to have been about 275 ppm in 1850 and is now approximately 345 ppm. An amount of dissolved inorganic carbon 60 times that present in the atmosphere is present in the ocean, which has acted as a sink for the "excess" CO_2.

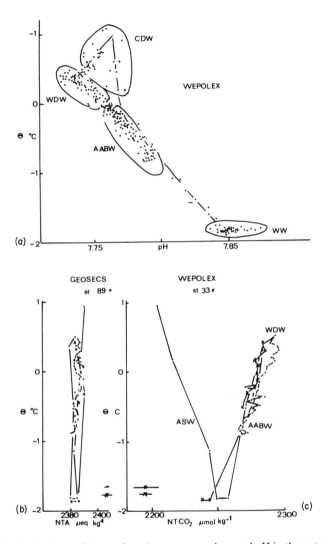

Figure 8.10 Distributions of inorganic carbon concentrations and pH in the water masses in the Southern Ocean (data from GEOSECS and WEPOLEX stations in the Weddell Sea during summer). (a) Plot of potential temperature (theta) versus pH, with the approximate mixing lines; pH is a good tracer for identification of the water masses in the Southern Ocean. (b) Plot of theta versus normalized total alkalinity (NTA); NTA is almost constant for the whole water masses. (c) Plot of theta versus normalized total inorganic carbon in the different water masses (ASW = Antarctic Surface Water; WDW = Warm Deep Water; CDW = Circumpolar Deep Water; AABW = Antarctic Bottom Water). From Poisson and Chen (1987).

Apparently, most of the net absorption of CO_2 by the oceans occurs in the polar regions; Chen and Drake (1986) stated that the surface wind-mixed layer of most of the oceans is more or less saturated with excess CO_2, whereas the deep oceans, except for regions of deep-water formation, are expected to contain relatively little industrial-age CO_2.

A direct method has been developed to estimate the oceanic penetration of excess CO_2 at individual stations of the Southern Ocean on the basis of carbonate data. The water masses of the Weddell Sea contain large amounts of inorganic compounds (Fig. 8.10a–c), pH values are <7.9 (i.e., clearly lower than the mean value for the world ocean), and normalized total inorganic carbon ($NTCO_2 = TCO_2 \times 35.00/S$) increases from about 2220 μmol kg^{-1} in surface waters to 2300 μmol kg^{-1} in deep and bottom waters. Both parameters are excellent tracers of water masses in the Southern Ocean, unlike normalized total alkalinity (NTA) and calcium, which are virtually constant throughout the water column (about 2385 and 10,230 μmol kg^{-1}, respectively). The concentrations of these parameters are comparable to those measured in the Arctic Ocean (see Section II,B). Using the measured TCO_2 values and correcting for the influences of $CaCO_3$ dissolution and organic carbon decomposition, Poisson and Chen (1987) calculated the excess CO_2 for individual stations in the Weddell Sea. The excess CO_2 (Fig. 8.11) compared with WDW is about 28 μmol kg^{-1}.

Unlike the deep layers of the northern hemisphere near the sources of anthropogenic CO_2 outgassings, AABW contains little anthropogenic CO_2 (6 ± 5 μmol kg^{-1}). As stated, the winter ice pack blocks the atmosphere–sea exchange of gases and also diminishes the downward wind mixing of surface properties. Thus the upwelled old deep water sinks back down as a result of cooling without absorbing much of the excess surface CO_2, and measurable CO_2 penetrates only to about 200 m (Fig. 8.11a), not much deeper than the homogeneous surface layer (Poisson and Chen, 1987).

2. Antarctic Intermediate Water

According to Deacon (1982), the Antarctic Intermediate Water (AAIW) is identifiable in the Polar Front zone as the core layer characterized by a temperature minimum extending to depths greater than 200 m. Wust (1936), Deacon (1933, 1937), Gordon et al. (1977a,b), and Molinelli (1981) have all concluded that this water mass results mainly from the isopycnal mixing of WW and the overlying waters when these sink at the Polar Front. In contrast to this proposed mechanism, McCartney (1977) explained the characteristics of this water mass almost entirely by local modifications of the subantarctic surface water, without any significant role played by Antarctic water masses. For the Indian sector, Jacobs and Georgi (1977), Le Jehan and Treguer (1983), and Le Corre and Minas (1983) characterized

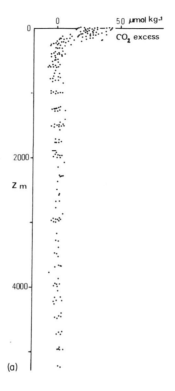

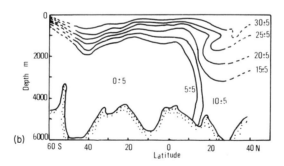

Figure 8.11 Excess (anthropogenic) CO_2 in the southern and world oceans. (a) Profile of excess CO_2 for stations located in the Weddell Sea (U.S.–USSR WEPOLEX Expedition, summer 1981). The WDW is taken as reference (mean value = zero); note that the penetration of anthropogenic CO_2 is limited to 200 m and the excess in AABW is very low. After Poisson and Chen (1987). (b) Distribution of excess CO_2 (micromoles per kilogram) in the western Atlantic Ocean (GEOSECS data). Although the data may be subject to possible systematic error in the polar oceans, this figure illustrates the slow penetration of industrial CO_2 in deep and bottom water masses of the southern hemisphere through the North Atlantic Deep Water flow. After Chen (1982).

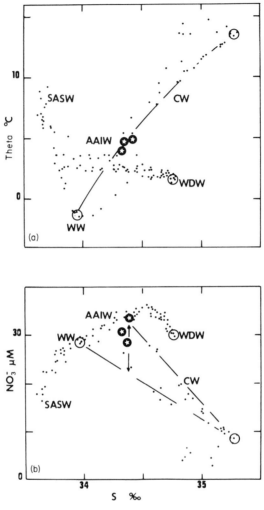

Figure 8.12 Formation of Antarctic Intermediate Water (AAIW). Both hydrographic parameters (potential temperature, salinity) and nutrients (nitrate, phosphate, silicate) are useful for identifying the processes that lead to the formation of AAIW, as illustrated by data from ANTIPROD 1 cruise (Indian sector of the Southern Ocean, summer 1977). (a and b) Diagrams drawn from data collected between the surface and S_{max} core layer reveal that AAIW cannot be formed in the subantarctic sector, as the hydrological and chemical characteristics of the AAIW are very different from those of the Subantarctic Summer Surface Water (SASW). A possible mechanism for the formation of AAIW is mixing of Antarctic Winter Water WW), shoaling down at the Polar Front, with Central Waters, present north of the Subtropical Convergence. Nevertheless, nonconservative anomalies for nitrate (b) might be due to local recycling (mineralization or organic matter settling down from the frontal area). Dots show possible mixing paths.

AAIW by temperatures of about 5°C; salinities of 34.40 per mil; and nitrate, phosphate, and silicate concentrations of 31, 1.94, and 29 μM, respectively. This layer has a dissolved oxygen concentration of approximately 210 μM and is thus markedly undersaturated (about 30%). Both theta–salinity and nutrient–salinity diagrams (Fig. 8.12) demonstrate that Antarctic surface waters influence the characteristics of AAIW. However, a simple bend between WW and Central Waters is insufficient to explain the excess nutrient and oxygen deficit in AAIW. The differences might be due either to nutrient recycling (the Polar Front zone is supposed to be a more productive area; see below) or to a small admixture of the oxygen minimum water mass situated beneath (see below). It is possible that both processes have some importance.

3. The Warm Deep Waters

The Circumpolar Deep Water (CDW) is the main core layer of the Southern Ocean system. South of the Polar Front this water mass is called the Warm Deep Water (WDW). The CDW/WDW, whose mean thickness is about 2300 m, is by far the most voluminous Antarctic water mass, with 81 million km^3, or nearly 58% of the Southern Ocean's volume. The residence time of CDW/WDW in the Southern Ocean is only about 50 years, which is quite short for a deep water mass but long in comparison with processes such as gas exchange with the atmosphere or decomposition of organic matter.

As for chemical distributions, two extremes are distinguishable within the WDW. The well-known salinity maximum of the North Atlantic Deep Water component coincides with a deep minimum in "NO" (Broecker, 1974) (see Fig. 8.13). The oxygen minimum layer (where nitrate and phosphate maxima are also detectable) is located 100–200 m beneath the temperature maximum layer. This oxygen minimum is apparently derived from oxygen-depleted deep water masses that lie beneath the productive coastal upwelling areas of the tropical Indian, Atlantic, and Pacific oceans (Fig. 8.14).

4. The Antarctic Bottom Waters

The Antarctic Bottom Waters (AABW) represent 28% of the total volume of the Southern Ocean (Carmack, 1977). It is now recognized that there are several varieties of bottom waters in the Antarctic basins. The bottom waters closest to the source regions reflect the influence of the near-surface component (i.e., cold, with higher dissolved oxygen and slightly diminished silicate concentrations; Fig. 8.15a–c). The variable salinity of the bottom waters near Antarctica corresponds to that of the local WDW component. As the bottom waters spread from source regions into deeper and more remote parts of the Antarctic basins, their chemical characteristics are altered

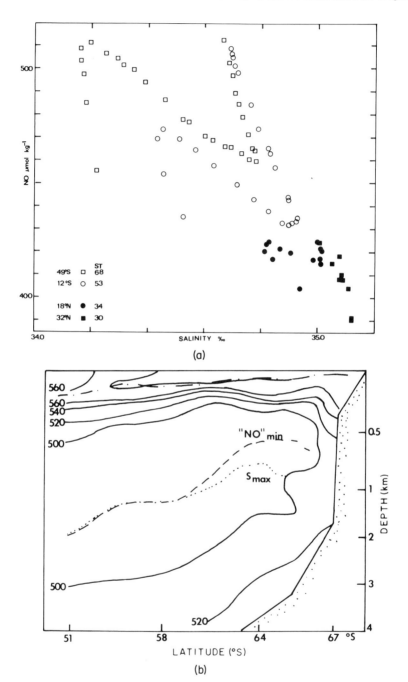

8 Chemical Oceanography

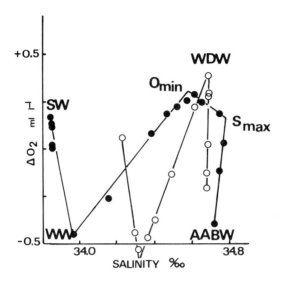

Figure 8.14 A characteristic oxygen minimum lies at the upper levels of the WDW. By using a multiple linear regression analysis (S, O_2, NO_3^-) Le Jehan and Treguer (1983) showed that this minimum is not caused by a biological phenomenon (it results from entrance of Indian Deep Water in the Indian sector of the Southern Ocean). The oxygen residual ΔO_2 theoretical value calculated by the regression analysis minus *in situ* concentration) versus salinity follows characteristic patterns with slope breaks at each water mass (WW = Winter Water; WDW composed of the salinity-maximum component and the minimum-oxygen layer; AABW = Antarctic Bottom Water). Both Indian and Atlantic components are present north of the Antarctic Divergence (filled circles) but a single one is observed south of the AD (open circles).

through interactions with the overlying water column and with the bottom sediments (Mantyla and Reid, 1983; Edmond *et al.*, 1979).

Formation of bottom water depends mainly on the presence of high-salinity water above the continental shelf (the so-called Western Shelf Water, WSW), and thus the main sources of bottom water in the Antarctic are in the Weddell and Ross seas. In comparison with the southeast Pacific sector, the Atlantic sector contains fresher, colder, and denser bottom waters. These bottom waters from the Atlantic sector are also richer in oxygen. Bottom

---

Figure 8.13 The main end member of the Warm Deep Water comes from North Atlantic Deep Water, which penetrates southward into the Southern Ocean. (a) This water mass, with a well-known salinity maximum, has a strong NO minimum (NO = O_2 + $9NO_3^-$ is a conservative tracer as shown by Broecker, 1974). This minimum spreads southward to the divergence area (about 65°S), as illustrated (b) for a transect through the Indian sector of the Southern Ocean, east of Kerguelen Plateau (NO in micromoles per kilogram). From Treguer *et al.* (1987).

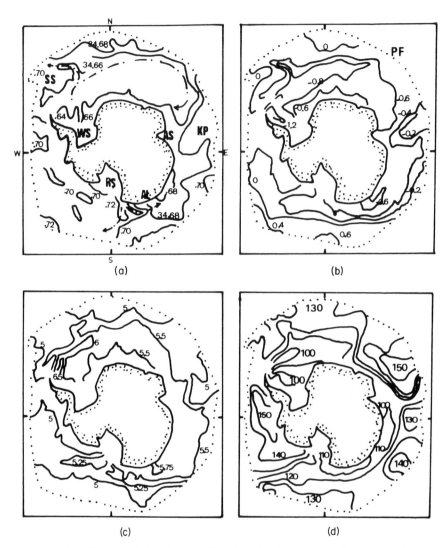

Figure 8.15 Different varieties of AABW flow along the abyssal slopes of Antarctica, the main sources being located in Weddell Sea (WS) and the Ross Sea (RS). Diagrams of (a) salinity (grams per kilogram), (b) potential temperature (degrees Celsius); and (c) oxygen (milliliters per liter are redrawn from Carmack (1977). (d) Diagram for silicate (micromolar; redrawn from Mantyla and Reid, 1983) shows the near-bottom value distributions south of the Polar Front. Arrows represent possible trajectories of the abyssal currents for WS and RS bottom waters.

waters of intermediate characteristics are found in the south Indian abyss (Table 8.3).

A three-stage process was proposed by Foster and Carmack (1976) (Fig. 8.16) to explain the formation of the bottom water in the Atlantic sector. This process goes through the intermediate formation of modified WDW by admixture of WW in the subsurface layer. According to Foster and Carmack's model, the composition of AABW is approximately $\frac{5}{8}$WDW + $\frac{2}{8}$WSW + $\frac{1}{8}$WW. It must be noted that if tracers such as oxygen (Fig. 8.16), nitrate, and phosphate are used, the agreement with the model is quite good, but that silicate shows nonconservative behavior (Edmond et al., 1979; Le Jehan and Treguer, 1983; Mantyla and Reid, 1983). This model was found to be consistent with data collected later on tritium distribution: Michel (1978) reported undetectable tritium levels in Weddell Sea bottom waters, and Weiss et al. (1979) estimated the flux of AABW to be approximately 4.5 Sv on the basis of tritium data.

Although the circulation patterns within the Ross Sea are similar to those within the Weddell Sea, Jacobs et al. (1970) proposed a somewhat different mechanism for the Ross Sea Bottom Water formation. Two varieties of AABW, characterized by relatively high oxygen concentrations (about 240 μM), might result from a mixture of WDW with either saline Ross Sea Shelf Water or the somewhat fresher Ice Shelf Water.

Minor local sources of bottom water have been identified all around Antarctica by their high oxygen and low silicate content. Gordon and Tchernia (1972) identified such a source near the Adélie Coast in the south Indian basin, and some evidence of local bottom-water formation in the Davis Sea was given by Treshnikov et al. (1973). Bottom-water formation near the Enderby Land–Prydz Bay coast has been suggested by Jacobs and Georgi (1977) and Treguer et al. (1987). L. Memery and A. Poisson (unpublished results) give support to this hypothesis using results from Freon 11 and Freon 12 profiles.

Table 8.3 Comparative Characteristics of the Weddell Sea and Ross Sea Bottom Waters (Depth over 3500 m)

	T (°C)	S (g/kg^{-1})	O_2 (ml/l^{-1})	NO_3^- (μM)	PO_4^{3-} (μM)	H_4SiO_4 (μM)
Weddell Sea	−0.88	34.65	6.1	29	2.2	99
Ross Sea	−0.5	34.72	5.25	n.d	n.d	100
Adélie Land	−0.6	34.68	5.75	n.d	n.d	110
Amery Sea	−0.4	34.68	5.6	33	2.2	120

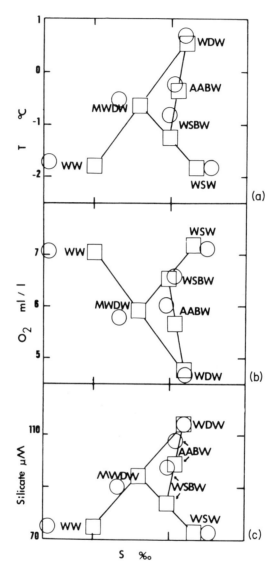

Figure 8.16 The Foster and Carmack (1976) model for the formation of AABW in the Weddell Sea area, applied to physical and chemical tracers. (a) Potential temperature versus salinity; (b) oxygen versus salinity; (c) silicate versus salinity.

5. Distributions of trace metals and dissolved organic matter

Only a few modern data (Boyle and Edmond, 1975; Orren and Monteiro, 1985; Bordin et al., 1987) are available to describe the trace-metal concentrations in the Southern Ocean. Although these data must be considered preliminary, they indicate that concentrations in surface water are similar to those in other oceanic regions (Jacques and Treguer, 1986). The profiles of Cu (Fig. 8.17a), Cd, and Zn in the Indian and Atlantic sectors increase with depth, as is typical of other oceanic regions. The Mn profiles also show an increase with depth and are different from Mn profiles in most other oceanic regions, which generally show a surface maximum resulting from aeolean inputs. The positive Mn gradient toward AABW (Fig. 8.17b) might result from the surface water component of AABW (see above). All other trace-metal profiles reveal the presence of a maximum at intermediate depths, more or less coinciding with the oxygen minimum. E. Suess (personal communication) also measured high Mn concentrations in the deep waters of Bransfield Strait and hypothesized that these might result from hydrothermal activity.

Only a few profiles of total dissolved organic nitrogen (DON) and dissolved organic phosphorus (DOP) are available in the Southern Ocean. Le Jehan and Treguer (1983, Fig. 8.18), found that DOP concentrations varied between 0.2 and 0.02 μM in a transect across the Polar Front, with a minimum in the WDW. Similar profiles are shown by Jacques and Treguer (1986) for DON south of the Polar Front (PF), with the WDW minimum being about 5 μM.

B. Elemental Cycles

In the previous section we stressed the linkages between chemical distributions and physical processes such as gas exchange with the atmosphere, deep circulation, and bottom-water formation. Seawater chemistry is also significantly linked to the dynamics of the Southern Ocean ecosystem through the distribution and cycling of biologically active chemical constituents. In the Southern Ocean some of the chemical distributions are very different from those elsewhere in the ocean, including the Arctic, with the result that the nature of the physical/chemical control of biological productivity may change qualitatively at the Polar Front. Also, the low surface temperatures of the Antarctic and subantarctic increase the solubility of carbon dioxide (as well as other gases) and decrease the dissolution rate of biogenic silica, making the Southern Ocean a region of major importance in the global CO_2 and silica cycles.

After a general overview of the distribution and cycling of biologically active chemical constituents within the Antarctic Circumpolar Current and

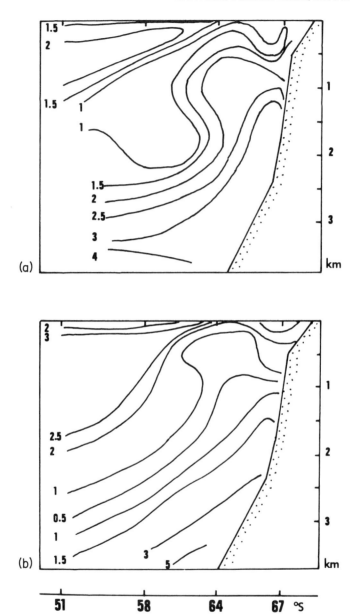

Figure 8.17 Distribution of trace metals in the Indian sector of the Southern Ocean: (a) copper and (b) manganese, both in nanomoles per kilogram. Redrawn after Bordin *et al.* (1987).

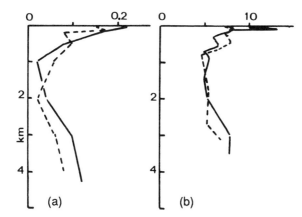

Figure 8.18 Profiles of total dissolved organic nitrogen (DON and phosphorus (DOP) in the Indian sector of the Southern Ocean (a) DOP in μM south (—) and north (- -) of Polar Front; (b) DON in μM south of Polar Front, 51° (—) and 58° (- -).

the resulting chemical uniqueness of the Southern Ocean as a habitat for life, we will consider these distributions and processes in several specific areas that appear to be of particular geochemical and ecological interest: the divergence waters, the Polar Front system, marginal ice zones, and shallow waters in the vicinity of islands.

1. Open-ocean areas of the Antarctic Circumpolar Current

As described in Section III,A,1, the Antarctic Divergence and the Circumpolar Current are characterized by massive upwelling of deep ocean water. The vertical nutrient transport associated with this upwelling makes the surface waters south of the Polar Front (an area equal to about 10% of the world ocean) a perpetually nutrient-replete environment for phytoplankton. Uptake by phytoplankton in summer results in clear seasonal cycles in the near-surface concentrations of nitrate, phosphate, and silicic acid (e.g., Jennings *et al.*, 1984), and depletion of nitrate and phosphate to below the detection limit of conventional colorimetric analyses has been reported within one very intense phytoplankton bloom in a marginal ice zone (Nelson and Smith, 1986; see subsection 2,a below). However, the open-ocean areas of the Antarctic Circumpolar Current (ACC) appear to provide an environment characterized by continuous nutrient sufficiency for the phytoplankton.

This generalized nutrient sufficiency within the surface waters results in a qualitative difference between the Southern Ocean and most other marine habitats with respect to the physical processes controlling phytoplankton

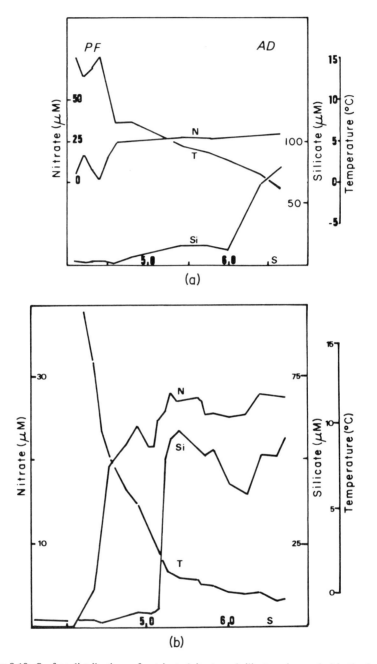

Figure 8.19 Surface distributions of nutrients (nitrate and silicate, micromolar) in the Southern Ocean during summer show striking variations when going northward from the Antarctic Divergence (AD) to the Polar Front (PF). (a) Indian sector (after Le Jehan and Treguer, 1983); (b) Atlantic sector (after van Bennekom et al., 1988).

productivity. It is clear that over most of the ocean, localized wind-driven upwellings, mesoscale eddies, oceanic fronts, and even passing storms can result in localized areas of greater primary productivity. The common property of all these mesoscale features and events in tropical, temperate, and arctic environments is that they lead to enhanced vertical transport of nutrients from intermediate depths to a nutrient-depleted surface layer. The ensuing release from nutrient limitation permits more rapid growth of the phytoplankton and increases both its biomass and productivity. Although localized upwellings, eddies, and fronts clearly occur in the Southern Ocean and storms are very common (see Chapter 5), it is not clear that they have anywhere near the biological significance there that they have in other parts of the ocean (see Chapter 9). The main reason for this lack of response is that intensified vertical transport of nutrients is of little importance when nutrient concentrations in the surface layer are already high enough to prevent nutrient limitation of phytoplankton growth.

In fact, as discussed below, most of the known areas of increased primary productivity in the Southern Ocean are associated with *diminished* vertical exchange of water. This pattern, which is nearly the opposite of that which prevails in most other oceanic regions, is consistent with a general condition in which light, rather than any nutrient, represents the main limiting resource for the phytoplankton. Thus, localized vertical stabilization of the water column improves the light regime for the phytoplankton by restricting the vertical extent of wind-driven mixing, and the concurrent decrease in vertical nutrient exchange has little or no negative impact because of the very high initial concentrations of all major nutrients in the euphotic zone. It appears to be a fair generalization to say that the basic upper-ocean processes governing phytoplankton productivity differ qualitatively south of the Polar Front from those in the rest of the ocean.

2. Localized environments

a. The Antarctic Divergence The dramatic silicate gradient occurring in the waters flowing northward from the Antarctic Divergence (AD; Fig. 8.19), compared with the much smaller decrease of total inorganic nitrogen and phosphorus when going from the AD to the Polar Front, is a striking feature of the Indian sector of the Southern Ocean (Le Jehan and Treguer, 1983; Le Corre and Minas, 1983). Analogous patterns are apparent in the other sectors, with some difference in latitude (see, for example, van Bennekom *et al.*, 1988). Although this gradient may be due to conservative processes, several arguments suggest the importance of biological processes leading to N/P uptake ratios that vary between 10 and 16 and N/Si uptake ratios between 0.2 and 13 in surface waters. Copin-Montegut and Copin-Montegut (1978) have calculated uptake C:N:Si:P ratios of 62:11:2.5:1 for Antarctic phytoplankton. In the Drake Passage, Sommer and Stabel (1986) measured ratios of residual concentrations of Si and N that varied between 4 and 10 during summer. High concentrations of ammonium (reaching 1 μM) have been measured in Antarctic surface waters (e.g., Le Jehan and Treguer, 1983; Nelson et al., 1989; Biggs, 1982) and originate either from excretion of the high trophic levels or from bacterial activity.

During summer, the surface waters are not particularly rich in biogenic particulate matter. In the Indian sector, Copin-Montegut and Copin-Montegut (1978) measured particulate organic carbon (POC) and particulate organic nitrogen (PON), respectively, to be approximately 12 and 2 μmol l^{-1}, and Treguer et al. (1988) showed that the concentrations of biogenic silica are lower than 2 μmol l^{-1}, with corresponding chlorophyll a concentrations lower than 0.5 μg l^{-1}. The low biomass may be explained by a high grazing pressure and/or by rapid settlement of highly silicified diatoms; Dunbar et al. (1985) and Tsunogai et al. (1986) measured settling rates as high as 200 m day^{-1} for some particles. The distribution of both dissolved and particulate matter in the AD area may be dramatically modified (Heywood and Priddle, 1987) by the presence of gyres generated by topographic anomalies (Daniault, 1984).

The primary production measured in the Antarctic Divergence system during summer is especially low (Holm-Hansen, 1985; Jacques and Treguer, 1986) in spite of the high nutrient levels and abundant light at the surface (Fig. 8.19). This so-called "Antarctic paradox" might be explained by unfavorable conditions for phytoplankton growth due to combined influences of physical factors (e.g., temperature and vertical mixing; Jacques and Treguer, 1986; Priddle et al., 1986). Trace metals and dissolved organic matter seem to play a minor role (Jacques et al., 1984); their surface concentrations (Fig. 8.14) are almost in the same range as those in the world ocean (Orren and Monteiro, 1985; Bordin et al., 1987). However, Martin and Fitzwater (1988) have suggested a possible limiting effect of iron because of low aeolien inputs into the Southern Ocean, and recent experiments by de Baar et al. (1989) tend to support this hypothesis. Finally, although the Southern Ocean as a whole is an upwelling ecosystem in which the nutrient concentrations are generally in excess of phytoplankton requirements, silicate concentrations drop below 10 μM near the Polar Front. The affinity constant (Michaelis–Menten half-saturation constant, K_s) of some Antarctic diatoms reaches unusually high values (over 12 μM), which led Jacques (1983) to suggest possible limitation of diatom growth by a deficiency of this element.

b. The Polar Front Two main hydrographic fronts characterize the Southern Ocean: the Subtropical Convergence (STC) and the Polar Front

(PF), which dramatically modify the circulation of the chemical constituents. Although these hydrographic discontinuities result from energy transfers from the atmosphere to the sea surface, their locations coincide with bottom topography. Daniault (1984) demonstrated the permanence of mesoscale cyclonic gyres (amplitude, 80 km; wavelength, 500 km; duration about 20 days) in the subantarctic or in the Antarctic domains, with the patches of kinetic energy over 900 $cm^2 s^{-2}$ being predominantly located at the STC. It is also well known that the location of the PF all around Antarctica is almost coincident with the main rifts in the Atlantic, Indian, and west Pacific sectors.

We have previously discussed the nutrient distributions, which show dramatic changes when crossing the STC and PF. Relatively little is known about the particulate matter distribution across the frontal zones, but it is now apparent that in the strong temperature gradients, relatively high biomass of phytoplankton can be found at least some of the time (Lutjeharms *et al.*, 1975). The gyres associated with these frontal zones (Daniault, 1984) may exert a favorable influence on phytoplankton growth in at least two ways that are largely independent of vertical nutrient transport. First, the duration of the phytoplankton growth coincides quite well with that of the mesoscale hydrographic structure, resulting in relatively stable mesoscale containment of the developing phytoplankton biomass (i.e., much less lateral dissipation). Second, significant upwelling rates within the gyre are able to retard the sinking of the biogenic particles, thus tending to increase biomass in the surface layer (Heywood and Priddle, 1987).

High sediment accumulation rates of biogenic silica between the Polar Front and the northern limit of seasonal sea ice (DeMaster, 1981) tend to support the idea of high primary production in the overlying surface waters. DeMaster (1981) estimated the average net accumulation of biogenic silica for this region to be $0.2-0.3$ mol Si m^{-2} yr^{-1}, which is equivalent to a total accumulation of $2-3 \times 10^{14}$ g SiO_2 yr^{-1}. According to Nelson and Gordon (1982), approximately 30–35% of the silica production is recycled in the photic layer in the ACC region in spring. These authors estimated from mass balance that an additional 40–50% of the surface-produced silica dissolves between 100 m and the bottom. Assuming that these values approximate the yearly average, we can estimate the average amount of biogenic silica produced in the surface layer south of the PF to be between 0.7 and 2 mol-Si m^{-2} yr^{-1}. The C:Si mole ratio of diatoms grown in culture without nutrient limitation averages 7.7 (Brzezinski, 1985), but this ratio may be as low as 2 within Antarctic diatom blooms (Nelson and Smith, 1986). Thus, the silica production rates calculated here correspond to annual primary productivity estimates (for diatoms) of $17-185$ g C m^{-2} yr^{-1}. These values tend to be higher than most estimates of the total annual productivity of open-ocean areas of the ACC (e.g., Holm-Hansen, 1985; Smith and Nelson, 1986), which generally range from 15 to 30 g C m^{-2} yr^{-1}. Van Bennekom *et al.* (1988), using calculations similar to those just presented for areas south of the Polar Front, estimated higher silica production rates in the PF zone itself. They calculated that in the PF north of the Weddell Sea, the observed accumulation of about 1 mol SiO$_2$ m^{-2} yr^{-1} implies a net production of 2 mol SiO$_2$ m^{-2} yr^{-1} (corresponding to 48-185 g C m^{-2} yr^{-1}) in the PF surface layer.

c. Marginal ice zones About 1.6×10^7 km^2, or over one-third of the surface area of the Southern Ocean, undergoes a seasonal cycle of advancing and retreating pack ice (e.g., Zwally *et al.*, 1983). With the exception of brief periods of rapidly advancing ice in autumn, the ice edge zone is a region of net ice melt at all times of year. The heat source causing the ice to melt is solar and atmospheric heating from above in spring and summer, when the ice edge is retreating and/or at very southerly locations, and heat transfer from relatively warm (+ 1 to + 2 °C) surface waters beneath the ice in winter, when the ice edge is near its maximum northward extent (see Chapter 5). This ice melt produces a lens of diminished salinity in the ice edge zone, typically extending 50-200 km seaward into ice-free or nearly ice-free waters (e.g., Smith and Nelson, 1985; Nelson *et al.*, 1987).

It is quite common for phytoplankton blooms to develop within the marginal ice zones of both northern and southern oceans (see Chapter 9). In the Southern Ocean ice edge systems where physical and biological data sets have been obtained together, a reasonably consistent pattern has emerged: the seaward extent of the bloom appears to be controlled by that of the meltwater lens, and biomass levels decrease sharply when the upper water column is no longer stabilized by low salinity from the melting ice (e.g., Smith and Nelson, 1985; Wilson *et al.*, 1986; Nelson *et al.*, 1987). Nutrient concentrations diminish within the ice edge bloom, and this depletion is generally far too pronounced to have resulted from simple dilution by meltwater (e.g., Nelson and Smith, 1986; Nelson *et al.*, 1987). Most of the localized nutrient depletion thus results from uptake by the phytoplankton, and two stations have been occupied within an intense diatom bloom near the Ross Sea ice edge at which nitrate and/or phosphate was depleted to colorimetrically undetectable levels (Nelson and Smith, 1986) (Fig. 8.20). However, nutrient sufficiency is by far the most prevalent condition within these blooms, even though they are relative nutrient minima in comparison with the surrounding surface waters (Fig. 8.21). The main process responsible for the development of ice edge phytoplankton blooms appears to be the vertical stabilization of the upper 30-80 m by meltwater (see Chapter 9). Thus, the observations of ice edge phytoplankton blooms in the Southern Ocean ap-

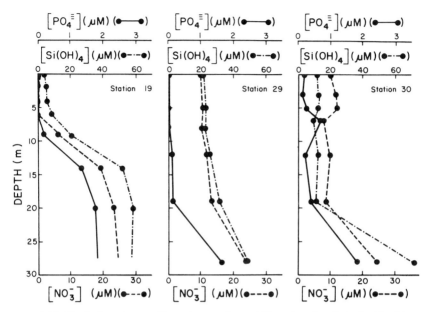

Figure 8.20 Vertical nutrient profiles at three stations within an ice edge phytoplankton bloom in the Ross Sea during January 1983.

pear to be consistent with the general conditions of nutrient sufficiency and light limitation for phytoplankton south of the Polar Front.

Martin and Fitzwater (1988) presented data indicating that phytoplankton growth in the subarctic Pacific may at times be limited by the availability of iron. They further suggested that, because of the absence of apparent sources, the Southern Ocean may typically be an iron-limited habitat for phytoplankton. Experiments conducted in the Weddell–Scotia Sea in the spring of 1989 support this hypothesis (de Baar et al., 1990). We are not aware of any studies to date that characterize the iron (or other trace metal) chemistry of the marginal ice zone or any of the other localized areas of high productivity in the Antarctic. However, no existing data appear to be inconsistent with the hypothesis of iron limitation in the Southern Ocean, and if surface waters affected by ice melt differ significantly from the general conditions of the region with respect to trace-metal/organic ligand relationships, these chemical properties of meltwater may also be significant in initiating and sustaining ice edge phytoplankton blooms.

d. Island effects Although there are few islands in the Southern Ocean, and thus they play a minor role in the biogeochemical cycles of the region as a whole, they represent a somewhat different kind of system. Specifically, the

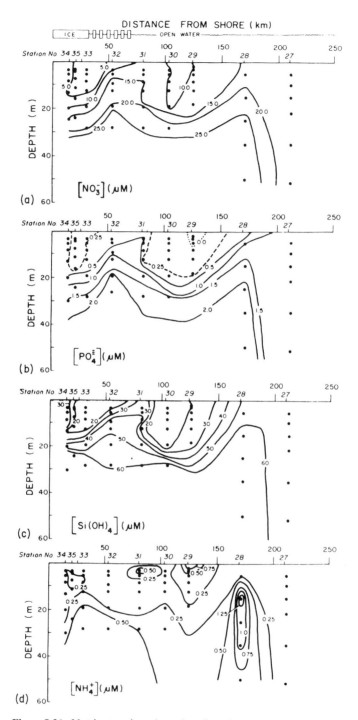

Figure 8.21 Nutrient sections through an ice edge phytoplankton bloom.

net flux of organic material between land and ocean appears to be in the opposite direction from that occurring in most other coastal regions. The soils in the temperate and northern polar zones are much richer in organic matter than those in the Antarctic. Thus, the marine food web of the Southern Ocean supports the terrestrial and coastal environments, either directly or indirectly (Myrcha *et al.*, 1985; Smith, 1985), while the reverse is true in most other coastal regions. Although the major nutrient concentrations of coastal waters in the Antarctic are similar to those of their oceanic surroundings (Treguer, 1987a), the coastal waters tend to be much more productive for both pelagic and benthic food chains. Le Jehan and Treguer (1983) measured chlorophyll *a* concentrations as high as 6 μg l^{-1} in the Kerguelen coastal waters during summer, 10 times as high as in the open ocean. These waters contained relatively high concentrations of ammonium (2 μM), urea (1 μM), and dissolved organic phosphorus (0.15 μM) (i.e., two to three times as high as in offshore waters; Fig. 8.22). High biomass concentration is thought to result mainly from a severe reduction of turbulence in those sheltered areas, leading to a stratified hydrographic structure and an increased residence time of phytoplankton cells in the photic layer. The role of trace metals and dissolved organic matter (DOM), which may be present at higher concentrations in nearshore environments, may also be significant (Treguer, 1987a). A more complete description of the variations of DOM, such as sugars and proteic material, in the coastal waters of King George Island was given by Dawson *et al.* (1985), who showed a direct relation between DOM (especially for sugars) and benthic macrophyte activity.

Coastal upwelling patterns were detected during the prevailing westerly winds by Le Jehan and Treguer (1983) and by Grindley and David (1985) for Kerguélan and Marion islands, respectively. Conversely, Allenson *et al.* (1985) found no evidence of upwelling in the lee of Prince Edwards, despite the occurrence of strong westerly winds. Heywood and Priddle (1987) suggest that islands represent significant topographic obstacles in the Antarctic Circumpolar Current and may play an indirect but important role in the generation of gyres, thereby creating favorable conditions for the accumulation of particulate matter in the photic layer.

Unlike the relatively small island areas in the Antarctic Circumpolar Current, the continental shelf areas adjacent to Antarctica are extensive. It is becoming apparent that they play a major role in regional, and even global, biogeochemical cycles (e.g., Ledford-Hoffman *et al.*, 1986). These nearshore processes in close proximity to the Antarctic continent are described in the next section, with special reference to the silicon cycle.

3. The biogeochemical cycling of silicon

Any discussion of biogeochemical cycles in the Southern Ocean must pay particular attention to the silica cycle, and any discussion of the global

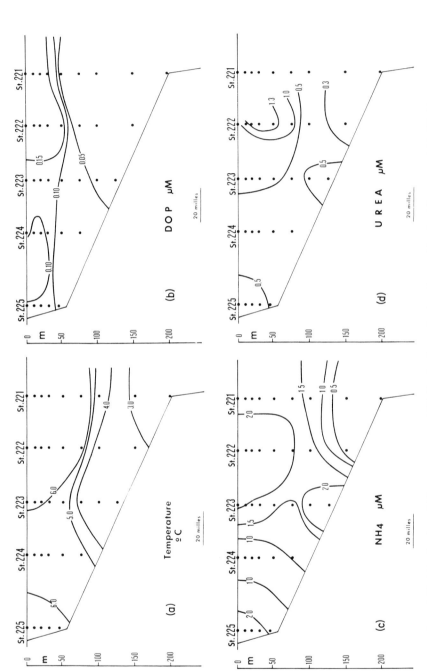

Figure 8.22 Distribution of (a) temperature and (b) dissolved organic phosphorus; After Lutjeharms *et al.* (1985); (c) ammonium, and (d) urea in coastal waters of the Kerguélen Islands.

oceanic silica cycle must pay particular attention to processes in the Southern Ocean. The reason is that the cycling of siliceous material in the Antarctic and subantarctic differs from that in the other major oceanic provinces in a way that causes the Southern Ocean to dominate the global oceanic silica budget. In this section we discuss these processes to the extent that they are presently understood. We use the Ross Sea as an example, not because we have any conclusive indication of how typical or atypical that system is with respect to the Southern Ocean as a whole, but because production and accumulation rates of both organic and siliceous particulate material have been measured there and very preliminary mass balances for siliceous and organic matter can be constructed.

a. Regional and global cycles of siliceous and organic matter Two global mass balances for silica, drawing on estimates of silica accumulation rates derived from radioisotope profiles in sediment cores, have estimated that over 75% of the accumulation of biogenic silica in marine sediments that is presently occurring worldwide is taking place south of the Polar Front (DeMaster, 1981; Ledford-Hoffman et al., 1986). The Southern Ocean is not nearly this important quantitatively in the global deposition pattern of organic matter, accounting for only about 5% of the total accumulation of organic carbon in marine sediments (Lisitzin, 1972; Holland, 1978). The absence of an unusually great surface source of biogenic material is also apparent in data on the total primary productivity in the region. Whereas the estimated annual primary productivity of the Southern Ocean has been estimated to be about 1×10^{15} g C yr^{-1} (Smith and Nelson, 1986), this still represents only about 5% of the global primary productivity of the ocean estimated by Ryther some 20 years ago (Ryther, 1969). It now appears likely that Ryther significantly underestimated the productivity of the major mid-ocean gyres (e.g., Jenkins and Goldman, 1985). Thus, the best present estimates are that the Southern Ocean is the site of $\leq 5\%$ of either production of organic matter in the upper ocean or accumulation of organic matter in marine sediments but is the site of about 75% of the global-scale removal of biogenic silica to the seabed.

This change in the relationship between the opal and organic matter cycles is also reflected in nutrient concentration profiles in the upper ocean: Zentara and Kamykowski (1981), examining all available nutrient profiles in a worldwide data base, found that only in the Antarctic and subantarctic are the vertical concentration gradients of silicate greater than those of nitrate in a way that causes a silicate versus nitrate regression line to extrapolate to zero silicate while large amounts of nitrate ($> 10\ \mu M$) remain. The implication of this relationship is that the Southern Ocean is the only major oceanic province in which the net rate of removal of silicate from the surface layer significantly exceeds that of nitrate. Thus both the rate of accumulation of siliceous sediments at the seafloor (DeMaster, 1981) and the dissolved nutrient distributions in the upper 1000 m (Zentara and Kamykowski, 1981) suggest that the vertical flux of opal out of the surface layer is unusually great in the Southern Ocean.

Production rates of biogenic silica in surface waters of the Southern Ocean (e.g., Nelson and Gordon, 1982) do not appear to be significantly higher than in the other marine environments in which they have been measured (e.g., Goering et al., 1973; Azam and Chisholm, 1976; Nelson and Goering, 1978; Paasche and Ostergren, 1980; Nelson et al., 1981; Banahan and Goering, 1986; Brzezinski, 1987). Therefore, it seems difficult to ascribe either the very high rates of siliceous sedimentation or the major regional anomaly in silicate–nitrate relationships in the upper ocean to unusually high rates of silica production by diatoms. However, both the sedimentation pattern and the nutrient profile anomalies are consistent with the sketchy picture now emerging from data on the net balance between production and dissolution of biogenic silica in near-surface waters. Previous studies in the subantarctic, in several coastal upwelling systems, and in Gulf Stream warm-core rings have shown that in some temperate and tropical surface waters the recycling of biogenic silica can be very intense. For example, rate measurements off the coast of northwest Africa in spring and in Gulf Stream warm-core rings in summer have indicated that 70–100% of the opal produced by siliceous phytoplankton was dissolving in the upper 60–100 m (Nelson and Goering, 1977; Brzezinski, 1987). The opposite end of the spectrum, as revealed in studies to date, is the Antarctic Circumpolar Current, where on average only 30–35% of the surface-produced opal was dissolving in the upper 100 m and estimates at individual stations were as low as 18% (Nelson and Gordon, 1982). Thus the major geographic differences in both accumulation of opal at the seafloor (DeMaster, 1981) and net removal of silicic acid from the upper 1000 m (Zentara and Kamykowski, 1981) may result from an unusually low rate of silica dissolution in the surface layer of the Southern Ocean.

The dissolution rate of biogenic silica in seawater is known to be highly temperature dependent from laboratory studies of both physically intact diatoms and acid-cleaned diatom frustules (e.g., Lewin, 1961; Hurd, 1972; Kamatani, 1982). If surface layer temperature is a major determinant of silica dissolution rates in the ocean, it would result in a significantly higher fraction of the surface-produced silica being transported to depth at polar and subpolar latitudes (and in temperate regions in winter) than in the tropics and subtropical gyres. Because the deep ocean is uniformly cold, these temperature-dependent differences in the intensity of near-surface recycling may be reflected in differences in opal flux throughout the water column and to the seabed.

Such a mechanism is consistent with the global and regional mass balance information summarized above and also with data from the subarctic Pacific (Honjo, 1984) showing the vertical flux of opal at a depth of 3800 m to be nearly as great as the flux to the seafloor estimated by DeMaster (1981) in the subantarctic (1.3 versus 1.7 mmol SiO_2 m^{-2} day^{-1}). These fluxes exceed the global mean delivery rate of silica to the seabed (e.g., Calvert, 1983) by more than a fact of ten. In addition to temperature, biological processes such as grazing and gravitational settling of fecal pellets almost certainly play a significant role in the transport and cycling of silica, and the relative importance of physical and biological mechanisms in regulating this cycle is not known, either regionally or globally. It is clear, however, that the silica cycle in the Antarctic and subantarctic differs from that in other parts of the ocean in a way that makes the Southern Ocean the main site of removal of silica from the ocean.

Two major differences between the Southern Ocean silica cycle and the rest of the oceanic silica cycle are logically possible: biogenic silica may be produced in higher proportion to organic carbon in the surface waters, and/or an unusually high fraction of that silica may be delivered to the seabed rather than dissolving within the water column. These explanations are not mutually exclusive, and both may be necessary to explain the anomalously great importance of the Southern Ocean in the global silica cycle. There is both indirect (Le Jehan and Treguer, 1983) and direct (Nelson and Smith, 1986; Nelson *et al.,* 1989) evidence that natural phytoplankton assemblages in the Southern Ocean may produce 2–5 times as much silica per unit carbon as they do elsewhere in the ocean. However, recent data from the Ross Sea (the only Southern Ocean location where all relevant rates have been estimated) point to the importance of differences in the regenerative portions of the silica and carbon cycles. Existing Ross Sea data allow crude mass balances for silica and carbon to be estimated, and these are discussed in the next section.

b. Preliminary mass balances for the Ross Sea Production rates of biogenic silica and organic carbon in the upper ocean and accumulation rates of both phases in the seabed have been measured in the western Ross Sea (e.g., El-Sayed *et al.,* 1983; Wilson *et al.,* 1986; Nelson and Smith, 1986; Ledford-Hoffman *et al.,* 1986). By expressing these data in the same rate units and making certain assumptions regarding the seasonal cycle of primary production, it is possible to construct very preliminary budgets for organic carbon and biogenic silica within the water-sediment column of the region. It should be borne in mind that most of the terms in these budgets are subject to considerable uncertainty and that the applicability of the conclusions to the Southern Ocean at large is unknown. However, if they are even approximately correct, they demonstrate both a major decoupling between the cycles of siliceous and organic matter in this particular portion of the Southern Ocean and the fact that this decoupling can result in very high silica accumulation rates in a region of modest primary productivity.

The annual cycle of primary production in the Ross Sea is strongly constrained by the seasonal cycle of solar irradiance. The region receives no direct sunlight from May through July, and the period of energetically significant subsurface irradiance is further diminished by extensive ice cover over the entire Ross Sea from early April through late September (Zwally et al., 1983). During most years an area of diminished ice cover begins to appear adjacent to the Ross Ice Shelf at approximately 175°E in October and develops into a large open-water polynya, still surrounded on all sides by significant ice cover, by December (Zwally et al., 1983). By January the western edge of this polynya has moved to within 100 km of the Victoria Land coast, and during late January and early February of 1983 (the one year during which these events were studied) there was a very intense diatom bloom about 200 km in east–west extent immediately offshore of the ice edge and spatially confined to a low-salinity meltwater lens approximately 25 m thick (Smith and Nelson, 1985). This bloom contained only moderately high chlorophyll concentrations (5–7 μg l^{-1}) but unusually high particulate organic carbon levels (40–50 μmol l^{-1}) and massive biogenic silica concentrations (30–40 μmol l^{-1}) (Smith and Nelson, 1985).

Primary productivity in summer at Ross Sea locations outside the ice edge bloom appears to average about 200 mg C m^{-2} day^{-1}, with very few individual observations exceeding 500 mg C m^{-2} day^{-1} (e.g., Holm-Hansen et al., 1977; El-Sayed et al., 1983). Within the 1983 diatom bloom in the western Ross Sea, the mean primary productivity was 960 mg C m^{-2} day^{-1} with a range of 310 to 1750 (Wilson et al., 1986). These rates, taken together with the presumption that productivity is close to zero from April through September because of light limitation, indicate that the ice edge diatom bloom represents a pulse of biogenic material that is highly significant in the annual primary productivity of the region (Table 8.4). It has been suggested that this is a common annual cycle in areas that experience a seasonal advance and retreat of pack ice. Jennings et al. (1984) estimated that events of this kind cause the annual primary productivity of the Weddell Sea to be about twice as great as indicated by productivity surveys that do not include their effects, and Smith and Nelson (1986) calculated that adequately accounting for the effect of ice edge blooms increases estimates of the overall annual primary productivity of the entire Southern Ocean by about 60%.

No data are available on rates of biogenic silica production in the Ross Sea at locations far removed from the ice edge. Within the ice edge diatom bloom, however, silica production rates averaged 38 mmol Si m^{-2} day^{-1}

Table 8.4 Estimated Annual Production of Organic Carbon and Biogenic Silica in the Southwestern Ross Sea[a]

	Carbon	Silica
Within bloom		
Mean daily production (mmol m^{-2} day^{-1})	80[b]	38[c]
Duration (days yr^{-1})	50[d]	50[d]
Annual bloom production (mol m^{-2} yr^{-1})	4.0	1.9
Outside bloom		
Mean daily production (mmol m^{-2} day^{-1})	17[e]	3[f]
Duration (days yr^{-1})	120	120
Annual nonbloom production (mol m^{-2} yr^{-1})	2.0	0.4
Total annual production (bloom + nonbloom; mol m^{-2} yr^{-1})	6.0	2.3
Percent of annual production attributable to bloom	67	83

[a] These calculations assume that all production occurs during a 170-day period in summer and that an ice edge phytoplankton bloom traverses the area, lasting 50 days at any one location.
[b] From Wilson et al. (1986).
[c] From Nelson and Smith (1986).
[d] From Smith and Nelson (1986).
[e] From El-Sayed et al. (1983).
[f] From Nelson and Gordon (1982); data are from 1000–1500 km to the north but are the only data available for Southern Ocean locations remote from the ice edge.

with a range of 7 to 93 (Nelson and Smith, 1986). These production rates are an order of magnitude higher than those measured in the Antarctic Circumpolar Current remote from ice edge effects (Nelson and Gordon, 1982) and comparable to those measured in highly productive diatom blooms associated with coastal upwelling (data from various sources, summarized by Nelson and Smith, 1986). We have thus calculated that the ice edge bloom represents a significant, and perhaps the dominant, source term in the annual silica budget of the western Ross Sea (Table 8.4).

The biogenic silica content of Ross Sea sediments is very low on banks and ridges and much greater in topographic depressions, due to bottom resuspension and redeposition (Tolstikov, 1966; Kellogg and Truesdale, 1979; J. B. Anderson et al., 1984). Within the basins, both the biogenic silica content and the accumulation rate of silica in the sediments increase from east to west (Ledford-Hoffman et al., 1986); accumulation rates range from a minimum of about 1 mg SiO_2 cm^{-2} yr^{-1} in Sulzberger Bay to a maximum of about 36 mg SiO_2 cm^{-2} yr^{-1} southwest of the Pennell Bank in the western Ross Sea. Expressed in the same units as silica production rates, these accumulation rates represent annual average daily rates ranging from 0.46 to 16 mmol Si m^{-2} day^{-1}. The average for all areas deeper than 500 m in the western Ross Sea has been estimated by Ledford-Hoffman et al. (1986) to be 28 mg SiO_2 cm^{-2} yr^{-1} (= 12.8 mmol Si m^{-2} day^{-1}).

These are very high rates. The global mean rate of siliceous sedimentation calculated from data used by Calvert (1983) in his global silica budget is 0.06 mmol Si m^{-2} day^{-1}, about two orders of magnitude less than the rate estimated above for the Ross Sea. Even in deep-water areas of the Antarctic Circumpolar Current, which apparently account for a large fraction of the total oceanic siliceous sedimentation, the highest accumulation rates averaged over any large area (the subantarctic Atlantic) were equivalent to 1.7 mmol Si m^{-2} day^{-1} (DeMaster, 1981), or about one-eighth of the mean rate for basins in the Ross Sea. The total amount of siliceous sedimentation on the entire Ross Sea shelf has now been estimated as 2.1×10^{13} g SiO_2 yr^{-1} (Ledford-Hoffman et al., 1986), which is not much less than the total rate of silica accumulation on all other continental shelves of the world combined and implies that continental shelves in the Antarctic may account for about 25% of the removal of silica from the world ocean.

Accumulation rates this high appear to require very high rates of opal production and flux to support them. A simple comparison of silica production and accumulation rates in the western Ross Sea (summarized in Tables 8.4 and 8.5) demonstrates this point. Ledford-Hoffman et al. (1986) estimated that the mean accumulation rate of biogenic silica in deep basins of the western Ross Sea was approximately 28 mg SiO_2 cm^{-2} yr^{-1}. Because this area of the Ross Sea has extensive shallow banks containing very little siliceous sediment (e.g., Kellogg and Truesdale, 1979), 50% of this number (14 mg SiO_2 cm^{-2} yr^{-1}) approximates an areawide average. This rate is equivalent to 2.3 mol Si m^{-2} yr^{-1}. Thus, at the mean rate of silica production we measured *within* an intense diatom bloom in the marginal ice zone (38 mmol Si m^{-2} day^{-1}) (Nelson and Smith, 1986), it would take 60 (i.e., 2.3/0.038) days of silica production to support the estimated annual sedimentation rate even if no silica dissolved within the water column. Our best present estimate of the duration of the bloom over any one location on the seabed is about 50 days (Smith and Nelson, 1986). Thus the mean annual accumulation rate of biogenic silica in the sediments of the region appears to be approximately equal to the annual production of silica in the marginal ice

Table 8.5 Estimated Annual Fluxes of Organic Carbon and Biogenic Silica in the Southwestern Ross Sea[a]

	Carbon	Silica
Surface-layer production (mol m^{-2} yr^{-1})	6.0[a]	2.3[a]
Accumulation in sediment (mol m^{-2} yr^{-1})		
Area-weighted mean	0.65[b]	2.3[c,d]
Range	0–2.0[b]	0–7.0[c]
Accumulation as percent of production		
Area-weighted mean	11	100[e]
Range	4–33	7–300

[a] From Table 8.1.
[b] Based on weight percent of carbon measurements (Dunbar et al., 1989a, plus unpublished data) and sediment accumulation rates (Ledford-Hoffman et al., 1986).
[c] From Ledford-Hoffman et al., (1986).
[d] Calculated assuming that deep basins (mean silica accumulation rate = 4.6 mol SiO_2 m^{-2} yr^{-1}) and banks (little or no accumulation of biogenic silica) each constitute 50% of the seafloor of the western Ross Sea (Tolstikov, 1966).
[e] The fact that the estimated rates of silica production and accumulation are exactly the same is fortuitous. The rates were measured on different time scales and in different areas. These values are present because (1) they are the best we can do with the data we now have and (2) they strongly suggest that at least a high fraction of the surface-produced silica is delivered to the seabed.

zone. Clearly if the various individual estimates are correct, at least one of two things must be true:

1. A very high fraction of the biogenic silica produced within the ice edge diatom bloom reaches the seabed, or
2. There are reasonably high rates of silica production in summer during times when no ice edge diatom bloom is present.

The annual production and accumulation estimates shown in Tables 8.4 and 8.5 should be viewed as highly simplified calculations to compare general magnitudes of processes. The annual production cycles of both carbon and silica appear to be dominated by diatom productivity in the vicinity of the receding ice edge, and the data on these rates were obtained in a single summer cruise. Moreover, although some data on primary productivity at

Ross Sea locations remote from the ice edge are available (e.g., El-Sayed et al., 1983), no parallel data on silica production exist; our estimate of silica production outside the ice edge zone is based on data from the Antarctic Circumpolar Current more than 1000 km to the north (Nelson and Gordon, 1982). Similarly, the accumulation-rate information derived from ^{210}Pb profiles (Ledford-Hoffman et al., 1986) represents maximum estimates because the amount of bioturbation affecting the vertical distribution of ^{210}Pb is unknown. Bioturbation appears to be minor on the basis of very sketchy data on plutonium profiles in the sediment (Leford-Hoffman et al., 1986). However, this cannot be confirmed without obtaining both ^{210}Pb and ^{14}C profiles from the same cores, and data of this kind are not yet available for the Ross Sea.

The species composition of diatoms in both the phytoplankton and the sediment appears to be consistent with the hypothesis that an ice edge diatom bloom of the kind observed in 1983 is a recurrent annual event that dominates the flux of opal to the sediments of the western Ross Sea. The 1983 diatom bloom was composed of a fairly diverse phytoplankton assemblage, in which Carbonell (1985) identified 50 diatom and 48 dinoflagellate species. It was, however, dominated numerically (> 80%) by a single species, the small pennate diatom *Nitzschia curta* (Carbonell, 1985; Wilson et al., 1986). If *N. curta* were typically the dominant diatom in ice edge phytoplankton blooms in the western Ross Sea and if these blooms were quantitatively important in the annual flux of silica to the sediment, one would expect a high proportion of *N. curta* in the sedimentary diatom assemblage. Biostratigraphic studies of Ross Sea sediments (Kellogg and Truesdale, 1979; Truesdale and Kellogg, 1979) show *N. curta* to be the dominant diatom in the sediments of the western Ross Sea at present and during most of the past 18,000 years. Moreover, the observed westward increase in the accumulation rate of siliceous sediments in the Ross Sea (Ledford-Hoffman et al., 1986) is paralleled by a westward increase in the relative abundance of *N. curta* in the surface sediment (Truesdale and Kellogg, 1979).

The hypothetical annual cycle of primary productivity, silica production, and silica flux that emerges from the observations detailed above is this: as the Ross Sea ice edge retreats westward seasonally, a diatom bloom dominated by *N. curta* follows it, intensifying with time until in its fully developed midsummer condition it represents a major seasonal pulse in the primary production of both organic and siliceous material. The organic matter is recycled to a large degree within the water column, but a high fraction of the biogenic silica is transported to the seabed. These processes cause both the accumulation rate and the *N. curta* content of siliceous sediments to increase from east to west. They also result in silica accumulation rates in the western Ross Sea that are too high to be supported by the surface productivity unless

(1) the ice-edge diatom bloom is a regular annual event and the major quantitative contributor of silica to the sediment, and (2) a high fraction of the biogenic silica produced near the receding ice edge is transported to the sediment without dissolving in the intervening water column.

References

Aagaard, K. & L. K. Coachman. 1975. Toward an ice-free ocean. *Eos* **56**: 484–486.
Aagaard, K., L. K. Coachman & E. C. Carmack. 1981. On the halocline of the Arctic Ocean. *Deep-Sea Res.* **28**: 529–545.
Aagaard, K. & P. Greisman. 1975. Toward new mass and heat budgets for the Arctic Ocean. *JGR, J. Geophys. Res.* **80**: 3821–3827.
Aagaard, K., J. H. Swift & E. C. Carmack. 1985. Thermohaline circulation in the Arctic Mediterranean seas. *J. Geophys. Res.* **90**: 4833–4846.
Allenson, B. R., B. Boden, L. Parker & C. Duncombe Rae. 1985. A contribution to the oceanology of the Prince Edward Islands. *In* "Antarctic Nutrient Cycles and Food Webs" (W. R. Siegfried, P. R. Condy & R. M. Laws, eds.), pp. 38–45. Springer-Verlag, Berlin.
Anderson, J. B., C. Brake & N. Meyers. 1984. Sedimentation on the Ross Sea continental shelf. *Mar. Geol.* **57**: 295–333.
Anderson, L. G. & D. Dyrssen. 1980. "Constituent Data for Leg 2 of the Ymer 80 Expedition," Rep. Chem. Seawater 24. Dept. Anal. Mar. Chem., Univ. Göteborg, Göteborg, Sweden.
———. 1981. Chemical constituents of the Arctic Ocean in the Svalbard area. *Oceanol. Acta* **4**: 305–311.
Anderson, L. G. & E. P. Jones. 1985. Measurements of total alkalinity, calcium, and sulphate in natural sea ice. *J. Geophys. Res.* **90**: 9194–9198.
———. 1986. Water masses and their chemical constituents in the western Nansen basin of the Arctic Ocean. *Oceanol. Acta* **9**: 277–283.
———. 1990. On the origin of the chemical and physical properties of the Arctic Ocean halocline north of Ellesmere Island. *Cont. Shelf Res.* (in press).
Anderson, L. G., D. Dyrssen, E. P. Jones & M. Lowings. 1983. Inputs and outputs of salt, fresh water, alkalinity, and silica in the Arctic Ocean. *Deep-Sea Res.* **30**: 87–94.
Anderson, L. G., E. P. Jones, R. Lindegren, B. Rudels & P.-I. Sehlstedt. 1988a. Nutrient regeneration in cold, high salinity water of the Arctic shelves. *Cont. Shelf Res.* **8**: 1345–1355.
Anderson, L. G., E. P. Jones, R. Lindegren, B. Rudels & P.-I. Sehlstedt. 1988b. On the chemistry of the cold, high salinity bottom waters of the Arctic Ocean shelves. *Cont. Shelf Res.* **8**: 1345–1355.
Anderson, L. G., E. P. Jones, K. P. Koltermann, P. Schlosser, J. H. Swift & W. R. Wallace. 1989. The first oceanographic section across the Nansen basin in the Arctic Ocean. *Deep-Sea Res.* **36**: 475–482.
Anderson, L. G., D. Dyrssen & E. P. Jones. 1990. An assessment of the transport of atmospheric CO_2 into the Arctic Ocean. *J. Geophys. Res.* (in press).
Armstrong, F. A. J. 1965. Silicon. *In* "Chemical Oceanography, Volume I" (J. P. Riley and G. Skirrow, eds.), pp. 409–432. Academic Press, New York.
Azam, F. & S. W. Chisholm. 1976. Silicic acid uptake and incorporation by natural marine phytoplankton populations. *Limnol. Oceanogr.* **21**: 427–433.
Bainbridge, A. E. 1980. *GEOSECS Atlantic Expedition Volume 2, Sections and Profiles*, U.S. Government Printing Office, Washington, D.C., 148 pp.

Banahan, S. & J. J. Goering. 1986. The production of biogenic silica and its accumulation on the southeastern Bering Sea shelf. *Cont. Shelf Res.* **5**: 199–213.

Biggs, D. C. 1982. Zooplankton excretion and NH_4 cycling in near surface waters of the Southern Ocean. 1. Ross Sea, austral summer 1977–1978. *Polar Biol.* **1**: 55–67.

Bordin, G., P. Appriou & P. Treguer. 1987. Distributions horizontales et verticales du cuivre, du manganèse et du cadmium dans le secteur indïen de l'Océan Antarctique. *Oceanol. Acta* **10**: 411–420.

Boyle, E. A. & J. M. Edmond. 1975. Copper in surface waters south of New Zealand. *Nature (London)* **253**: 107–109.

Broecker, W. S. 1974. "NO", a conservative-mass tracer. *Earth Planet. Sci. Lett.* **23**: 100–107.

Brzezinski, M. A. 1985. The Si:C:N ratio of marine diatoms: Interspecific variability and the effect of some environmental variables. *J. Phycol.* **21**: 347–357.

———. 1987. Physiological and environmental factors affecting diatom species competition in a Gulf Stream warm-core ring. Ph.D. Dissertation, Oregon State Univ., Corvallis.

Callahan, J. E. 1972. The structure and circulation of deep water in the Antarctic. *Deep-Sea Res.* **19**: 563–575.

Calvert, S. E. 1983. Sedimentary geochemistry of silicon. *In* "Silicon Geochemistry and Biogeochemistry" (S. R. Aston, ed.), pp. 143–186. Academic Press, London.

Carbonell, M. C. 1985. Phytoplankton of an ice-edge bloom in the Ross Sea, with special reference to the elemental composition of Antarctic diatoms. M.S. Thesis, Oregon State Univ., Corvallis.

Carmack, E. C. 1977. Water characteristics in the Southern Ocean south of the Polar Front. *Deep-Sea Res.* **24**(Suppl.): 15–41.

Chen, C. T. 1982. On the distribution of the anthropogenic CO_2 in the Atlantic and Southern Ocean. *Deep-Sea Res.* **29**: 563–580.

———. 1984. Carbonate chemistry of the Weddell Sea. U.S. Dep. Energy Rep. DOE/EV/10611–10614.

Chen, C. T. & E. Drake. 1986. Carbon dioxide increase in the atmosphere and oceans and possible effects on climate. *Annu. Rev. Earth Planet. Sci.* **14**: 201–235.

Chen, C. T. & A. Poisson. 1984. Excess carbon dioxide in the Weddell Sea. *Antarct. J. 1984 Rev.*, pp. 74–75.

Coachman, L. K. & C. A. Barnes. 1962. Surface water in the Eurasian basin of the Arctic Ocean. *Arctic* **15**: 251–277.

Codispoti, L. A. & F. A. Richards. 1968. Micronutrient distributions in the East Siberian and Laptev seas during summer 1963. *Arctic* **21**: 67–83.

———. 1971. Oxygen supersaturation in the Chukchi and East Siberian seas. *Deep-Sea Res.* **18**: 341–351.

Collos, Y. 1982. Regimes transitoires dans l'assimilation de l'azote par le phytoplancton marin. Theses d'Etat es Sciences. pp. 123. Universite d'Aix-Marseille.

Copin-Montegut, C. and G. Copin-Montegut. 1978. The chemistry of particulate matter from the South Indian and Antarctic oceans. *Deep-Sea Res.* **25**: 911–931.

Craig, H. R. 1983. *GEOSECS Pacific Expedition Vol. 2, Sections and Profiles,* U.S. Government Printing Office, Washington, D.C.

Daniault, N. 1984. Apport des connaissances spatiales à la connaissance des courants de surface. Applications à l'Océan Antarctique. Ph.D. Dissertation, Univ. Bretagne Occidentale, Brest.

Danielsson, L.-G. & S. Westerlund. 1983. Trace metals in the Arctic Ocean. *In* "Trace Metals in Sea Water" (C. S. Wong, E. Boyle, K. W. Bruland, J. D. Burton & E. D. Goldberg, eds.), pp. 85–96. Plenum, New York.

Dawson, R., W. Schramm & M. Bolter. 1985. Factors influencing the production, decomposition and distribution of organic and inorganic matter in Admiralty Bay, King George

Island. *In* "Antarctic Nutrient Cycles and Food Webs" (W. R. Siegfried, P. R. Condy & R. M. Laws, eds.), pp. 109–114. Springer-Verlag, Berlin.

Deacon, G. E. R. 1933. A general account of the hydrology of the South Atlantic Ocean. *'Discovery' Rep.* **7**: 171–238.

———. 1937. The hydrology of the Southern Ocean. *'Discovery' Rep.* **15**: 1–124.

———. 1982. Oxygen in Antarctic water. *Deep-Sea Res.* **11**: 1369–1371.

de Barr, H. J. W., A. G. Buma, R. F. Nolting, G. C. Cadee, G. Jacques & P. Treguer. 1990. On iron limitation of the Southern Ocean: experimental observations in the Weddel and Scotia Seas. *Mar. Ecol.: Prog. Ser.* (in press).

DeMaster, D. J. 1981. The supply and accumulation of silica in the marine environment. *Geochim. Cosmochim. Acta* **45**: 1715–1732.

Dunbar, R. B., A. J. Macpherson & G. Wefer. 1985. Water column particulate flux and seafloor deposits in the Bransfield Strait and southern Ross Sea, Antarctica. *Antarct. J. U.S.* **19**: 98–100.

Dunbar, R. B., A. R. Leventer & W. L. Stockton. 1989. Biogenic sedimentation in McMurdo Sound, Antarctica. *Mar. Geol.* **85**: 155–179.

Dyrssen, D. 1977. The chemistry of plankton production and decomposition in seawater. *In* "Ocean Sound Scattering Prediction" (N. R. Anderson & B. J. Zahuranc, eds.), pp. 65–84. Plenum, New York.

Edmond, J. M., S. S. Jacobs, A. L. Gordon, A. W. Mantyla & R. F. Weiss. 1979. Water column anomalies in dissolved silica over opaline pelagic sediments and the origin of the deep silica maximum. *J. Geophys. Res.* **84**: 7809–7826.

El-Sayed, S. Z., D. C. Biggs & O. Holm-Hansen. 1983. Phytoplankton standing crop, primary productivity and near-surface nitrogenous nutrient fields in the Ross Sea, Antarctica. *Deep-Sea Res.* **30**: 871–886.

Foster, T. D. & E. C. Carmack. 1976. Frontal zone mixing and Antarctic Bottom Water formation in the Weddell Sea. *Deep-Sea Res.* **23**: 301–307.

Glibert, P. M. 1982. Regional studies of daily, seasonal, and size fractions variability in ammonium remineralization. *Mar. Biol.* **70**: 209–222.

Glibert, P. M., D. C. Biggs & J. J. McCarthy. 1982. Utilization of ammonium and nitrate during austral summer in the Scotia Sea. *Deep-Sea Res.* **29**: 837–850.

Goering, J. J., D. M. Nelson & J. A. Carter. 1973. Silicic acid uptake by natural population of marine phytoplankton. *Deep-Sea Res.* **20**: 777–789.

Gordon, A. L. & E. M. Molinelli. 1982. "Southern Ocean Atlas: Thermohaline Chemical Distributions and the Atlas Data Set." Columbia Univ. Press, New York.

Gordon, A. L. & P. Tchernia. 1972. Waters of the continental margin off Adélie Coast, Antarctica. *In* "Antarctic Oceanology. II. The Australian–New Zealand Sector" (D. E. Hayes, ed.), Antarct. Res. 19, pp. 59–69. Am. Geophys. Union, Washington, D.C.

Gordon, A. L., H. W. Taylor & D. T. Georgi. 1977a. Antarctic oceanography. *In* "Polar Oceans" (M. Dunbar, ed.), pp. 45–76. Arctic Inst. North Am., Calgary, Alberta, Canada.

———. 1977b. Antarctic Polar Front in the western Scotia Sea—summer 1975. *J. Phys. Oceanogr.* **7**: 309–328.

Gordon, A. L., C. T. A. Chen & W. G. Metcalf. 1984. Winter mixed layer entrainment of Weddell Deep Water. *J. Geophys. Res.* **89**: 637–640.

Gorshkov, S. G. 1983. "World Ocean Atlas," Vol. 3. Pergamon Press, Oxford. 190 pp.

Grindley, J. R. & P. David. 1985. Nutrient upwelling and its effects in the lee of Marion Island. *In* "Antarctic Nutrient Cycles and Food Webs" (W. R. Siegfried, P. R. Condy & R. M. Laws, eds.), pp. 46–51. Springer-Verlag, Berlin.

Heywood, R. B. & J. Priddle. 1987. Retention of phytoplankton by an eddy. *Cont. Shelf Res.* **7**: 937–955.

Holland, H. D. 1978. "The Chemistry of the Atmosphere and Oceans." Wiley, New York.

Holm-Hansen, O. 1985. Nutrient cycles in Antarctic marine ecosystems. *In* "Antarctic Nutrient Cycles and Food Webs" (W. R. Siegfried, P. R. Condy & R. M. Laws, eds.), pp. 61–10. Springer-Verlag, Berlin.

Holm-Hansen, O., S. Z. El-Sayed, G. S. Franceschini & R. L. Cuhel. 1977. Primary production and factors controlling phytoplankton growth in the Southern Ocean. *In* "Adaptations Within Antarctic Ecosystems" (G. Llano, ed.), pp. 11–50. Gulf Publ. Co., Houston, Texas.

Honjo, S. 1984. Study of ocean fluxes in time and space by bottom-tethered sediment trap arrays: A recommendation. *In* "Global Ocean Flux Study Workshop" pp. 306–324. National Research Council, Washington, D.C.

Hurd, D. C. 1972. Factors affecting solution rate of biogenic opal in seawater. *Earth Planet. Sci. Lett.* **15**: 411–417.

Jacobs, S. S. & D. T. Georgi. 1977. Observations on the southwest Indian/Antarctic Ocean. *Deep-Sea Res.* **24**(Suppl.): 43–84.

Jacobs, S. S., F. A. Amos & P. M. Bruchhausen. 1970. Ross Sea oceanography and Antarctic Bottom Water formation. *Deep-Sea Res.* **17**: 935–962.

Jacques, G. 1983. Some ecophysiological aspects of the Antarctic phytoplankton. *Polar Biol.* **2**: 27–33.

Jacques, G. & P. Treguer. 1986. "Les Ecosystems Pelagiques Marins," Collect. Ecol. No. 19. Masson, Paris.

Jacques, G., M. Fiala & L. Oriol. 1984. Demonstration à partir de tests biologiques de l'effet negligeable des éléments traces sur la croissance du phytoplancton Antarctique. *C.R. Seances Acad. Sci.* **298**: 527–530.

Jenkins, W. J. & J. C. Goldman. 1985. Seasonal oxygen cycling and primary production in the Sargasso Sea. *J. Mar. Res.* **43**: 465–491.

Jennings, J. C., Jr., L. I. Gordon & D. M. Nelson. 1984. Nutrient depletion indicates high primary productivity in the Weddell Sea. *Nature (London)* **308**: 51–54.

Jones, E. P. & L. G. Anderson. 1986. On the origin of the chemical properties of the Arctic Ocean halocline. *J. Geophys. Res.* **91**: 10759–10767.

Jones, E. P., A. R. Coote & E. M. Levy. 1983. Effect of sea ice meltwater on the alkalinity of seawater. *J. Mar. Res.* **41**: 43–52.

Jones, E. P., D. Dyrssen & A. R. Coote. 1984. Nutrient regeneration in deep Baffin Bay with consequences for measurements of the conservative tracer NO and fossil fuel CO_2 in the oceans. *Can. J. Fish. Aquat. Sci.* **41**: 30–35.

Kamatani, A. 1982. Dissolution rates of silica from diatoms decomposing at various temperatures. *Mar. Biol. (Berlin)* **53**: 29–35.

Kellogg, T. B. & R. S. Truesdale. 1979. Late quaternary paleoecology and paleoclimatology of the Ross Sea: The diatom record. *Mar. Micropaleontol.* **4**: 137–158.

Kinney, P., M. E. Arhelger & D. C. Burrell. 1970. Chemical characteristics of water masses in the Amerasian basin of the Arctic Ocean. *J. Geophys. Res.* **75**: 4097–4104.

Krysell, M. & D. W. R. Wallace. 1988. Arctic Ocean ventilation studied using carbon tetrachloride and other anthropogenic halocarbon tracers. *Science* **242**: 746–749.

Le Corre, P. & H. J. Minas. 1983. Distribution et évolution des éléments nutritifs dans le secteur indïen de l'Océan Antarctique en fin de période estivale. *Oceanol. Acta* **6**: 365–381.

Ledford-Hoffman, P. A., D. J. DeMaster & C. A. Nittrouer. 1986. Biogenic silica accumulation in the Ross Sea and the importance of Antarctic shelf deposits in the marine silica budget. *Geochim. Cosmochim. Acta* **50**: 2099–2110.

Le Jehan, S. & P. Treguer. 1983. Uptake and regeneration Si:N:P ratios in the Indian sector of the Southern Ocean. *Polar Biol.* **2**: 127–136.

Le Jehan, S. & P. Treguer. 1985. The distribution of inorganic nitrogen, phosphorus, silicon and dissolved organic matter in surface and deep waters of the Southern Ocean. *In* "Antarctic

Nutrient cycles and food webs" (W. R. Siegfried, P. R. Condy, and R. M. Laws, eds.), pp. 22–29. Springer-Verlag Press, Berlin.

Levitus, S. 1982. "Climatologic Atlas of the World Ocean," NOAA Prof. Pap. 13. Natl. Oceanic Atmos. Admin., Washington, D.C.

Lewin, J. C. 1961. The dissolution rate of silica from diatom walls. *Geochim. Cosmochim. Acta* **21**: 182–195.

Lisitzin, A. P. 1972. Sedimentation in the World Ocean. *Spec. Publ.—Soc. Econ. Paleontol. Mineral.* **17**: 1–218.

Livingston, H. D. 1988. The use of Cs and Sr isotopes as tracers in the Arctic Mediterranean seas. *Philos. Trans. R. Soc. London, Ser. A* **325**: 161–176.

Livingston, H. D., S. L. Kupferman, V. T. Bowen & R. M. Moore. 1984. Vertical profile of artificial radionuclide concentrations in the central Arctic Ocean. *Geochim. Cosmochim. Acta* **48**: 2195–2203.

Livingston, H. D., J. H. Swift & H. G. Ostlund. 1985. Artificial radionuclide tracer supply to the Denmark Strait overflow between 1972 and 1981. *J. Geophys. Res.* **90**: 6971–6982.

Lutjeharms, J. R. E., N. M. Walters & B. R. Allanson. 1985. Oceanic frontal systems and biological enhancement. *In* "Antarctic Nutrient Cycles and Food Webs" (W. R. Siegfried, P. R. Condy & R. M. Laws, eds.), pp. 11–21. Springer-Verlag, Berlin.

Lyakhin, Y. I. 1970. Saturation of water of the Sea of Okhotsk with calcium carbonate. *Oceanology (Engl. Transl.)* **10**: 789–795.

Macdonald, R. W., C. S. Wong & P. E. Erickson. 1987. The distribution of nutrients in the southeastern Beaufort Sea: Implications for water circulation and primary production. *J. Geophys. Res.* **92**: 2939–2952.

Mantyla, A. W. & J. Reid. 1983. Abyssal characteristics of the world ocean waters. *Deep-Sea Res.* **30**: 805–833.

Mart, L., N. W. Nurnberg & D. Dyrssen. 1984. Trace metal levels in the eastern Arctic Ocean. *Sci. Total Environ.* **39**: 1–14.

Martin, J. H. & S. E. Fitzwater. 1988. Iron deficiency limits phytoplankton growth in the north-east Pacific subarctic. *Nature (London)* **331**: 341–343.

McCartney, J. J. 1977. Subantarctic mode water. *Deep-Sea Res.* **34**(Suppl.): 103–119.

Melling, H. & E. L. Lewis. 1982. Shelf drainage flows in the Beaufort Sea and their effect on the Arctic Ocean pycnocline. *Deep-Sea Res.* **20**: 967–985.

Mel'nikov, I. A. & G. L. Pavlov. 1978. Characteristics of organic carbon distribution in the waters and ice of the Arctic basin. *Oceanology (Engl. Transl.)* **18**: 163–167.

Michel, R. L. 1978. Tritium distributions in Weddell Sea water masses. *J. Geophys. Res.* **83**: 6192–6198.

———. 1984. Oceanographic structure of the eastern Scotia Sea. II. Chemical oceanography. *Deep-Sea Res.* **31**: 1157–1168.

Midttun, L. 1985. Formation of dense bottom water in the Barents Sea. *Deep-Sea Res.* **32**: 1233–1241.

Molinelli, E. J. 1981. The Antarctic influence on Antarctic Intermediate Water. *J. Mar. Res.* **39**: 139–145.

Moore, R. M. 1981. Oceanographic distributions of zinc, cadmium, copper and aluminum in waters of the central Arctic Ocean. *Geochim. Cosmochim. Acta* **45**: 2475–2482.

Moore, R. M. & N. J. Smith. 1986. Disequilibria between ^{226}Ra, ^{210}Pb and ^{210}Po in the Arctic Ocean and the implications for chemical modification of the Pacific water flow. *Earth Planet. Sci. Lett.* **77**: 285–292.

Moore, R. M., M. G. Lowings & F. C. Tan. 1983. Geochemical profiles in the central Arctic Ocean: Their relation to freezing and shallow circulation. *J. Geophys. Res.* **88**: 2667–2674.

Myrcha, A., S. J. Pietr & A. Tratur. 1985. The role of pygoscelid penguin rookeries in nutrient cycles at Admiralty Bay, King George island. *In* "Antarctic Nutrient Cycles and Food Webs" (W. R. Siegfried, P. R. Condy & R. M. Laws, eds.), pp. 156–162. Springer-Verlag, Berlin.

Nelson, D. M. & J. J. Goering. 1977. Near-surface silica dissolution in the upwelling region off northwest Africa. *Deep-Sea Res.* **24**: 65–73.

———. 1978. Assimilation of silicic acid by phytoplankton in the Baja California and northwest Africa upwelling systems. *Limnol. Oceanogr.* **23**: 508–517.

Nelson, D. M. & L. I. Gordon. 1982. Production and pelagic dissolution of biogenic silica in the Southern Ocean. *Geochim. Cosmochim. Acta* **46**: 491–501.

Nelson, D. M. & W. O. Smith, Jr. 1986. Phytoplankton bloom dynamics of the Ross Sea ice edge. II. Mesoscale cycling of nitrogen and silicon. *Deep-Sea Res.* **33**: 1389–1412.

Nelson, D. M., J. J. Goering & D. W. Boisseau. 1981. Consumption and regeneration of silicic acid in three coastal upwelling systems. *In* "Coastal Upwelling" (F. A. Richards, ed.), pp. 242–256. Am. Geophys. Union, Washington, D.C.

Nelson, D. M., W. O. Smith, Jr., L. I. Gordon & B. A. Huber. 1987. Spring distributions of density, nutrients and phytoplankton biomass in the ice edge zone of the Weddell–Scotia Sea. *J. Geophys. Res.* **92**: 7181–7190.

Nelson, D. M., W. O. Smith, Jr., R. D. Muench, L. I. Gordon, C. W. Sullivan & D. M. Husby. 1989. Particulate matter and nutrient distributions in the ice-edge zone of the Weddell Sea: Relationship to hydrography during late summer. *Deep-Sea Res.* **36**: 191–209.

Nikiforev, Ye.G., Y. V. Belysheva & N. I. Blinov. 1966. The structure of water masses in the eastern part of the Arctic basin. *Oceanology (Engl. Transl.)* **6**: 59–64.

Olson, R. I. 1980. Nitrate and ammonium uptake in Antarctic waters. *Limnol. Oceanogr.* **25**: 1064–1074.

Orren, M. J. & P. M. S. Monteiro. 1985. Trace elements geochemistry in the Southern Ocean. *In* "Antarctic Nutrient Cycles and Food Webs" (W. R. Siegfried, P. R. Condy & R. M. Laws, eds.), pp. 30–37. Springer-Verlag, Berlin.

Ostlund, H. G. 1982. The residence time of the freshwater component in the Arctic Ocean. *J. Geophys. Res.* **87**: 2035–2043.

Ostlund, H. G. & G. Hut. 1984. Arctic Ocean water mass balance from isotope data. *J. Geophys. Res.* **89**: 6373–6381.

Ostlund, H. G., R. Oleson & R. Brescher. 1980. "GEOSECS Indian Ocean Radiocarbon and Tritium Results." Univ. of Miami, Coral Gables, Florida.

Ostlund, H. G., Z. Top & V. E. Lee. 1982. Isotope dating of waters at Fram III. *Geophys. Res. Lett.* **9**: 1117–1119.

Ostlund, H. G., H. Craig, W. S. Broecker & D. Spencer. 1987a. "GEOSECS Atlantic, Pacific and Indian Ocean Expeditions," Vol 7. U.S. Govt. Printing Office, Washington, D.C.

Ostlund, H. G., G. Possnert & J. H. Swift. 1987b. Ventilation rate of the deep Arctic Ocean from carbon-14 data. *J. Geophys. Res.* **92**: 3769–3777.

Paasche, E. & I. Ostergren. 1980. The annual cycle of diatom growth and silica production in the inner Oslofjord. *Limnol. Oceanogr.* **25**: 481–494.

Poisson, A. & C. T. Chen. 1987. Why is there little anthropogenic CO_2 in the Antarctic Bottom Water? *Deep-Sea Res.* **34**: 1255–1275.

Priddle, J., I. Hawes & J. C. Ellis-Evans. 1986. Antarctic aquatic ecosystems as habitats for phytoplankton. *Biol. Rev. Cambridge Philos. Soc.* **61**: 199–238.

Probyn, T. A. & S. J. Painting. 1985. Nitrogen uptake by size-fractionated phytoplankton in Antarctic surface waters. *Limnol. Oceanogr.* **30**: 1327–1332.

Redfield, A. C., B. H. Ketchum & F. A. Richards. 1963. The influence of organisms on the composition of seawater. *In* "The Sea" (M. N. Hill, ed.), Vol. 2, pp. 26–67. Wiley (Interscience), New York.

Reeburgh, W. S. & M. Springer-Young. 1983. New measurements of sulfate and chlorinity in natural sea ice. *J. Geophys. Res.* **88**: 2959–2966.

Rudels, B. 1986. The theta-S relations in the northern seas: Implications for the deep circulation. *Polar Res.* **4**: 133–159.
Ryther, J. H. 1969. Photosynthesis and fish production in the sea. *Science* **166**: 72–76.
Schlosser, P. 1986. Helium: A new tracer in Antarctic oceanography. *Nature (London)* **321**: 233–235.
Schlosser, P. & F. Sarano. 1987. Concentrations en sels nutritifs de l'eau de surface dans le secteur indïen de l'Océan Austral (Campagne APSARA 2-ANTIPROD 3). *In* "Characteristiques biologiques, chimiques, et sedimentologiques du secteur indïen de l'Ocean Austral (Plateau des Kerguelen)" (M. Fontugne & M. Fiala, eds.), Rapp. Camp. Mer TAAF 84–01: 87–103.
Schlosser, P., W. Roether & G. Rohardt. 1987. Helium-3 balance of the upper layers of the northwestern Weddell Sea. *Deep-Sea Res.* **34**: 365–377.
Simon, V. 1986. Le systeme assimilation-regeneration des sels nutritifs de l'eau de surface dans le secteur indien de l'Ocean Austral. *Mar. Biol.* **92**: 431–442.
Simon, V. & F. Sarano. 1987. Concentratioons en sels nutritifs de l'eau de surface dans le secteur indien de l;Ocean Austral (Campagne APSARA 2- ANTIPROD 3). *In* "Characteristiques biologiques. chimiques, et sedimentologiques du secteur indien de l'Ocean Austral (Plateau des Kerguelen)" (M. Fontugne, M. Fiala, eds.). Rapp. Camp. Mer TAAF 84–01: 87–103.
Slawyk, G. 1979. ^{13}C and ^{15}N uptake by phytoplankton in the antarctic upwelling area: results from the Antiprod 1 cruise in the Indian Ocean sector. *Austr. J. Mar. Fresh. Res.* **30**: 431–448.
Smethie, W. M., Jr., D. W. Chipman, J. H. Swift & K. P. Koltermann. 1988. Chlorofluoromethanes in the Arctic Mediterranean seas: Evidence for formation of bottom water in the Eurasian basin and deep water exchange through Fram Strait. *Deep-Sea Res.* **35**: 347–369.
Smith, J. N., K. M. Ellis & E. P. Jones. 1990. Cs-137 transport into the Arctic Ocean through Fram Strait. *J. Geophys. Res.* **95**: 1693–1701.
Smith, R. I. L. 1985. Nutrient cycling in relation to biological productivity in Antarctic and subantarctic terrestrial and freshwater ecosystems. *In* "Antarctic Nutrient Cycles and Food Webs" (W. R. Siegfried, P. R. Condy & R. M. Laws, eds.), pp. 138–155. Springer-Verlag, Berlin.
Smith, W. O., Jr. & D. M. Nelson. 1985. Phytoplankton bloom produced by a receding ice edge in the Ross Sea: Spatial coherence with the density field. *Science* **227**: 163–166.
―――. 1986. The importance of ice-edge phytoplankton production in the Southern Ocean. *BioScience* 251–257.
Sommer, U. & H. H. Stabel. 1986. Near surface nutrient and phytoplankton distribution in the Drake Passage during early December. *Polar Biol.* **6**: 107–110.
Subba Rao, D. V. & T. Platt. 1984. Primary production of Arctic waters. *Polar Biol.* **3**: 191–201.
Swift, J. H., T. Takahashi & H. D. Livingston. 1983. The contribution of the Greenland and Barents seas to the deep water of the Arctic Ocean. *J. Geophys. Res.* **88**: 5981–5986.
Takahashi, T., W. S. Broecker & S. Langer. 1985. Redfield ratio based on chemical data from isopycnal surfaces. *J. Geophys. Res.* **90**: 6907–6924.
Tan, F. C. & P. M. Strain. 1980. The distribution of sea ice meltwater in the eastern Canadian Arctic. *J. Geophys. Res.* **85**: 1925–1932.
Tan, F. C., D. Dyrssen & P. M. Strain. 1983. Sea ice meltwater and excess alkalinity in the East Greenland Current. *Oceanol. Acta* **6**: 283–288.
Tolstikov, E. I. 1966. "Atlas Antarktika, Sovietskaia antarkticheskaia ekspeditsia." Bur. Geogr. Cartogr., Moscow, USSR (in Russian).
Treguer, P. 1987a. Comparative physical and chemical characteristics of coastal waters from Kerguélen and Heard islands and their oceanic surroundings during summer, relation to

phytoplankton growth. *J. CNFRA:* 17-26. Special issue, SCAR/SCOR. Group of specialists on Southern Ocean ecosystems and their living resources, Paris 25 Juin, 1985.

———. 1987b. Resultats Apasara 2. *In* "Charactéristiques biologiques, chimiques, et sedimentologiques du secteur indïen de l'Ocean Austral (Plateau des Kerguelen)" (M. Fontugne & M. Fiala, eds.), Rapp. Camp. Mer TAAF 84-01.

Treguer, P., P. Appriou, G. Bordin, P. Cuhet, S. Le Jehan & P. Souchu. 1987. Caractéristiques physiques et chimiques des masses d'eaux du secteur indïen de l'Ocean Austral, au sud du Front Polaire. *In* "Caractéristiques biologiques, chimiques et sedimentologiques du secteur indïen de l'Ocean Austral (Plateau des Kerguelen)" (M. Fontugne & M. Fiala, eds.), Rapp. Camp. Mer TAAF, 84-01: 54-74.

Treguer, P., A. Kamatani & S. Gueneley. 1988. Biogenic silica and particulate organic matter from the Indian sector of the Southern Ocean. *Mar. Chem.* **23**: 167-180.

Treguer, P., D. M. Nelson, S. Gueneley, C. Zeyons, J. Morran & A. Buma. 1990. The distribution of biogenic and lithogenic silica and the composition of particulate matter in the Scotia Sea and Drake Passage during autumn 1987. *Deep-Sea Res.* (in press).

Treshnikov, A. F., A. A. Girs, G. I. Baranov & V. A. Yefimov. 1973. "Preliminary Programme of the Polar Experiments for the South Polar Region." Arctic and Antarctic Research Institute, Leningrad, USSR. 55 pp.

Truesdale, R. S. & T. B. Kellogg. 1979. Ross Sea diatoms: Modern assemblage distributions and their relationship to ecologic, oceanographic and sedimentary conditions. *Mar. Micropaleontol.* **4**: 13-31.

Tsunogai, S., S. Noriki, K. Haradaka, T. Kurosaki, Y. Wanatabe & M. Maedaa. 1986. Large but variable particulate flux in the Antarctic Ocean and its significance for the chemistry of Antarctic water. *J. Oceanogr. Soc. Jpn.* **42**: 83-90.

van Bennekom, A. J., G. W. Berger, S. J. van der Gaast & R. T. P. de Vries. 1988. Primary productivity and the silica in the Southern Ocean (Atlantic sector). *In* "The Polar Ocean: the Antarctic: present and past" (E. Olansson, ed.). *Palaeogeogr. Palaeoclimatol., Palaeoecol., Spec. Issue* **67**: 19-30.

Wallace, D. W. R. & R. M. Moore. 1985. Vertical profiles of CCl_3F (F-11) and CCl_2F_2 (F-12) in the central Arctic Ocean basin. *J. Geophys. Res.* **90**: 1155-1166.

Wallace, D. W. R., E. P. Jones & R. M. Moore. 1987. Ventilation of the Arctic Ocean cold halocline: Rates of diapycnal and isopycnal transport, oxygen utilization and primary production inferred using chlorofluoromethane distributions. *Deep-Sea Res.* **34**: 1957-1979.

Weiss, R. F. & J. L. Bullister. 1984. Chlorofluoromethanes in the Southern Ocean: The ventilation in the Weddell Gyre. *Eos* **65**: 915 (abstr.).

Weiss, R. F., H. G. Ostlund & H. Craig. 1979. Geochemical studies of the Weddell Sea. *Deep-Sea Res.* **26**: 1093-1120.

Wilson, D. L., W. O. Smith, Jr. & D. M. Nelson. 1986. Phytoplankton bloom dynamics of the western Ross Sea ice edge. I. Primary productivity and species-specific production. *Deep-Sea Res.* **33**: 1375-1387.

Wust, G. 1936. Schichtung und Zirkulation des Atlantischen Ozeans. Stratosp., senschaftl. *Ergeb. Dtsch. Atl. Exp. Meteorol.* **6**: 1925-1927.

Yeats, P. A. 1988. Manganese, nickel, zinc and cadmium distributions at the Fram 3 and CESAR ice camps in the Arctic Ocean. *Oceanol. Acta* **11**: 383-388.

Zentara, S. J. & D. Kamykowski. 1981. Geographic variations in the relationship between silicic acid and nitrate in the South Pacific Ocean. *Deep-Sea Res.* **28**: 455-465.

Zwally, H. J., J. C. Comisco, C. L. Parkinson, W. J. Campbell, F. D. Carsey & P. Gloersen. 1983. "Antarctic Sea Ice 1973-1976: Satellite Passive Microwave—Observations," NASA, Spec. Publ. 459. Natl. Aeron. Space Admin., Washington, D.C.

9 Polar Phytoplankton

Walker O. Smith, Jr.
Botany Department and Graduate Program in Ecology
University of Tennessee
Knoxville, Tennessee

Egil Sakshaug
Trondhjem Biological Station, The Museum
University of Trondheim
Trondheim, Norway

I. Introduction 477
   A. Patterns of Primary Productivity in Polar Systems 478
II. Pelagic Production 480
   A. Environmental Factors Influencing Phytoplankton Growth 480
   B. Primary Production Cycles 492
   C. Photosynthesis and Growth Versus Irradiance 499
   D. Nutrient Uptake Relationships 505
   E. Elemental Ratios 509
III. Processes Affecting Phytoplankton Distribution 511
IV. Ice-Algal Production 512
   A. Types of Sea Ice Communities 512
   B. Environmental Factors Influencing Growth of Ice Algae 513
   C. Ice-Algal Productivity and Irradiance–Nutrient Relationships 514
V. Coupling of Ice Community with Water Column 515
VI. Conclusions 516
   References 517

I. Introduction

Polar ecosystems are often characterized by an extreme range of environmental conditions, yet they also have a set of physical and chemical parameters which are relatively constant. The entire set of environmental variables creates a unique habitat, and each condition has a direct impact on phytoplankton growth, accumulation, and distribution. Some environmental parameters undergo large variations on various time scales; for example, solar radiation (including both absolute photon flux and photoperiod) changes on a daily, weekly, and seasonal basis. On the other hand, water temperatures are relatively constant in most polar systems, varying much less than in most temperate and tropical systems. Thus the range and scales of temporal change in polar oceans encompass those in other oceanic realms, but the special combination of conditions results in the environment which we recognize as being polar. This chapter will attempt to contrast and compare algal biomass, growth, and distributions in the Arctic and Antarctic and give particular attention to the environmental conditions which modify growth and result in the unusual characteristics observed in polar algal communities.

We define polar systems both oceanographically and geographically. The Southern Ocean is usually defined by that body of water south of the Antarctic Convergence (Deacon, 1982); by this definition, the boundaries of the Southern Ocean change in both space and time. The Arctic Ocean is an oceanic basin, but the polar waters of the Arctic also include the marginal seas of the Arctic Ocean proper (e.g., Bering, Greenland, Kara, Barents, East Siberian, Chuckchi, and Beaufort seas). These regions undergo substantial variations in ice cover over the course of a year (see Fig. B, page xv) and hence extreme variations in phytoplankton biomass and growth. We will treat these regions collectively as being included in polar systems.

A. Patterns of Primary Productivity of Polar Systems

Primary productivity in polar regions is generally considered to be characterized by a single pulse which follows the rapid increase in solar radiation. Nutrients are reduced during this period (Fig. 9.1). Zooplankton generally lag significantly behind this increase in phytoplankton biomass, but exceptions to this pattern within the zooplankton community exist (see Chapter 10). Variations within polar systems also exist and are induced by the depth of the water column, the presence of ice and meltwater input to surface waters, and latitude. These factors modify the timing and duration of the pulse as well as its magnitude, and they can have an important impact not only on phytoplankton growth but also on the local trophic dynamics. With the advent of satellite sensors which can detect variations in ocean color, it has become possible to observe phytoplankton biomass over large areas in polar seas (Fig. 9.2; color plate facing page 486). Although this depiction exhibits spatial and temporal bias (due to the low number of satellite overflights and large cloud cover in polar regions; see Chapter 3), the data reinforce the pattern generally observed in field work; i.e., phytoplankton concentrations in the shallow, marginal seas of the Arctic can reach high levels during the summer, whereas the Antarctic is characterized by low biomass and blooms which are highly restricted in space.

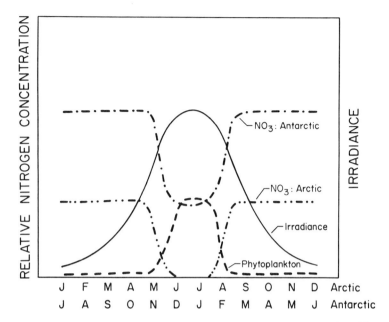

Figure 9.1 Idealized annual cycle of solar irradiance, nitrate, and phytoplankton biomass in polar regions.

Another important source of primary productivity which is not covered by the conceptual model presented in Fig. 9.1 is that of ice algae. Growth of ice algae usually precedes that of phytoplankton in time and is much more restricted both seasonally and spatially (Fig. 9.3). However, those attributes are exactly what make ice-algal production a potentially important source of food for herbivores.

Historically, polar productivity has been thought to be primarily diatomaceous; furthermore, polar phytoplankton have been characterized as being large (greater than 20 μm; e.g., Hart, 1934). However, we now know that other sizes and groups are important as agents of biogenic production (Hewes et al., 1985); furthermore, it is very clear that during certain periods small plankton dominate carbon and nitrogen dynamics (e.g., Fay, 1974; Probyn and Painting, 1985; Hewes et al., 1985; Koike et al., 1986; W. O. Smith et al., 1987). Although we will not concentrate on the taxonomic description of polar plankton, it is recognized that the phytoplankton assemblages in polar regions are extremely diverse with regard to both taxonomy and size.

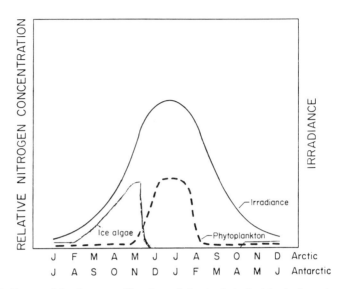

Figure 9.3 Temporal development of ice algae relative to phytoplankton in the water column.

II. Pelagic Production

A. Environmental Factors Influencing Phytoplankton Growth

As in other oceans, the phytoplankton biomass at one point in space and time is the difference between growth and loss processes. The growth rate is determined by physical and chemical factors such as temperature, light (including variations induced by ice cover and the vertical movements of water), nutrient supply ("new" nutrients brought to the euphotic zone from depth by vertical transport and "regenerated" nutrients produced by the activities of heterotrophic organisms), organic growth factors, and heavy metals. The losses are mainly due to grazing by micro- and macrozooplankton and to passive sinking, both of which in turn affect the light regime by altering the water column attenuation by particulate matter. The three environmental features which appear to exert the greatest controls over phytoplankton growth are low temperatures, seasonally dependent or permanent occurrence of sea ice, and extreme seasonal variations in the light regime, and it is the unique combination of these factors that makes the spatial and temporal patterns of phytoplankton growth in polar areas different from those in other oceans.

1. Temperature

Water temperatures of polar oceans may range from the freezing point of seawater ($-1.8°C$) to $4-6°C$ in waters advected from the Atlantic into the Barents and Greenland seas. Because temperature generally places an upper limit on metabolic processes, it directly controls phytoplankton growth rate, which is the difference between gross photosynthesis and respiration (plus any extracellular production). The Q_{10} value for respiration is apparently higher (2.3–12; Tilzer and Dubinsky, 1987) than for photosynthesis (1.4–2.2; Neori and Holm-Hansen, 1982; Tilzer and Dubinsky, 1987). As a result, nocturnal respiration may constitute only a few percent of photosynthesis at temperatures less than $0°C$, which in turn would result in net growth even when days are short. The marked temperature effect on photosynthesis (Tilzer *et al.*, 1986; Tilzer and Dubinsky, 1987) has been suggested to result from a temperature dependence of maximum quantum yields. C. W. Sullivan (unpublished) also found increases in the rates of light-saturated photosynthesis of ice algae over a small range of temperatures. Li (1985) found that Arctic phytoplankton had temperature optima which were much higher than *in situ* temperatures (ca. $10°C$ greater); furthermore, activation energies for photosynthesis and ribulose -1,5-bisphosphate carboxylase (RuBPC) activity are similar (Li *et al.*, 1984), indicating that RuBPC may represent an important rate-limiting step in photosynthesis of polar phytoplankton. Thus, Li's data indicate that high-latitude phytoplankton, by virtue of the fact that the temperature at which they live is far from their optimal temperature, are not sensitive to slight temperature increases in their habitat.

Eppley (1972) investigated the relationship between temperature and phytoplankton growth rate by compiling all available laboratory data on phytoplankton growth at various temperatures and optimal nutrient levels. Although the number of species was limited and the minimum temperature used during culturing was $2°C$, the relationship has been widely used to predict maximum rates of phytoplankton growth in polar regions. The equation that describes the growth rate–temperature relationship is

$$\log_{10} \mu = 0.0275T - 0.070 \qquad (9.1)$$

where μ is the specific growth rate in doublings (dbl) per day and T is the temperature in degrees Celsius. Based on this equation, the maximum growth rate at -1.8 and $3°C$ would be 0.76 and 1.03 dbl day^{-1}, respectively, which corresponds to a Q_{10} value of 1.88. Table 9.1 shows actual growth rates measured for Arctic and Antarctic phytoplankton. The average values range from 0.19 to 0.75 dbl day^{-1}. It is noteworthy that none of the averages surpass the predicted rates, not even values which represent optimum irradiances (Sakshaug and Holm-Hansen, 1986; Gilstad, 1987). The

Table 9.1 Summary of Observed Growth Rates of Natural Phytoplankton Assemblages in Polar Regions

Study area	Season	μ (doublings per day)	Reference
Arctic			
Baffin Bay	Summer 1978	0.31 (0.01–0.76)	Harrison et al. (1982)
Greenland Sea	Summer 1984	0.71 (0.36–1.45)	W. O. Smith et al. (1987)
Greenland Sea	Summer 1987	0.38 (0.03–1.55)	Culver and Smith (1989)
Barents Sea[a]		0.75 (0.63–0.93)	Gilstad (1987)
Antarctic			
Ross Sea	Summer 1972	0.01–0.33	Holm-Hansen et al. (1977)
Ross Sea	Summer 1983	0.19 (0.1–0.5)	Wilson et al. (1986)
Scotia Sea	Summer 1981	0.30–0.76	Rönner et al. (1983)
Weddell Sea	Summer 1977	0.71 (0.37–1.30)	El-Sayed and Taguchi (1981)
Weddell Sea	Spring 1983	0.53 (0.07–2.50)	W. O. Smith and Nelson (1990)
Weddell Sea	Autumn 1986	0.59 (0.14–2.58)	W. O. Smith and Nelson (1990)
Weddell Sea[b]	Summer 1985	0.75 (0.38–1.33)	Spies (1987)
Ice edge[c]	Summer 1985	0.42 (0.37–0.49)	Sakshaug and Holm-Hansen (1986)

[a] Laboratory cultures at −0.5°C, maximum rates for 11 diatom species.
[b] Shipboard cultures at −1°C, 34 cultures of diatoms in unenriched seawater.
[c] Shipboard cultures at −1 to 4°C with unenriched seawater.

observed variation in growth rates of natural phytoplankton is in some cases large (Table 9.1), particularly when growth rate has been calculated on the basis of ^{14}C uptake measurements, and it is likely that some of the highest values in Table 9.1, especially those involving natural populations, reflect unbalanced short-term features rather than a steady-state growth rate. Conversely, some of the lower field estimates may be a result of light or nutrient limitation. Species differences, however, also play an important role (Jacques, 1983; Gilstad, 1987; Spies, 1987). It appears that Eppley's function realistically approximates maximum *in situ* rates of growth for multispecies assemblages of polar phytoplankton and that the major role of temperature in polar regions is to set an upper limit for growth of phytoplankton, which is then modified by other environmental factors such as light and nutrients.

2. Irradiance

Irradiance, because of its extreme seasonal variation, is considered to be a major environmental parameter controlling phytoplankton growth in polar regions. At 60° latitude daylength ranges from about 6 h in midwinter to about 19 h in midsummer, and beyond 66.7° there is at least some period of total darkness and of continuous light (Fig. 9.4). At the poles the dark period lasts for nearly 6 months, and at extreme latitudes the transition from total darkness to continuous light is abrupt (40–120 days from 82° to 70°; Holm-Hansen et al., 1977).

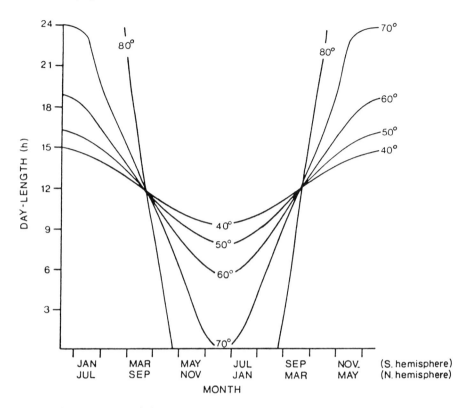

Figure 9.4 Seasonal variation in daylength at various latitudes. From Sakshaug and Holm-Hansen (1984).

Daylength has a pronounced effect on algal growth rate. Gilstad (1987) found that the growth rate of diatoms increased with irradiance and reached a maximum at 20-40 μmol photons m^{-2} s^{-1} (= μEin m^{-2} s^{-1}), after which growth rate became independent of irradiance. However, growth rate was three to six times greater when the photoperiod was 19-24 h/day rather than 4 h/day for the same irradiance. A similar daylength response was noted for *Skeletonema costatum* isolated from Trondheimsfjord (63°N; Sakshaug and Andresen, 1986).

The amount of incident radiation is controlled by atmospheric conditions such as atmospheric turbidity, relative atmospheric thickness, and the diffuse radiation component (Kremer and Nixon, 1978). Equally important in controlling the amount of radiation entering the surface waters of polar regions is the solar angle. The maximum solar angle during the daily cycle can be calculated based on the latitude and the magnitude of refraction and

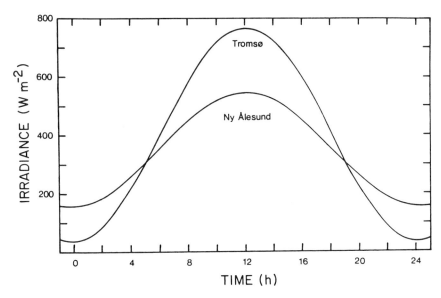

Figure 9.5 Diurnal variation in irradiance on clear midsummer days at Tromsø (70°N) and Ny Ålesund, Svanlbard (79°N). From Sakshaug and Holm-Hansen (1984).

reflection through the surface waters by Beer's and Snell's laws (Jerlov, 1968). Examples of diurnal cycles at high latitudes in midsummer are shown in Fig. 9.5. Reflectance of incident irradiance depends on the angle of incidence and the roughness of the sea (Fig. 9.6). In polar regions the incidence angle of direct sunlight at 60° latitude will always be greater than 37° and greater than 67° at the poles. Thus, reflection losses will generally be pronounced, especially during the early spring and late autumn, when the solar angle is extreme. Qualitatively, low solar angles result in greater attenuation of shorter wavelengths than of longer ones, and hence the atmosphere removes the wavelengths which are most effective in driving photosynthesis. The loss of radiation at the sea surface in polar regions is variable, as in all oceanic systems, but quantitatively greater than in temperate and tropical areas.

In addition to reflectance, light is greatly attenuated above the sea surface by clouds and fog, which are regular features of polar oceans. This stochastic attenuation can range from 40 to 90%, whereas the reflectance loss of diffuse radiation may be less than 5–7% (Preisendorfer, 1957). Long-term measurements of scalar irradiance at 0.5 m in Trondheimsfjord (63°N) at noon have ranged from 30 μmol m^{-2} s^{-1} in midwinter to about 1500 μmol m^{-2} s^{-1} in midsummer (Hegseth and Sakshaug, 1983), although maximum values at 0.5 m in the Barents Sea rarely reach 1000 μmol m^{-2} s^{-1} (E. N. Hegseth,

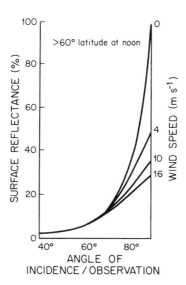

Figure 9.6 Reflectance of surface irradiance as a function of the angle of incidence in polar regions at various wind speeds. Redrawn from Kirk (1983).

unpublished). It is obvious that the light environment in the surface waters of polar systems is highly variable on all time scales.

Light is greatly attenuated by sea ice and its condition (i.e., its thickness and degree of snow cover and the presence of ice algae, brine pockets, and air bubbles). Snow is extremely opaque to light; attenuation coefficients range from 16 to 45 m^{-1} (Palmisano et al., 1986a), which implies that a 10-cm-thick snow layer will reduce the incidence irradiance to 1.1–20% of surface irradiance and a 50-cm-thick layer will reduce incidence light to 0.01–3%. Sea ice itself is considerably more transparent, with attenuation coefficients of 1.5–1.6 m^{-1} (Maykut and Grenfell, 1975; Palmisano et al., 1986a). One meter of ice will attenuate about 80% of incident radiation; furthermore, the spectral composition of the light will be greatly modified (Sullivan et al., 1984). However, the major impact of ice and snow is the marked reduction of the quantity of photosynthetically active radiation at the surface of the water column.

The vertical diffuse attenuation coefficient of polar waters is also variable. Strongly pigmented particles (e.g., phytoplankton and phytodetritus) account for 80% of this variation (Tilzer et al., 1986). In waters with extremely low biomass, such as those under ice and in winter, attenuation coefficients for photosynthetically active radiation (PAR, 400–700 nm) are from 0.036 to 0.07 m^{-1}, as they are for temperate and tropical waters. Effects of dis-

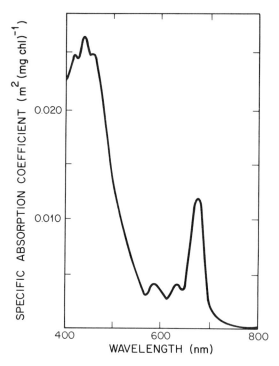

Figure 9.7 Specific absorption coefficient of an Arctic *Phaeocystis* bloom in the Barents Sea in May 1987. Data of E. Sakshaug (unpublished).

solved organic matter on light attenuation are generally constant and limited, except in localized river plumes of the Arctic during the summer.

Phytoplankton absorbs strongly in the blue, blue-green, and red parts of the electromagnetic spectrum (Fig. 9.7). Averaged for PAR, the specific absorption coefficient for phytoplankton has been reported to vary between 0.001 and 0.030 m² (mg chlorophyll a)$^{-1}$ (Kirk, 1975; Bannister, 1979; Bricaud *et al.*, 1983; Mitchell and Kiefer, 1988), and an average specific absorption coefficient for several species and photoadaptive states may be between 0.005 and 0.015, depending on species and photoadaptive state (Maske and Haardt, 1987; SooHoo *et al.*, 1987). Multiplied by the concentration of chlorophyll and chlorophyll-like pigments (e.g., phaeopigments), this figure yields the contribution to the total diffuse attenuation coefficient due to absorption by particles containing chlorophyll or its derivatives. Thus the total attenuation coefficient of a water column containing 40 µg l^{-1} chlorophyll and its degradation products includes a contribution from pigmented particles of 0.4–0.52 m^{-1} and will reduce the 1% isolume from

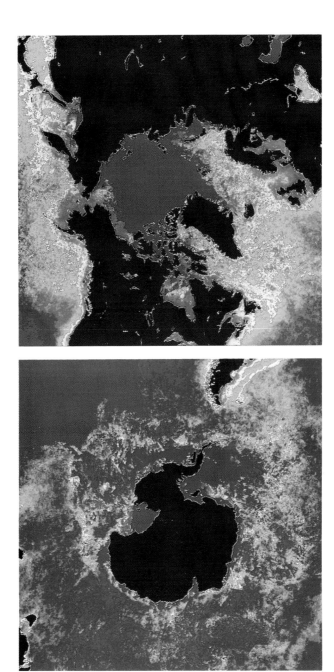

Figure 9.2 Spatial distribution of phytoplankton biomass as determined from ocean color using the Coastal Zone Colar Scanner on the Nimbus-7 satellite. Violet indicates the lowest concentrations and red the highest; gray represents ice-covered waters or regions where no data were obtained. In general, the Arctic is characterized by large accumulations of phytoplankton in the shallow regions, whereas the Antarctic is much more oligotrophic with spatially restricted phytoplankton blooms. Image courtesy of G. Feldman, Goddard Space Center/NASA.

about 115 m in the clearest water to less than 10 m. Peak chlorophyll plus phaeopigment concentrations of more than 40 μg l^{-1} have been observed in the nutrient-rich waters of the Bering Sea and in the Antarctic (Iverson *et al.*, 1979; Whitaker, 1982); thus, self-shading by algae can be pronounced in polar waters. In reality, self-shading is even more pronounced than these calculations suggest, because the photosynthetically usable blue-to-blue-green part of the spectrum is attenuated more by algae than the overall PAR, and mainly green to yellow wavelengths remain at depth. The usable red part of the spectrum is also strongly attenuated by the water itself (the attenuation coefficient at 650 nm is about 0.4 m^{-1}), and as a result these wavelengths have little effect below 10 m. In the Antarctic and the part of the Arctic influenced by Atlantic waters, the chlorophyll level is generally less than 10 μg l^{-1}, and the impact of self-shading becomes correspondingly smaller.

Another factor which controls the availability of irradiance to phytoplankton in polar regions is vertical mixing. During vertical transport within water, phytoplankton experience a large irradiance gradient, and hence with deeper mixing the time–depth integrated irradiance available to phytoplankton is reduced. Sverdrup (1953) introduced the concept of "critical depth" to describe mathematically the onset of net production and a spring bloom in northern temperate areas. Critical depth is the depth at which the integrated (through depth and time) gross photosynthesis equals the integrated respiration by phytoplankton. The critical depth is a function of the atmospheric light regime, the transmission across the air–water interface, and the total vertical attenuation coefficient of the water. If the depth of mixing is less than the critical depth, integrated gross photosynthesis is larger than integrated respiration, and net production will result. However, to be of general use, this model must balance gross photosynthesis against all losses, e.g., passive sinking and grazing. Furthermore, the effects of self-shading and spectral composition of irradiance are now understood well enough to be realistically incorporated into such models. Nonetheless, given that grazing and sinking losses as well as self-shading effects are generally considered to be minor when net production becomes positive, Sverdrup's conceptual model remains crucial to understanding phytoplankton growth in well-mixed water columns.

Open ocean areas in polar regions can exhibit extremely deep mixed layers, at times greater than 1000 m (O. M. Johannessen *et al.*, unpublished results). However, it has been observed that blooms usually do not occur in polar regions when the mixed layer is greater than 40–50 m (Sakshaug and Holm-Hansen, 1984), although blooms of the prymnesiophyte *Phaeocystis pouchetii* (Hariot) Lagerheim have been observed in mixed layers of 80 m (E. Sakshaug, unpublished). A critical depth of 80 m is far less than expected from the original Sverdrup model if calculations are made using realistic initial conditions. For example, assuming clear waters (attenuation coefficients from 0.04 to 0.075 m^{-1}) and irradiances of 50-700 μmol m^{-2} s^{-1} at the surface (March through June in the northern hemisphere), the initial average irradiance in the water column before the onset of self-shading should be sufficient for growth even in mixed layers of 750 m or more. Clearly, this does not occur. Despite the conceptual utility of Sverdrup's approach, it cannot be used explicitly in polar regions without inclusion of more detailed spectral and physiological information.

Physiological constraints also exist within Sverdrup's model, in that phytoplankton are able to adapt their photosynthetic machinery within an irradiance gradient on finite time scales. The time required for photoadaptation, according to studies of cellular changes in chlorophyll, appears to be variable, ranging from 4 h to a number of days, and is positively related to growth rate (Falkowski, 1983; Gallegos et al., 1983). If the mixing rate is large relative to the rate of photoadaptation, algal properties, including physiological parameters, tend to be evenly distributed within the water column; conversely, if the mixing rate is much less than the rate of adaptation, a gradient in photosynthetic capacity (and ultimately phytoplankton biomass if the mixed layer remains shallow) should become evident (Harris, 1980; Lewis et al., 1984). The latter is most apparent during stratification and after the development of a chlorophyll maximum layer at depth. Some data are available from the Arctic which indicate that the photoadaptational response is asymmetrical (i.e., adaptation from high to low irradiance takes different amounts of time than adaptation from low to high irradiance; Gallegos et al., 1983). Data for photoadaptational rates are sparse, and models yield conflicting results about the symmetry of the response (Harris, 1980; Falkowski, 1984; Lewis et al., 1984; Post et al., 1985; Geider and Platt, 1986; Cullen and Lewis, 1988). If adaptation from low to high irradiance takes less time than the reverse process, the average photoadaptation status would correspond, after some cycles, to an irradiance higher than the average irradiance in the mixed layer, whereas the opposite asymmetry would maximize shade adaptation.

An additional consideration with regard to the photoadaptation of polar phytoplankton is the response during rapid transfers within the water column. E. N. Hegseth (unpublished) has observed that Arctic diatoms can grow at irradiances as low as 10 μmol m^{-2} s^{-1}, but not if directly transferred from 450 μmol m^{-2} s^{-1}. This particular light gradient would correspond to a downward transport of 95 m in clear oceanic waters and only 38 m in waters with an attenuation coefficient of 0.1 m^{-1}. Consequently, rapid (and deep) mixing may be analogous to sudden transfer of algae to darkness, which would arrest the photoadaptational response (Sakshaug et al., 1987). Gallegos et al. (1983) also found that Arctic phytoplankton adapted slowly (complete photoadaptation taking from 2 to 6 weeks) when moved from high to low irradiances and much more rapidly (ca. 6 h) when moved from low to high irradiances. Consequently, during mixing, surface phytoplankton would retain the photoadaptational characteristics they had previously and exhibit low photosynthetic efficiency at low light (uptake of carbon relative to carbon biomass) and elevated respiration rates. Generally, respiration rates become low when algae become adapted to shade (Falkowski and Owens, 1980; Cosper, 1982). This may be an even more important modification in a critical depth – mixed layer model than the symmetry of the time course of photoadaptation. Nevertheless, it is clear that our present understanding of vertical mixing and photoadaptational rates is insufficient to produce realistic models of phytoplankton growth in polar regions.

3. Nutrients

Significant differences exist in nutrient concentrations among the various polar regions (see Chapter 8). In general, nutrient levels in the Southern Ocean are elevated as a result of two factors: (1) a large-scale divergence which supplies preformed (or "new") nutrients to the surface waters and (2) low rates of phytoplankton removal. Arctic waters, with the exception of Bering Sea water, exhibit lower nutrient concentrations. Bering Sea water can be traced in the Arctic Ocean basin for some distance by its nutrient characteristics (Codispoti, 1979), but the low surface nutrient concentrations found throughout the Arctic Ocean are maintained by the physical features which produce the permanent halocline (see Chapter 8). In all polar water masses the atomic N/P ratio of maximum nutrient concentrations is close to the Redfield ratio (Sakshaug, 1989).

Nutrients are generally depleted after a phyotoplankton bloom in the Arctic, even in waters of Pacific origin (Hameedi, 1978; Alexander and Niebauer, 1981; Heimdal, 1983; S. L. Smith *et al.*, 1985; Rey and Loeng, 1985; Fig. 8.2), although in some portions of the Arctic nutrients remain after the spring bloom (Subba Rao and Platt, 1984). In contrast, nutrients rarely become depleted in the Southern Ocean. In the remaining part of the Arctic growing season, the length of which depends on the initiation date and duration of the spring phytoplankton bloom, the surface layer is characterized by a distinct pycnocline, which generally occurs in the upper 25 m (e.g., Bering Sea: Sambrotto *et al.*, 1984; Barents Sea: Rey and Loeng, 1985; Chukchi Sea: Hameedi, 1978; Fram Strait: S. L. Smith *et al.*, 1985; W. O. Smith *et al.*, 1987). This stratification remains unless physical forces (such as intense surface wind action; shear forces between currents, banks, and islands; vertical motions induced by upwelling or eddies; or tidally induced mixing in shallow waters) produce turbulence to weaken or dissipate the stability. During this stratified period, primary productivity is low and based on nutrients regenerated within the surface layer (Harrison et al., 1982). Any disruption of the pycnocline introduces preformed nutrients again into the euphotic zone and stimulates phytoplankton growth. As a result, the annual productivity of Arctic regions near banks and in shallow waters is greater due to intermittent nutrient addition. In the regions where no mechanism for positive nutrient flux has been observed (e.g., northern Barents Sea), the stratification remains throughout the summer until the formation of new ice in the autumn (Rey and Loeng, 1985).

Although enhanced vertical transport of preformed nutrients into the euphotic zone can stimulate primary productivity, it also creates an unfavorable light environment for phytoplankton growth. Riley (1963) suggested that optimal productivity will occur under alternating conditions of turbulence and stability, which is clearly the case in upwelling regions (Barber and Smith, 1981). Because the low Arctic (55–70°N) is exposed to more or less regular passages of low-pressure systems, wind-induced turbulence may increase the flux of nutrients into the euphotic zone and stimulate productivity. However, wind is generally an inefficient mechanism for transferring energy into the deeper portion of the surface layer, and other mechanisms such as upwelling along ice edges (see Chapter 7) may be more important in the annual productivity cycle of polar regions.

In the Antarctic the question is not whether nutrients are limiting, but why nutrients are never reduced to limiting conditions. Usually this is explained by light limitation via deep vertical mixing (Sakshaug and Holm-Hansen, 1984). Postbloom nutrient concentrations nearly always remain elevated (Jennings et al., 1984); furthermore, the distribution of phytoplankton biomass is often related to the residence time of phytoplankton in the upper layers (Heywood and Priddle, 1987). Recently, it has been suggested that micronutrients (e.g., trace metals) may limit phytoplankton growth in nutrient-rich regions such as the Antarctic (Martin and Fitzwater, 1988; Martin and Gordon, 1988). Experiments in the sub-Arctic Pacific using ultraclean techniques have shown significant growth stimulation upon the addition of iron, and it was proposed that phytoplankton growth in that region was regulated by the rate of atmospheric input of iron. It was further hypothesized that nutrients are not removed to depletion in the waters off the continental shelf of the Antarctic because of low input rates of atmospheric iron and the lack of other sources (e.g., from sediments). Martin and Fitzwater (1988) also interpret the presence of ice edge and island blooms as being related to greater inputs of iron from ice melt and resuspended sediments. Although these ideas have not been experimentally tested, such a hypothesis would explain not only the geographic variations observed in Antarctic productivity but also their temporal nature. Complete exhaustion of nutrients has been found in cultures of natural waters (Sakshaug and

Holm-Hansen, 1986; W. O. Smith, unpublished), and Whitaker (1982) found no growth stimulation on addition of trace metals, both alone and together with chelators, but these results may have been influenced by trace metal contamination during sampling. These studies were also conducted in shallow, continental shelf regions. The results of Martin and Fitzwater (1988) and Martin and Gordon (1988) may alter the interpretation of previous productivity results from the Antarctic and can be resolved only by a rigorous examination of trace metal–nutrient–light interactions in the field.

4. Organic factors

Although little is known concerning the role of organic material in regulating phytoplankton growth in polar oceans, a few studies have measured concentrations of ecologically significant vitamins (Carlucci and Cuhel, 1977; Fiala and Oriol, 1984). The former study suggested that low levels of vitamin B_{12} may have limited microplankton growth in the Southern Ocean; the latter investigation showed a slight stimulation of phytoplankton growth by vitamin B_{12} despite the fact that it was not a requirement for growth of two species of Antarctic diatoms. Whitaker (1982), however, found no stimulation of photosynthesis by vitamins. No studies comparable to these have been conducted in Arctic waters. The presence or absence of organic ligands in polar waters is similarly unknown. Whitaker (1982) found no stimulation of photosynthesis on addition of chelators to natural populations, although ultraclean techniques were not used. Whether the concentrations of dissolved organic material are at a level such that the complexation of toxic trace metals is complete is unknown.

It has been suggested that the Southern Ocean is a site of export of large amounts of dissolved organic matter to tropical regions (e.g., Sorokin, 1971). This hypothesis was effectively refuted by Banse (1974), and it is now thought that polar regions have low levels of dissolved organic material. Based on carbon and nutrient budgets, Karl *et al.* (1990) hypothesized that there is a production of significant amounts of carbon-rich DOC within the summer bloom in the Bransfield Strait, but direct confirmation is lacking. No information is available on the fate or seasonal dynamics of this carbon pool.

The use of dissolved organic matter by algae in polar regions has long been suggested as a mechanism by which cells retain viability over the long periods of reduced light. The use of organic substrates in the dark (heterotrophy) or in the light (photoheterotrophy) was investigated in the Arctic by Horner and Alexander (1972), who were unable to detect any significant uptake by ice algae, and Bunt and Lee (1972) found no effect of organics on the survival of various ice-algal species. However, three species of diatoms were found to obtain from 0.003 to 0.3% of their total carbon requirements via organic uptake (Palmisano and Sullivan, 1983), and Rivkin and Putt (1987a) found that amino acids were actively incorporated into cellular substrates (e.g., protein) by diatoms. In the latter study substrate levels provided for uptake were low, and hence it was suggested that organic matter removal might be ecologically important. However, none of the experiments in these studies were conducted when the ambient dissolved organic pool concentration was known, and hence the potential for experimental bias remains. Until *in situ* dissolved organic carbon concentrations are measured in concert with kinetic experiments, the role of heterotrophy in algal energetics will remain unresolved.

B. Primary Production Cycles

1. Open water and marginal ice zones

The pelagic realm of polar regions is generally characterized by a single pulse of "new" production or a bloom. The prebloom environmental conditions include deep vertical mixing, maximum nutrient concentrations, low incident irradiance, and low levels of phytoplankton biomass. When the light regime becomes adequate (by increased insolation and/or vertical stratification), net production occurs and phytoplankton biomass increases (Fig. 9.1). The duration of the bloom in polar regions appears to be controlled by either nutrient levels or the availability of light. For example, in the Barents Sea and Baffin Bay a marked pycnocline often forms in the upper 25 m near the ice edge and nutrients can be depleted within this stratified region in 2 weeks (Rey and Loeng, 1985). In the Antarctic, stratification is less pronounced and nutrients are rarely depleted. Phytoplankton growth occurs over a longer time period and apparently is limited by light effects. Maximum concentrations of chlorophyll within a bloom range from 10 to 15 μg l^{-1} in the Barents Sea (Rey and Loeng, 1985) to more than 40 μg l^{-1} in the Bering Sea (Iverson *et al.*, 1979). Chlorophyll concentrations greater than 30 μg l^{-1} have been observed at the ice edge in the Weddell Sea (W. O. Smith and Nelson, 1986), but usually peak standing stocks are 10 μg l^{-1} or less (El-Sayed, 1984). One observation of a chlorophyll level of 180 μg l^{-1} at the surface has been made (El-Sayed, 1971) but undoubtedly represents accumulation via frazil ice concentration (e.g., Garrison *et al.*, 1983). Postbloom chlorophyll levels in both the Arctic and Antarctic are less than 1.0 μg l^{-1}, and values less than 0.1 μg l^{-1} are not uncommon in both regions.

The seasonally ice-covered regions of polar oceans always have a well defined region where the open ocean grades into the ice, known as the marginal ice zone. These areas have been known for years as sites of enhanced biological activity at all trophic levels (e.g., Hart, 1934; Bradstreet and Cross, 1982; Ainley *et al.,* 1986; W. O. Smith, 1987). It has been suggested for a number of years that the enhanced productivity at the ice edge is due to increased stratification imparted by the melting ice (Gran, 1931; Marshall, 1957), and a number of investigations have shown elevated phytoplankton standing stocks in the vicinity of the ice edge (McRoy and Goering, 1976; El-Sayed and Taguchi, 1981). More detailed investigations of the temporal and spatial dynamics of these features have led to the recognition that the blooms in the marginal ice zone are a quantitatively significant feature in both the Arctic and Antarctic (Alexander and Niebauer, 1981; W. O. Smith and Nelson, 1985, 1986; Rey and Loeng, 1985; S. L. Smith *et al.,* 1985; Nelson *et al.,* 1987; W. O. Smith *et al.,* 1987). When the ice melts, nutrient-rich waters are exposed to high light levels, and strong vertical stratification provides optimal conditions for phytoplankton growth (Fig. 9.8). As the pack ice retreats, the bloom is initiated near the ice edge and is dissipated at some distance (via either light or nutrient effects), so that the bloom appears to trail the receding ice edge. The spatial extent (normal to the ice edge) depends on the physical dynamics of each ice edge, which vary considerably. For example, the phytoplankton bloom in the Bering Sea extends about 50 km, as do those of the Barents Sea and Fram Strait, whereas those in the Ross and Weddell seas extend about 250 km from the ice edge (Alexander and Niebauer, 1981; Rey and Loeng, 1985; S. L. Smith *et al.,* 1985; W. O. Smith and Nelson, 1985; Nelson *et al.,* 1987).

The conceptual model presented in Fig. 9.8 suggests that an ice-edge bloom should occur earlier at lower-latitude ice edges, and this is in fact observed. At the maximum (and most southerly) ice extent in the Barents Sea, a bloom is initiated in April, considerably earlier than in the ice-free but turbulent Norwegian Sea farther south (Rey *et al.,* 1987). Increased concentrations of phytoplankton do not occur until August off Ellesmere Island (82°N; Appolonio, 1980). In the Antarctic, blooms begin at least by November at the northern extent of the ice (ca. 60°S; Hart, 1934; Nelson *et al.,* 1987) and continue into March farther south (at the Filchner Ice Shelf; El-Sayed and Taguchi, 1981). However, these trends may be greatly modified by other processes: water mass history, geographic variations of meltwater input, and advective effects (Rey and Loeng, 1985; Nelson *et al.,* 1987; Sullivan *et al.,* 1988; Comsio *et al.,* 1990).

The Fram Strait marginal ice zone (MIZ) is different from other Arctic ice-edge systems due to its greater water column depth. As a result, the processes which replenish nutrients on the continental shelf systems (e.g., upwelling: Alexander and Niebauer, 1981; tidal mixing: Niebauer and Alexander, 1985) are apparently not as frequent in the Fram MIZ during the growing season (although upwelling has been observed in February along the ice edge; Buckley *et al.,* 1979). On the other hand, in the Fram Strait area

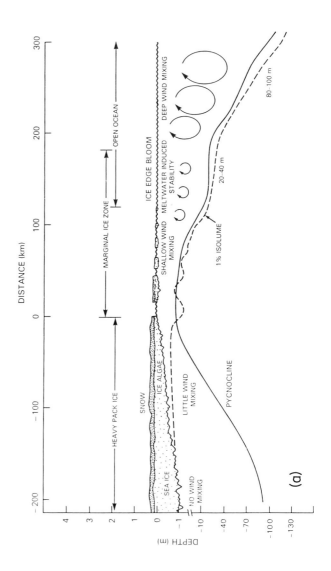

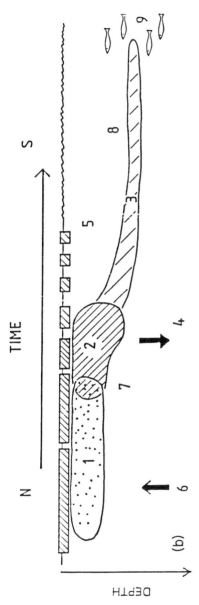

Figure 9.8 Conceptual model of the development of phytoplankton blooms in the marginal ice zone of (a) the Antarctic (from Sullivan et al., 1988) and (b) the Barents Sea (from Rey et al., 1987). The numbers in (b) correspond to the following stages: 1, prebloom conditions; 2, ice-edge phytoplankton bloom; 3, postbloom deep chlorophyll maximum; 4, sedimentation of phytoplanton; 5, nutrient-depleted postbloom surface layer; 6, upward migration of zooplankton; 7, spawning of zooplankton; 8, growth and development of zooplankton; 9, capelin feeding and migration.

eddies are commonly generated by a number of mechanisms (see Chapters 3 and 7), and a biological enhancement associated with eddies has been observed (S. L. Smith et al., 1985). However, the relationship between nutrient enhancement, light availability, and vertical stability within these eddies is far from clear (Niebauer and Smith, 1989), although analogies with other eddy-like features in the Southern Ocean exist (Heywood and Priddle, 1987).

2. Primary productivity

For many years the Southern Ocean was considered to be highly productive (El-Sayed, 1984). However, we now know that this conclusion resulted from a strong seasonal and spatial bias in sampling. Furthermore, large standing stocks of higher trophic levels (see Chapter 11 and 12) were observed, and it was assumed that this food web was supported by elevated primary productivity. Further investigations have shown exactly the opposite: the Southern Ocean is largely oligotrophic (low biomass and productivity) with localized pulses of enhanced productivity (Hasle, 1969; Holm-Hansen et al., 1977; Sakshaug and Holm-Hansen, 1984). The Southern Ocean is the site of large numbers of pelagic organisms (whales, birds), and extensive accumulations of euphausids have been observed (Holm-Hansen and Huntley, 1984); however, the degree to which large-scale rather than small- and mesoscale productivity patterns support this elevated higher trophic level biomass is unclear. Also, the Antarctic has substantial deposits of diatomaceous sediments; again, the early assumption that such deposits reflected large surface productivity appears to be in error (Nelson and Smith, 1986; Ledford-Hoffman et al., 1986). Compilations of productivity have been made previously (e.g., El-Sayed and Turner, 1977; El-Sayed, 1984), and most workers now consider the productivity of pelagic waters of the Southern Ocean to be about 134 mg C m^{-2} day^{-1} or 16 g C m^{-2} yr^{-1} (Holm-Hansen et al., 1977).

The Arctic Ocean has long been considered the most oligotrophic ocean basin of all as a result of its perennial ice cover and short growing season (Platt and Subba Rao, 1975). Subba Rao and Platt (1984) estimate the primary productivity of the shelf waters to be approximately 225 mg C m^{-2} day^{-1} or 27 g C m^{-2} yr^{-1} and the open ocean water productivity to be 75 mg C m^{-2} day^{-1} or 9 g C m^{-2} yr^{-1}. Thus the productivities of the Arctic and Antarctic are not greatly different based on a comparison of mean values. Maximum values for the Arctic and Antarctic are also similar (4.89 g C m^{-2} day^{-1} in the Canadian Arctic: Hsiao and Trucco, 1980, in Subba Rao and Platt, 1984; 3.62 g C m^{-2} day^{-1} near Deception Island, Antarctica: Burkholder and Mandelli, 1965), and maxima occur in shallow, protected areas. Primary productivity rates of greater than 12 g C m^{-2} day^{-1} have been measured in the Chukchi Sea when chlorophyll concentrations exceeded 40

μg l^{-1} (C. P. McRoy, D. Hansell, A. Springer, J. J. Goering, J. J. Walsh & T. Witledge, unpublished). Both regions appear to be quantitatively similar in their rates of biogenic production, although the relative importance of light and nutrients to the annual productivity cycle in both polar oceans is uncertain.

McRoy and Goering (1976) found that the productivity at the ice edge of the Bering Sea was 89 mg C m^{-2} day^{-1}, which, although not large, was 15 times that under the ice or in open water. Alexander (in W. O. Smith, 1987) found that ice edge productivity increased from 0.32 g C m^{-2} day^{-1} in March to 6.60 g C m^{-2} day^{-1} in June. Nitrate was nearly depleted during the same interval (Alexander and Niebauer, 1981), which implies that the mean productivity of the ice-edge bloom during that period is approximately 1.1 g C m^{-2} day^{-1}. The Bering Sea ice edge retreats rapidly (Parkinson *et al.*, 1987); therefore its impact at one location is minor, despite the large absolute production during the bloom. However, the total primary production associated with ice-edge blooms in the Bering Sea is significant, providing up to 50% of the annual production of the inner-shelf region (W. O. Smith, 1987). Productivity of the ice edge in the Barents Sea is also tightly coupled to the meltwater-induced vertical stability (Rey and Loeng, 1985), and the magnitude of productivity in the marginal ice zone appears to be eightfold greater than outside.

Primary productivity of ice-edge blooms in the Southern Ocean is also much greater than that under ice or outside the marginal ice zone. In the Ross Sea the mean productivity in the ice-edge bloom was 962 mg C m^{-2} day^{-1}, which is over seven times greater than at a location not influenced by ice (Wilson *et al.*, 1986). In the Weddell Sea during austral spring, productivity averaged 562 mg C m^{-2} day^{-1}, a fivefold increase over that of open ocean waters (W. O. Smith and Nelson, 1990). W. O. Smith and Nelson (1986) used a simple model of ice-edge bloom genesis and calculated that the ice edge contributes nearly 40% of the annual production south of the Antarctic Convergence (and that present estimates of the total production of the Southern Ocean are low by ca. 60% if the effect of the ice edge is not incorporated). The model was extended to investigate interannual variability in production resulting from observed variations in ice cover (W. O. Smith *et al.*, 1988). The production maximum occurred in 1973–1974 and the minimum in 1980–1981, and the maximum was over 50% greater than that predicted for the minimum. Mean productivity of the region influenced by pack ice was calculated to be 333.2×10^{12} g C yr^{-1}.

Few studies have investigated the rates of primary productivity during the winter/spring transition, which is characterized by extremely low incident irradiance levels and low solar angles (relative to a tangent to the earth's surface). Brightman (1987) observed extremely low primary productivity values in Fram Strait during March–April (slightly over 10 mg C m^{-2}

day^{-1}), and Brightman and Smith (1989) found productivity to be about 7 mg C m^{-1} day^{-1} in the Bransfield Strait region during June, the period of minimal solar irradiance. Productivity during these periods is clearly minor and further emphasizes the oligotrophic nature of polar regions.

3. Space–time variations in primary productivity

Large variations in primary productivity occur as a result of temporal and spatial variability in polar regions, and these variations can be attributed to factors such as latitude, nutrients, and depth. El Sayed and Weber (1982) concluded that the temporal variations found were insignificant when compared to geographic variations, which implies that there are small- or mesoscale processes operating which strongly influence primary productivity and obscure any large-scale trends.

Evidence for strong diel periodicity in Antarctic phytoplankton has been reported (Rivkin and Putt, 1987a). This periodicity varies temporally, with photosynthetic maxima occurring at midday during the austral spring and at midnight during the austral summer under continuous irradiance. Concern was expressed about the effect such a pattern might have on measurements of polar productivity, since some measurements are short-term incubations and centered around solar noon. Whether this effect is uniform throughout polar oceans is unknown.

Few biological studies have addressed the question of interannual variability in phytoplankton standing stocks in polar regions, whereas upwelling areas and temperate systems have been relatively well studied (Barber and Smith, 1981; Cushing, 1981). W. O. Smith et al. (1987) noted significant differences in nitrate and chlorophyll concentrations within the Fram Strait marginal ice zone, but it was impossible to show any cause–effect relationship with any physical process. Substantial interannual differences are known to occur in large-scale ice concentrations (Zwally et al., 1983; Parkinson et al., 1987), and in some locations these have been tied to global perturbations (Niebauer, 1984, 1988). W. O. Smith et al. (1988) modeled the effects of observed variations of ice cover and concentration on the contribution of ice-edge phytoplankton blooms to Antarctic productivity and concluded that such variations could give rise to interannual productivity variations of 50%. They further speculated that such variations will greatly affect growth and survivorship at all trophic levels and have a strong influence on the structure of the Southern Ocean food web.

Recently, remote sensing of biological parameters has been applied to polar regions (Maynard and Clark, 1987; Sullivan et al., 1988; Comiso et al., 1990; see Chapter 3). Satellite sensing of pigments has shown that significant large-scale variability exists in polar regions. At this time there is no operable color sensor in space, but as future satellites are deployed, the assessment of phytoplankton biomass and its relationship with ice and other physical parameters will undoubtedly be greatly clarified.

C. Photosynthesis and Growth Versus Irradiance

The irradiance regime affects phytoplankton growth mainly through variation in irradiance and photoperiod. The algae in turn respond by adaptation, provided that the change in light is not too large or abrupt. Photoadaptation involves alterations of the properties of the photosynthetic apparatus of the cells, characteristics of respiration, as well as overall chemical composition, and the net effects is to minimize the impact of variations in light regime on the growth rate. Some polar phytoplankton species can grow at a nearly constant growth rate for a 4- to 25-fold variation in irradiance provided that there is time to adapt to each irradiance (Sakshaug and Holm-Hansen, 1986; Gilstad, 1987; Sakshaug, 1989). At irradiances below the optimum, the growth rate of adapted cultures may be linearly related to the logarithm of the irradiance (Falkowski and Owens, 1980; Sakshaug and Andresen, 1986). Compensation for variations in daylength is less pronounced; the growth rate is almost linearly proportional to daylength for short days but varies negligibly between 19 and 24 h of daylength (Sakshaug and Andresen, 1986; Gilstad, 1987).

The photosynthetic response (P^B) is usually expressed as a function of irradiance (I). Platt *et al.* (1980) derived an expression for this relationship:

$$P^B = P_S^B(1 - e^{-\alpha I/P_S^B})\, e^{-\beta I/P_S^B} \qquad (9.2)$$

in which P^B is the photosynthesis (milligrams of carbon per milligram of chlorophyll per hour) at irradiance I (micromoles of photons per square meter per second), P_S^B the maximum photosynthetic rate, α the slope of the curve at zero irradiance (units those of P^B over I) and β a term describing photoinhibition (units those of α). The irradiance at which photosynthesis becomes saturated is I_k, or

$$I_k = P_S^B/\alpha \qquad (9.3)$$

I_k is conventionally used as an index of photoadaptation, and the coefficients α and β also vary with the physiological state. If a population is moved from one irradiance to another and given sufficient time, I_k, α, and β will all change to new values. Ideally, data for P versus I curves should therefore be generated rapidly to preclude adaptation, which in practice means that experiments should be less than 4 h (Gallegos *et al.*, 1983), so that the derived parameters are an instantaneous measure of the *in situ* photosynthetic state.

Irradiance has been measured in a number of manners in the past as well.

Table 9.2 Range of Photosynthesis/Irradiance Parameters Determined in Polar Waters

Study area	Season	Depth	α^a	β^b	I_k^c	P_S^{Bd}	Reference
Arctic							
Lancaster Sound	Summer	Surface	0.008–0.011	0.172–0.249	275–321	0.90–0.95	Gallegos et al. (1983)
		Pycnocline	0.006–0.012	0.097–0.616	95–317	0.24–1.09	
Baffin Bay	Summer	Surface	0.006–0.008	0.097–0.165	489–713	0.76–1.50	Gallegos et al. (1983)
		Pycnocline	0.001–0.056	0.086–0.820	154–443	0.15–1.58	
Baffin Bay	Summer	Surface	0.009–0.100	0.021–0.236	88–174	0.95–2.97	Platt et al. (1982)
		Pycnocline	0.003–0.017	0.451–3.412	43–177	0.71–2.60	
Scott Inlet	Autumn	Surface	0.009	0.408	177	1.05	Platt et al. (1980)
		Pycnocline	0.008	0.472	79	0.81	
Greenland Sea	Winter	Surface	0.019–1.119	0.148–20.38	6–71	1.42–6.55	Brightman (1987)
		Subsurface	0.048–0.422	1.733–10.34	19–104	6.06–8.83	
Greenland Sea	Summer	Surface	0.006–0.134	0.226–10.89	41–118	0.43–7.20	W. O. Smith, Jr. (unpublished)
		Pycnocline	0.009–0.017	0.489–7.049	41–66	0.44–2.63	
Resolute Bay	Spring	Ice algae	0.007–0.094	0.01–0.44	0.5–4.3	0.03–0.08	Cota (1985)

								Reference
Antarctic								
McMurdo Sound	Summer	Surface	0.025–0.124	0.763–10.1	47–144	4.57–9.55		Palmisano et al. (1986a)
Bransfield St.	Winter	Surface	0.006–0.042	0.063–8.522	37–177	0.23–3.68		Brightman and Smith (1989)
		Subsurface	0.010–0.066	0.011–1.984	32– 48	0.53–2.50		
McMurdo Sound	Spring	Surface	0.015–0.029	ND[e]	50– 80	0.90–9.3		Rivkin and Putt (1987b)
		Ice Algae	0.015–0.019	ND	12– 25	0.80–1.9		
		Benthic Diatoms	0.08 –0.10	ND	6	8.9–34.0		
Scotia Sea	Autumn	Surface	0.009–0.045	ND	280–570	1.5–3.9		Sakshaug and Holm-Hansen (1986)
		Subsurface	0.010–0.049	ND	160–550	1.9–4.4		
Bransfield St.–Scotia Sea	Spring	Surface	0.004–0.011	ND	44– 88	0.56–2.66		Tilzer et al. (1986)
		Subsurface	0.005–0.009	ND	56– 95	1.54–2.47		
McMurdo Sound	Summer	Ice algae	0.011–0.043	0.072–0.394	3.4–8.5	0.08–0.23		Palmisano et al. (1986b)
Kerguelen Is.	Autumn	Surface	0.07–0.37[f]	ND	3.0–26[f]	0.13–2.23		Jacques (1983)

[a] mg C (mg Chl)$^{-1}$ h^{-1} (μmol m^{-2} s^{-1})$^{-1}$.
[b] mg C · 10^{-3} (mg Chl)$^{-1}$ h^{-1} (μmol m^{-2} s^{-1})$^{-1}$.
[c] μmol m^{-2} s^{-1}.
[d] mg C (mg Chl)$^{-1}$ h^{-1}.
[e] ND, Not determined.
[f] Expressed in klux ($\approx$17.5 μmol m^{-2} s^{-1}).

Irradiance (I, in units of energy) is measured with a cosine-corrected sensor, whereas scalar irradiance (E_0) is measured with a spherical, noncorrected sensor. Earlier work (e.g., Platt et al., 1982) generally used energy units, whereas more recent work has used scalar irradiance values (e.g., Palmisano et al., 1986b). The latter more closely measures the *in situ* environment of a cell, but in this chapter we will try to use the units which were originally used.

Polar phytoplankton exhibit substantial variation in photosynthetic parameters in different communities (Table 9.2). However, a pattern for Arctic phytoplankton relative to sampling depth has become apparent from studies in the Canadian Arctic (Platt et al., 1982) (Fig. 9.9). First, low-light-adapted phytoplankton have a value of P_S^B which is close to or somewhat lower than those of high-light-adapted populations. In addition, P_S^B is closely related to temperature (Harrison and Platt, 1986). Second, α values of low-light-adapted phytoplankton are close to or somewhat lower than those of high-light-adapted ones. Third, β is larger in low-light-adapted populations, so they are more susceptible to inhibition of photosynthesis at high irradiances.

Antarctic phytoplankton also exhibit adaptations in their photosynthetic responses. As in Arctic phytoplankton, P_S^B and α are highly variable and generally low, although Palmisano et al. (1986b) reported anomalously high photosynthetic parameters for *Phaeocystis*. The *Phaeocystis* populations were advected from open water and beneath the ice, and the highest values of α were observed in populations which presumably had been under the ice the

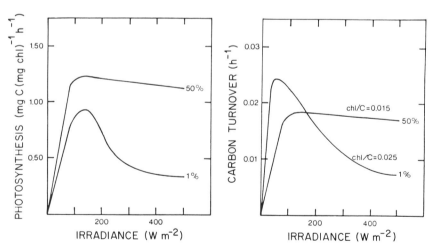

Figure 9.9 Photosynthesis versus irradiance responses in summer in Baffin Bay at 50 and 1% light depths. Photosynthetic rate is normalized to chlorophyll (left) and carbon (right). From Platt et al. (1982).

longest. This trend differs from that described for Arctic phytoplankton (Fig. 9.9) and for phytoplankton in general (Prezelin, 1981). In addition, presently available data (Table 9.2) indicate substantial differences among species [e.g., low α and high I_k values for *Odontella* indicate that high irradiances (low mixed-layers depths) are necessary for this species to bloom; Sakshaug and Holm-Hansen, 1986].

Few data are available from polar regions to discern the relationship between the irradiance at P_s^B (I_m of Platt *et al.*, 1980) and the irradiance to which the population is adapted. In the Canadian Arctic, Platt *et al.* (1982) found that for populations from the 50% isolume the I_m/I_k ratio was 5, and for populations from the 1% isolume the ratio was 3.2. Data for the diatom *Skeletonema costatum* indicate that the irradiance at P_s^B is from 3 (high-light adapted) to 50 (low-light adapted) times higher than the irradiance in which the cultures were grown (Falkowski and Owens, 1980). Thus *Skeletonema* can tolerate peak irradiances well above the one to which it is adapted. At high irradiances photoinhibition will occur, but for some species such irradiances are above the ambient light levels in polar waters. Typically, photoinhibition increases with exposure for up to 1 to 4 h (Lewis and Smith, 1983; Sakshaug and Andresen, 1986).

In order to calculate growth rates, it is necessary to convert the chlorophyll-specific photosynthetic response to a carbon-specific response. To do this requires an accurate estimate of the carbon/chlorophyll (C/Chl) ratio. Low-light-adapted algae usually have a lower C/Chl ratio than those adapted to high-light conditions. As a result, α (carbon-specific) becomes higher for low-light-adapted cells than for high-light-adapted populations (Fig. 9.9). Results for Antarctic phytoplankton indicate that P versus I curves for samples of different adaptive status become similar after conversion to carbon-specific rates (Palmisano *et al.*, 1986b; Sakshaug and Holm-Hansen, 1986).

The algal C/Chl ratio is a highly variable parameter and responds to changes in all environmental factors which affect the growth rate (temperature, daylength, irradiances, nutrients, etc.). The lowest ratio observed is about 20 (w/w), although 10-fold higher ratios are possible (Sakshaug, 1989). In principle, if the light regimen is kept constant, the C/Chl ratio should be closely related to the growth rate when growth is controlled by nutrients or temperature. This implies that the C/Chl ratio will become high when nutrient deficiency is also pronounced or temperature is low. However, the variation in the C/Chl ratio is correlated with variations in daylength and irradiance; for a constant irradiance the product of daylength and the C/Chl ratio may be inversely proportional to the growth rate, whereas for variable irradiance and constant daylength the variation in the C/Chl ratio and the growth rate is nonlinear (Sakshaug *et al.*, 1989).

Cultures of polar or subpolar phytoplankton at variable irradiance (when other factors are constant) indicate that the minimum C/Chl ratio appears when the cultures are grown at the lower part of the optimum irradiance range and that high photon fluxes, particularly in the inhibitory range, suppress the chlorophyll content of the cells markedly (C/Chl ratios by weight range from 20 to 50 at low irradiances and from 100 to 200 at high irradiances). Also, at low irradiances the C/Chl ratio may be significantly higher than it is near optimum growth. The latter tendency is very obvious in *Skeletonema* and is due to a marked increase in cellular carbon (and nitrogen) in extremely low-light-adapted cells, which possibly represents a storage pool of small, nitrogen-rich organic compounds (Sakshaug and Andresen, 1986).

The single greatest obstacle to the estimation of C/Chl ratios in natural phytoplankton populations (as well as the estimation of *in situ* growth rates) is the direct measurement of carbon in viable, active algal cells. Attempted corrections are mainly useful as an average for a large number of samples but may still lead to underestimates of the C/Chl ratio. Errors in this estimate are probably minimized in samples from blooms. For almost pure blooms of *Skeletonema* in the Trondheimsfjord near midsummer, C/Chl ratios in the mixed surface layer were about 67, in contrast to a ratio of 23 in the pycnocline; for the same species during the spring bloom in early April, the ratio was 32 (Sakshaug and Andresen, 1986). Thus both a vertical and seasonal gradient in irradiance and photoperiod occurred and was similar to that observed in culture. A pronounced difference between phytoplankton in the upper mixed layer and in the well-defined summer pycnocline has also been observe in Baffin Bay and the Barents Sea (Platt *et al.*, 1982; Sakshaug, 1989).

In the Southern Ocean, Hewes *et al.* (1989) have determined the C/Chl ratio in summer by estimation of phytoplankton carbon by measurement of cellular plasma volume (Utermohl and FTF technique; Hewes and Holm-Hansen, 1983). They found that the C/Chl ratio became lower as phytoplankton standing stocks increased. This was interpreted to be the result of marked self-shading in blooms. In a number of studies of marginal ice zones, elevated carbon/chlorophyll ratios have been observed (an average ratio of 118 in the Ross Sea during the austral summer and 87.4 in the Weddell Sea during austral spring; W. O. Smith and Nelson, 1985; Nelson *et al.*, 1989). Similar variations were not observed in the Fram Strait, where the average C/Chl was 43 (W. O. Smith *et al.*, 1987).

It is of particular relevance at high latitudes that irradiance and daylength are not equivalent with respect to growth rate. For example, 4 h of 200 μmol m^{-2} s^{-1} will not yield the same growth rate as 8 h of 100 μmol m^{-2} s^{-1}, despite the fact that the integrated irradiance for the two is identical (Sakshaug *et al.*, 1989). This may in part result from the fact that daylength regulation of the C/Chl ratio is mediated by a mechanism which is different from that of regulation by irradiance (Post *et al.*, 1985).

The models which relate growth to light regime have so far been restricted to steady-state situations (Bannister, 1979; Kiefer and Mitchell, 1983; Sakshaug *et al.*, 1989). Even if adaptations take place during vertical transport and within the diurnal cycle (Rivkin and Putt, 1987c), they may describe growth rate fairly adequately by integration of the variations in irradiance during the light cycle providing a realistic "average" set of adaptational coefficients (Gallegos and Platt, 1982). Estimating the C/Chl ratio in the field may be a greater problem.

D. Nutrient Uptake Relationships

Despite the fact that nutrients in Arctic waters reach undetectable levels during summer months and therefore are potentially limiting to phytoplankton growth, few direct measurements of nutrient uptake have been made. Relatively few direct measurements in the Southern Ocean have been made as well, although more observations are available for the Antarctic than the Arctic. Furthermore, the measurements which have been made to date have not indicated a general pattern, unlike experiments in temperate waters (McCarthy, 1981), and clearly a great deal of work remains to be done on nutrient dynamics in polar waters.

The uptake of nutrients has been investigated in polar regions using isotopic tracer techniques (Table 9.3) and is assumed to follow Monod kinetics, as has been found in temperate and tropical waters (e.g., McCarthy, 1981):

$$V = V_{max} [S/(K_s + S)] \qquad (9.4)$$

where V is the uptake velocity, V_{max} the maximum uptake velocity, S the substrate (or nutrient) concentration, and K_s the half-saturation constant for the substrate, which is equivalent to the velocity at which $V = V_{max}/2$.

Müller-Karger and Alexander (1987) found that the uptake of inorganic nitrogen in the marginal ice zone averaged 0.007 and 0.021 h^{-1} for ammonium and nitrate, respectively, for the mixed layer. These rates were sufficient during the 12-day study to preclude any limitation of phytoplankton growth by nutrients. W. O. Smith and Kattner (1989) found uptake rates in the Fram Strait region to be moderate (mean rates for nitrate and ammonium uptake at the surface of 0.0037 and 0.0028 h^{-1}, respectively), despite the relatively low ambient concentrations encountered. From the data presented in Harrison *et al.* (1982), nitrate and ammonium uptake rates can be approximated and are about 0.0029 and 0.0021 h^{-1} (if a 24-h photoperiod

Table 9.3 Ranges of Specific Uptake Rates of Various Nutrients as Determined Using Natural Assemblages in Polar Regions

Region	Season	Substrate	V (h^{-1})	Reference
Arctic				
Bering Sea	Summer	Silicic acid	0.0016–0.0075	Banahan and Goering (1986)
Bering Sea	Summer	Nitrate	0.0044–0.0113	Müller-Karger and Alexander (1987)
		Ammonium	0.0108–0.0358	
Fram Strait	Summer	Nitrate	0.0019–0.0092	W. O. Smith and Kattner (1989)
		Ammonium	0.0009–0.0068	
Barents Sea	Summer	Nitrate	0.0001–0.0093	Kristiansen and Lund (1989)
		Ammonium	0.0006–0.0076	
Baffin Bay	Summer	Nitrate	0.0008–0.0063	Harrison et al. (1982)
		Ammonium	0.0007–0.0060	
Antarctic				
Ross Sea	Summer	Nitrate	0.0008–0.0043	Nelson and Smith (1986)
		Ammonium	0.0012–0.0083	
		Silicic acid	0.0006–0.0066	
Scotia Sea	Autumn	Nitrate	0.0003–0.0014	Gilbert et al. (1982)
		Ammonium	0.0007–0.0038	
Scotia Sea	Spring	Nitrate	0.0016–0.0021	Olson (1980)
		Ammonium	0.0021–0.0031	
Scotia Sea	Summer	Nitrate	0.0001–0.0025	Rönner et al. (1983)
		Ammonium	0.0011–0.0034	
Scotia Sea	Autumn	Nitrate	0.0003–0.0014	Koike et al. (1986)
		Ammonium	0.0022–0.0059	

and a chlorophyll/particulate nitrogen ratio of 5.0 are assumed). Again, low ambient inorganic nitrogen concentrations were encountered, but uptake rates during sampling were not greatly depressed by *in situ* nutrient levels. The specific uptake rates from these studies are similar, which is surprising given the potential effects of detrital contamination during incubation, and may indicate the effect of temperature on controlling maximum uptake rates in polar systems.

Only a few studies have investigated the uptake of urea in Arctic waters. Harrison et al. (1985) investigated the uptake of urea relative to nitrate and ammonium and found that, on average, urea uptake accounted for 32% of the total (nitrate, ammonium plus urea) phytoplankton nitrogen demand. Urea concentrations commonly ranged from 0.1 to 2.0 μmol l^{-1}, and hence at a majority of stations urea was the largest pool of dissolved nitrogen. Ammonium and nitrate were the generally preferred substrates; in addition, the urea often was metabolized rapidly, with the carbon units being respired as CO_2 and the ammonium incorporated into cellular material. No light dependence was noted, although a general relationship between primary productivity and urea uptake was noted. Despite the fact that other studies have indicated that urea uptake is primarily via phytoplankton, the study of Harrison et al. (1985) was inconclusive in demonstrating the agent of urea removal. Hansell and Goering (1989) measured urea uptake in the Bering and Chukchi seas, where concentrations at some stations reached 6 μmol l^{-1}. During the periods when urea reached these concentrations, it was an extremely important source of nitrogen for phytoplankton metabolism. The source of the urea apparently was benthic and ultimately arose from the previous year's bloom which had sunk to the benthos.

The Southern Ocean is characterized by extremely high ambient nutrient concentrations, and hence *a priori* it might be expected that such waters would be dominated by nitrate removal. However, a number of studies have shown that ammonium is a quantitatively important source of nitrogen for phytoplankton growth. For example, Rönner et al. (1983) found that surface ammonium and nitrate uptake ratios averaged 0.0022 and 0.0006 h^{-1}, respectively (calculated by integrating the three points given per station and from a daily rate by division by 24). Similarly, Olson (1980), Glibert et al. (1982), and Koike et al. (1986) found ammonium uptake rates that were as large as or larger than nitrate uptake rates (Table 9.3). Considerable variation exists in the geographic and temporal settings of these studies, and this is reflected in the wide variation in results.

In contrast to these studies, Nelson and Smith (1986) found that a substantial portion of the phytoplankton nitrogen demand was satisfied by nitrate in the waters of the Ross Sea. Mean depth-integrated rates of removal of nitrate and ammonium were 0.0023 and 0.0036 h^{-1}, respectively. Glibert et al. (1982) found that nitrate uptake was less sensitive to light, and Rönner et al. (1983) found that the ammonium uptake rate in the dark ranged from 13 to 75% of the rates in the light. Nitrate uptake in dark bottles ranged from 27 to 100% of the light uptake rates, and the percentage was greater at the bottom of the euphotic zone. It has been speculated that this effect is the result of heterotrophic incorporation of nitrate (Glibert *et al.,* 1982), but experimental verification remains unavailable. W. O. Smith and Nelson (1986) also found that nitrate uptake was not as tightly coupled to light intensity as had been found in temperate waters (McCarthy, 1981), whereas the reverse was true for ammonium uptake.

No direct estimates of half-saturation constants for either nitrate or ammonium are available for polar populations. Rönner et al. (1983) conducted kinetic experiments but found that at all concentrations tested (with the lower range of additions being 10% of ambient concentrations) nitrate and ammonium uptake was saturated. Nitrate concentrations were always greater than 15 µmol l⁻¹, and ammonium concentrations at the stations where these experiments were run were always greater than 0.9 µmol l⁻¹. This suggests that the K_s value for ammonium is not greater than 0.5 µmol l⁻¹. Glibert et al. (1982) reported that ammonium uptake was saturated at concentrations of 5–10 µmol l⁻¹, but their data showed considerable variability. Present data are insufficient to demonstrate whether the nitrogen uptake properties of polar phytoplankton are different from those of phytoplankton from other oceans.

Uptake of silicic acid into biogenic silica by diatoms is of particular interest because of the large range of silicic acid concentrations found in polar regions and because of the presence of diatom-dominated phytoplankton assemblages in polar regions as well as the large deposits of diatomaceous oozes found in Arctic and Antarctic sediments. Banahan and Goering (1986) found that in the Bering Sea, specific uptake velocities of silicic acid ranged from 0.002 to 0.015 h⁻¹ in surface waters and were similar to values from the Ross Sea reported by Nelson and Gordon (1982) and Nelson and Smith (1986) (Table 9.3). Jacques (1983) reported K_s values for silicic acid in cultures of diatoms to range from 12 to 22 µmol l⁻¹; these values are extremely high, nearly an order of magnitude higher than those found in temperate species. If these kinetic constants can be extrapolated to natural populations, they suggest a possibility that in some diatom blooms (e.g., W. O. Smith and Nelson, 1985) growth can become limited by ambient silicic acid concentrations. Further work using natural populations is needed to confirm this suggestion.

An ecologically important parameter of nitrogen uptake is the f ratio, which is the ratio of nitrate uptake to total nitrogen uptake (Eppley and Peterson, 1979) and is a measure of the production available for export. Such material generally is exported to depth via passively sinking particles (whole cells, fecal material) and represents a loss of nitrogen from the euphotic zone. In areas where surface nitrogen concentrations and primary production are extremely low, f ratios are generally also low (ca. 0.1); conversely, where inorganic nitrogen fluxes support elevated primary production, f ratios are high, approaching 0.5 (Eppley and Peterson, 1979). Variations also occur within the water column, with values being greatest at depths with substantial nitrate flux and sufficient light and lowest near the surface in nitrate-depleted waters (Lewis et al., 1986). Some studies of polar waters have shown low f ratios, indicating that a large portion of the primary production was being supported by ammonium; however, an equal number of studies have reported large f ratios. When all data available from polar regions are plotted, no consistent trend is found (Fig. 9.10), implying that polar primary productivity is not limited by nutrient supply, as the Eppley and Peterson relationship requires.

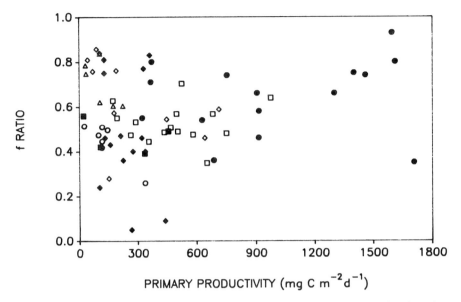

Figure 9.10 f ratios taken from various studies and seasons in polar regions as a function of independently derived primary productivity measurements. Data from Glibert et al. (1982) and El-Sayed and Taguchi (1981) (O), Nelson and Smith (1986) and Wilson et al. (1986) (●), W. O. Smith and Nelson (1990) [1983 spring: (□), 1986 autumn [△], W. O. Smith and Kattner (1989) and W. O. Smith et al. (1987) (◇), Harrison et al. (1982) (◆) and Olson (1980) (■).

E. Elemental Ratios

The ratios of various elemental constituents (e.g., carbon, nitrogen, silica, ATP, chlorophyll, protein) to each other have been used as indices of phytoplankton activity and physiological status. For example, high carbon/nitrogen ratios are often indicative of nitrogen deficiency, high carbon/phosphorus ratios can indicate phosphorus deficiency, and low ATP/cell ratios can indicate a variety of physiological stresses (e.g., nutrient deficiency; Parsons et al., 1984). Such ratios have been used in the field to evaluate the phytoplankton nutrient status (e.g., Sakshaug and Olsen, 1986; Paasche and Erga, 1988), although temperature and light may also affect the various ratios. In addition, species differences strongly influence community composition. For example, the N/P ratio of 16 (atoms) of nutrient-saturated mixed communities is the average of values produced by the represented species, which may range from 8 *(Skeletonema)* to 23 (*Amphidinium carterae*; Sakshaug et al., 1983). It appears that the major cause of these variations is the variation in cellular phosphorus, although some species with highly cellulosic cell walls (e.g., *Ceratium*) have high C/N ratios (Sakshaug et al., 1984). The major difficulty in the interpretation of elemental ratios in natural assemblages is the same as that in the interpretation of C/Chl ratios: to remove interference of detritus and heterotrophic contributions to the elemental pool.

Studies of nutrient-saturated cultures of Antarctic phytoplankton have shown that ratios between nitrogen, carbon, phosphorus, and ATP did not vary significantly as a result of adaptation to different light regimes (Sakshaug and Holm-Hansen, 1986). The C/N, N/P, and C/P ratios averaged 7.1, 15.1, and 107.2 (all by atoms), respectively, which are close to the Redfield ratios. The C/ATP ratio averaged 263.2, which is near the ratio originally proposed (Holm-Hansen, 1970), although we now know that the ratio shows extreme variations (e.g., Weiler and Karl, 1979). Field data from the Antarctic indicate that ratios of carbon, nitrogen, and phosphorus vary little from the expected Redfield ratios.

Nutrient limitation and signs of nutrient deficiency might be expected in surface waters of the Arctic after the major phytoplankton bloom. Data indicate relatively high C/N ratios at some stations (e.g., from the Barents Sea; Kristiansen and Lund, 1989), but no distinct physiological sign of nutrient deficiency during low-nutrient periods has been observed (Harrison *et al.*, 1982). The summer period may be characterized by a low standing stock of algae and a high grazing rate and therefore presumably high fluxes of ammonium. As a result, it is likely that algal biomass is limited by nutrients but that the algal growth rate is not limited by nutrient supply, a situation similar to that in central gyres (e.g., Goldman *et al.*, 1979). However, this is different from the situation in the Antarctic, where inorganic nutrients never limit phytoplankton biomass, although nutrient interactions influence growth rates.

The ratio of biogenic silica to other elements exhibits considerable variation in both culture and natural populations (Paasche, 1980; Brzezinski, 1985). The mean atomic silica/carbon ratio is 0.13 and is affected only slightly (an overall threefold variation) by temperature, light, nutrients, photoperiod, and species. The Si/N atomic ratios averaged 1.05 and showed variations similar to those of Si/C (Brzezinski, 1985). However, in the Antarctic elevated levels of Si/C and Si/N have been observed. For example, Copin-Montegut and Copin-Montegut (1978) reported Si/N ratios of 2.0–2.4, which correspond to Si/C ratios of 0.31–0.37. E. Sakshaug, O. Holm-Hansen & C. Hewes (unpublished) found mean Si/C ratios of 0.26, with maximum values of 0.42. It was also found that Si/N ratios based on nutrient anomalies were 3.1–3.9, corresponding to Si/C ratios of 0.5–0.6. W. O. Smith and Nelson (1985) and Nelson *et al.* (1989) also found extremely high silica/carbon ratios (0.62 in the Ross Sea during austral summer, 0.54 in the Weddell Sea during austral spring). Furthermore, the elemental ratio observed in the Ross Sea was nearly the same as the molar ratio of silica to carbon uptake (Nelson and Smith, 1986) and hence was clearly a result of growth rather than a detrital influence. Whether these ratios are an adaptation to a particular environmental parameter of marginal ice zones is unknown.

III. Processes Affecting Phytoplankton Distribution

A number of processes that affect phytoplankton distribution in polar regions are independent of those that influence growth. These include physical processes (e.g., fronts, eddies, interleaving) and biological processes (sinking, grazing). Although physical effects are of primary importance, such processes initially redistribute phytoplankton passively and secondarily influence their growth via nutrient and/or stratification effects. We will not deal with these processes directly but recognize that these oceanographic processes are extremely important in distributing biogenic material in the polar ocean's surface layers.

Sinking of phytoplankton cells has been investigated widely in temperate and tropical regimes, but only recently has the process been investigated in polar seas. Phytoplankton assemblages in the Weddell Sea sank approximately 1.0 m day^{-1}, which is faster than other assemblages measured in other regions (Johnson and Smith, 1986). Sinking was not a function of physical factors (e.g., density, viscosity) but was controlled by factors which influenced biological processes (light, nutrients, taxonomy of assemblage). In the Arctic, sinking rates were much lower (ca. 0.1 m day^{-1}) and appeared to be a function of the taxa present (flagellates rather than diatoms; Culver and Smith, 1989). In a shallow (60 m) Alaskan embayment, rates of phytoplankton sinking were moderate; in addition, the material sank from the euphotic zone and nearly quantitatively appeared in sediment traps (Laws *et al.*, 1988). Smetacek (1985) suggested that polar diatoms form large, mucilaginous aggregations which suddenly and rapidly sink from the euphotic zone; however, such occurrences have never been recorded. Finally, some diatoms form resting spores which sink from the euphotic zone. These resting spores have been observed in the Barents Sea after an ice-edge bloom (Rey and Skjoldal, 1987) and presumably are vertically mixed into the euphotic zone during the prebloom phase in the next year and thus seed the water column. Although these studies far from demonstrate the importance of passive sinking in polar systems, they do indicate the potential significance of the loss process and the quantitatively important link with the benthos.

Further evidence of the quantitative importance of sinking of intact phytoplankton cells is provided by sediment traps. Dunbar *et al.* (1985) found the material collected in the Ross Sea continental shelf area to consist mostly of intact diatom cells. Wefer *et al.* (1988) suggested that the large numbers of intact frustules found in the sediments of Bransfield Strait indicate that direct sedimentation of cells was a more important loss process than incorporation into fecal material through grazing. Large numbers of resting spores have also been found in sediment traps (von Bodungen *et al.,* 1986), although it is unknown what the environmental trigger for sporulation might be.

Grazing of phytoplankton is treated more completely in the following chapter (Chapter 10), but it is important to emphasize that herbivory by pelagic crustaceans (copepods, euphausiids) does not appear to be as important as passive sinking in polar regions (as far as controlling phytoplankton biomass and distribution is concerned). This seems especially true in shallower regions and marginal ice zones, where there are large amounts of new production which apparently is not used within the euphotic zone by herbivores. This material apparently sinks from the surface layer and supports a large and active benthic community (see W. O. Smith, 1987). There are deeper regions, however, in which grazing is relatively more important. For example, van Bodungen *et al.* (1986) found that in the productive waters of the Antarctic Peninsula and Bransfield Strait, grazing was negligible as a loss process; conversely, in the unproductive regions of Drake Passage and Scotia Sea, grazing was a quantitatively important loss process.

IV. Ice-Algal Production

A. Types of Sea Ice Communities

The major distinguishing feature of polar regions—ice—is also the site of a unique community: the sea ice community. This group of organisms includes autotrophic and heterotrophic members (see Garrison *et al.,* 1986), including large herbivores and a mixed assemblage of nektonic species (Ainley *et al.,* 1986). The composition of sea ice communities is largely dependent on the physical composition of the ice. Algae (and other associated organisms) grow to high densities on the undersurface of the ice, both attached to ice crystals and in the interstitial brine pockets. The algae can be in narrow bands or dispersed throughout the ice (Horner, 1985), and concentrations of algae vary widely on all space and time scales.

The biomass and composition of ice algae are well documented (Horner, 1976, 1985; Ackley *et al.,* 1979; Garrison *et al.,* 1986). Diatoms generally are the most abundant form in all sea ice communities, and chlorophyll concentrations of more than 300 mg m^{-2} in Antarctic ice (Palmisano and Sullivan,

1983) and 100 mg m^{-2} in the Arctic (R. E. H. Smith *et al.*, 1988) have been reported. In both of these studies the algae were concentrated at the ice–seawater interface. R. E. H. Smith *et al.* (1988) modeled the yields attainable by the relationship among irradiance penetrating ice, chlorophyll concentrations, and the spectral extinction coefficient of chlorophyll. The maximum biomass represents the standing stock beyond which any further increase would result in self-shading and negative net production in the portion of the assemblage below the compensation point. They found that the low irradiances under sea ice often limited growth rates but that observed standing stocks of ice algae are generally less than the maximum attainable biomass. Lower yields were further explained by decreases in net production at inhibitory irradiances and by grazing by amphipods.

B. Environmental Factors Influencing Growth of Ice Algae

The environmental parameters which influence the growth of ice algae are no different from those which affect phytoplankton: light, nutrients, micronutrients, etc. However, the distribution and flux of such factors are greatly different because the growth takes place at an interface and the vertical dimension is greatly reduced, producing a nearly two-dimensional system. The presence of ice greatly modifies the attenuation of light and the flux of nutrients, and hence the entire character of growth and accumulation is often different in sea ice communities.

Temperature is nearly constant in ice communities, and the overall growth response is similar to that which describes phytoplankton [Eq. (9.1)]. C. W. Sullivan (unpublished) has found a marked temperature effect on some photosynthetic parameters of ice algae but, in general, temperature is not considered to be a major factor regulating the growth, metabolism, and distribution of ice communities. Similarly, micronutrients (metals, organics) have not been investigated as controlling factors of ice-algal growth.

Light, unlike temperature, is highly variable within ice communities and greatly influences growth of ice algae on a variety of time scales. Ice communities are strongly shape adapted (Cota, 1985), with elevated α and P_S^B values (Table 9.2). Because of this, their growth can be initiated early in the spring, since they are maintained in an environment with relatively constant (albeit low) light by their presence on the ice (Fig. 9.3). Spatial variations in light occur due to snow cover and ice thickness (e.g., Palmisano and Sullivan, 1983; R. E. H. Smith *et al.*, 1988), and temporal variations are induced by short-term events of melting and snowfall. In general, irradiance at the ice–seawater interface is reduced to about 1% of that incident at the surface of the ice (Maykut and Grenfell, 1975; Sullivan *et al.*, 1984) and is even less if heavy snow cover is present. In addition, the spectral qualities of the light are altered during passage through ice, with the longer and shorter wavelengths being preferentially removed (Sullivan et al., 1984).

Nutrients are often greatly reduced at the ice–seawater interface (e.g., Cota et al., 1987), and hence it is probable that nutrients limit ice-algal growth. Because turbulent mixing within the water column is generally reduced during spring and is nearly absent directly under the ice, the flux of nutrients from the water column to the ice-algal community is minimal. Cota et al. (1987) found that nutrient demands of ice algae were substantial and that vertical mixing due to tidal flow was a major force in supplying nutrients to the ice-algal population. Nutrient fluxes were maximal during spring and neap tides, as were nutrient demands. Similar results have also been obtained in the Antarctic (Cota and Sullivan, 1990), where the vertical gradients in nutrients were much less but ambient concentrations of nutrients in the vicinity of extensive ice-algal growth were reduced. Thus the potential for nutrient limitation of ice-algal growth, particularly during periods of nearly constant surface irradiance, is great.

C. Ice-Algal Productivity and Irradiance–Nutrient Relationships

Ice-algal productivity is difficult to measure directly, since the environment is heterogeneous, cells are unevenly distributed, and even distribution of isotopic tracers within the ice cannot be ensured. Nonetheless, a number of attempts have been made to assess ice-algal productivity using conventional ^{14}C techniques. For example, chambers have been attached to the underside of ice to which radioisotope has been added (Clasby et al., 1973; Horner and Schrader, 1982). Palmisano et al. (1985) sectioned the ice into thin slices and perfused the slices with a uniformly radiolabeled seawater solution, and others have slowly melted ice samples and used isotope techniques with the resultant water samples (Palmisano et al., 1986b; Rivkin and Putt, 1987b). Such approaches suffer from artificial manipulation of samples, physiological damage due to osmotic shock (Garrison and Buck, 1986), and potential disruption of biotic interactions within the ice community, but they do provide estimates of the input of organic matter from this community.

Horner and Shrader (1982), using *in situ* techniques, found ice-algal productivity to range from 0.02 to 2.60 mg C m^{-2} h^{-1} from April through June, and Clasby et al. (1973) found a range from 0.3 to 7.67 mg C m^{-2} h^{-1}. These and other (Matheke and Horner, 1974) measurements were used by Subba Rao and Platt (1984) to estimate that ice-algal productivity to be about 10 g C m^{-2} yr^{-1}, similar to the estimates of Alexander (1974). Palmisano and Sullivan (1983) estimated productivity in McMurdo Sound from changes in algal biomass; their estimate of 4.1 g C m^{-2} is conservative, since losses due to grazing were not included. For the Antarctic Peninsula during winter, an ice-algal productivity of 35.0 mg C m^{-2} day^{-1} has been estimated (Kottmeier and Sullivan, 1987); if a growing season of 120 days is assumed, this estimate is similar to that of Subba Rao and Platt (1984). R. E. H. Smith et al. (1988) calculated the ice-algal production of Resolute Passage to be 5–23 g C m^{-2} yr^{-1}, a value approximately equal to that of the plankton of the region.

Photosynthesis/irradiance measurements have been made on a variety of ice-algal assemblages. Cota (1985) found that ice algae were extremely shade adapted. Mean P_S^B and α values of 0.052 mg C (mg chlorophyll)$^{-1}$ h^{-1} and 0.039 mg C (mg chlorophyll)$^{-1}$ h^{-1} (μmol m^{-2} s^{-1})$^{-1}$, respectively, were determined (Table 9.2). Similar α values were observed by R. E. H. Smith et al. (1987), who also found a correlation of assimilation number (P_S^B) with tidal frequency which they attributed to tidally induced nutrient flux. Antarctic ice algae also have been found to exhibit light-limited photosynthetic rates similar to those in the Arctic (Palmisano et al., 1985).

Cota et al. (1987) modeled the nutrient demand by Arctic ice algae and found that about 265 mg N m^{-2} day^{-1} was required to satisfy the growth of the assemblage. Growth rates calculated from these data were low, with a maximum doubling time of 0.26 doublings per day. No direct measurements of nutrient uptake by ice algae are as yet available.

V. Coupling of Ice Community with Water Column

It has long been recognized that significant taxonomic overlap exists between ice-algal communities and phytoplankton assemblages in the water column, and it has also been recognized that the phytoplankton of marginal ice zones is often different from that of the open ocean. Yet it remains unclear to what degree ice algae released during ice melt remain active in the water column or what their residence time might be. In the Beaufort and Bering seas the ice edge and ice-algal blooms are separated in time and have distinct taxonomic compositions (Clasby et al., 1973; Schandelmeier and Alexander, 1981; Horner and Schrader, 1982), whereas in the Ross and Weddell seas considerable taxonomic similarity exists between the two (W. O. Smith and Nelson, 1985; Garrison et al., 1987). Within the Antarctic Peninsula there was a great deal of overlap between ice-algal species and those present in the water column, and the community changed rapidly after the complete disappearance of ice (Krebs, 1983), which suggests that there is a constant interchange between ice and water in this region. W. O. Smith and

Nelson (1986) noted that in a phytoplankton bloom associated with a receding ice edge there should be a change in species composition away from the ice as the bloom intensifies (i.e., the greatest taxonomic similarity between ice and water assemblages will occur under the ice and decrease with distance from the ice). This apparently was observed in the Weddel Sea (Fryxell and Kendrick, 1988). Because there is such a disparity among marginal ice zones with regard to ice-algal biomass, rates of ice ablation, vertical mixing, and *in situ* growth rates, it is likely that the extent of "seeding" by ice algae will vary with the dynamics of each system.

Particle flux rates from ice undoubtedly are controlled by the rate of ablation, which in turn is dependent on the specific local conditions. In the Beaufort Sea, downward particle flux from ice began in early spring and continued at low and uniform rates until ice breakup (Carey, 1987). The flux rate was equal to approximately 1% of the rate of ice-algal production. Fecal material from ice-algal grazers consisted of a small portion of the flux and occurred predominately in late spring. The ice-algal contribution to the benthic carbon requirement in this shallow region was small but available earlier than that from the water column.

VI. Conclusions

Much remains to be learned about the physiology and ecology of polar phytoplankton. It appears that the overall productivities of the Arctic and Antarctic are similar, yet each environment is unique. In particular, the controls of phytoplankton growth remain paradoxical. We do not have a clear understanding of the seasonal cycles of production and flux in polar regions, and much remains to be learned about the partitioning of productivity within polar ecosystems. How phytoplankton "survive" the polar winter remains in the realm of speculation, as does the importance of phytoplankton processes in polar regions to global processes such as carbon dioxide budgets. Finally, much remains to be learned concerning the teleconnections of polar regions to the temperate and tropical seas. Significant oceanographic events occur with dramatic biological effects (e.g., massive mortality of birds in the Antarctic during 1983–1984; Heywood *et al.,* 1985), yet the causes of these occurrences can presently only be speculated on. Without a doubt, studies of polar phytoplankton processes will continue to receive increased attention, and these investigations in future decades will contribute much to our knowledge of the structure and function of polar ecosystems.

Acknowledgments

We thank W. G. Harrison and G. Cota for helpful reviews during the manuscript preparation, as well as our many colleagues and students who have worked with us under adverse polar conditions. We are also grateful for the support from the Office of Naval Research, the National Science Foundation, the Norwegian Research Council for Science and the Humanities, the Norwegian Fisheries Research Council, and the Ministry of the Environment.

References

Ackley, S. F., K. R. Buck & S. Taguchi. 1979. Standing crop of algae in the sea ice of the Weddell Sea region. *Deep-Sea Res.* **26**: 269–282.

Ainley, D. G., W. R. Fraser, C. W. Sullivan, J. J. Torres, T. L. Hopkins & W. O. Smith. 1986. Antarctic mesopelagic micronekton: Evidence from seabirds that pack ice affects community structure. *Science* **232**: 847–849.

Alexander, V. 1974. Primary productivity regimes of the nearshore Beaufort Sea, with reference to the potential role of ice biota. *In* "The Coast and Shelf of the Beaufort Sea" (J. C. Reed & J. E. Sater, eds.), pp. 609–632. Arctic Inst. North Am., Arlington, Virginia.

Alexander, V. & H. J. Neibauer. 1981. Oceanography of the eastern Bering Sea ice-edge in spring. *Limnol. Oceanogr.* **26**: 1111–1125.

Appollonio, S. 1980. Primary production in Dumbell Bay in the Arctic Ocean. *Mar. Biol. (Berlin)* **61**: 41–51.

Banahan, S. & J. J. Goering. 1986. The production of biogenic silica and its accumulation on the southeastern Bering Sea shelf. *Cont. Shelf Res.* **5**: 199–213.

Bannister, T. T. 1979. Quantitative description of steady state, nutrient-saturated algal growth, including adaptation. *Limnol. Oceanogr.* **24**: 76–96.

Banse, K. 1974. On the role of bacterioplankton in the tropical ocean. *Mar. Biol. (Berlin)* **24**: 1–5.

Barber, R. T. & R. L. Smith. 1981. Coastal upwelling ecosystems. *In* "Analysis of Marine Ecosytems" (A. R. Longhurst, ed.), pp. 31–68. Academic Press, New York.

Bradstreet, M. S. W. & W. E. Cross. 1982. Trophic relationships at high Arctic ice edges. *Arctic* **35**: 1–12.

Bricaud, A., A. Morel & L. Prieur. 1983. Optical efficiency factors of some phytoplankton. *Limnol. Oceanogr.* **28**: 816–832.

Brightman, R. I. 1987. Photosynthesis–irradiance relationships of polar phytoplankton populations during winter conditions. M. S. Thesis, Univ. of Tennessee, Knoxville.

Brightman, R. I. & W. O. Smith, Jr. 1989. Photosynthesis–irradiance relationships of Antarctic phytoplankton during austral winter. *Mar. Ecol.: Prog. Ser.* **53**: 143–151.

Brzezinski, M. A. 1985. The Si:C:N ratio of marine diatoms: Interspecific variability and the effect of some environmental variables. *J. Phycol.* **21**: 347–357.

Buckley, J. R., R. Gammelsrød, J. A. Johannessen, O. M. Johannessen & L. P. Røed. 1979. Upwelling: Oceanic structure at the edge of the Arctic ice pack in winter. *Science* **203**: 165–167.

Bunt, J. S. & C. C. Lee. 1972. Data on the composition and dark survival of four sea-ice microalgae. *Limnol. Oceanogr.* **17**: 458–461.

Burkholder, P. R. & E. F. Mandelli. 1965. Carbon assimilation of marine phytoplankton in Antarctica. *Proc. Natl. Acad. Sci. U. S. A.* **54**: 437–444.

Carey, A. G., Jr. 1987. Particle flux beneath fast ice in the shallow southwestern Beaufort Sea, Arctic Ocean. *Mar. Ecol.: Prog. Ser.* **40**: 247–257.

Carlucci, A. F. & R. L. Cuhel. 1977. Vitamins in the southern polar seas: Distribution and significance of dissolved and particulate vitamin B_{12}, thiamine, and biotin in the southern Indian Ocean. *In* "Adaptations Within Antarctic Ecosystems" (G. A. Llano, ed.), pp. 115–128. Gulf Publ. Co., Houston, Texas.

Clasby, R. C., R. Horner & V. Alexander. 1973. An in situ method for measuring primary productivity of Arctic sea ice algae. *J. Fish Res. Board Can.* **30**: 835–838.

Codispoti, L. A. 1979. Arctic Ocean processes in relation to the dissolved silicon content of the Atlantic. *Mar. Sci. Commun.* **5**: 361–381.

Comiso, J. C., N. G. Maynard, W. O. Smith, Jr. & C. W. Sullivan. 1990. Satellite ocean color studies of Antarctic ice edges in summer/autumn. *J. Geophys. Res.* **95** (in press).

Copin-Montegut, C. & G. Copin-Montegut. 1978. The chemistry of particulate matter from the southern Indian and Antarctic oceans. *Deep-Sea Res.* **25**: 911–931.

Cosper, E. 1982. Effects of diurnal fluctuations in *Skeletonema costatum* (Grev.) Cleve (Bacillariophyceae) in a cyclostat. *J. Exp. Mar. Biol. Ecol.* **65**: 229–239.

Cota, G. F. 1985. Photoadaptation of high Arctic ice algae. *Nature (London)* **315**: 219–222.

Cota, G. F. & C. W. Sullivan. 1990. Photoadaptation, growth and production of bottom ice algae in the Antarctic. *J. Physiol.* (in press).

Cota, G. F., S. J. Prinsenberg, E. B. Bennett, J. W. Loder, M. R. Lewis, J. L. Anning, N. Watson & L. R. Harris. 1987. Nutrient fluxes during extended blooms of Arctic ice algae. *J. Geophys. Res.* **92**: 1951–1962.

Cullen, J. J. & M. R. Lewis. 1988. The kinetics of algal photoadaptation in the context of vertical mixing. *J. Plankton Res.* **10**: 1039–1063.

Culver, M. E. & W. O. Smith, Jr. 1989. The effects of environmental factors on sinking rates of marine phytoplankton. *J. Phycol.* **25**: 262–270.

Cushing, D. H. 1981. Temporal variability in production systems. *In* "Analysis of Marine Ecosystems" (A. R. Longhurst, ed.), pp. 443–472. Academic Press, New York.

Deacon, G. E. R. 1982. Physical and biological zonation in the Southern Ocean. *Deep-Sea Res.* **29**: 1–15.

Dunbar, R. B., J. B. Anderson & E. W. Domack. 1985. Oceanographic influences on sedimentation along the Antarctic continental shelf. *Antarct. Res. Ser.* **43**: 291–312.

El-Sayed, S. Z. 1971. Observations on phytoplankton bloom in the Weddell Sea. *In* "Biology of the Antarctic Seas" (G. Llano & I. Wallen, eds.), pp. 301–312. Am. Geophys. Union, Washington, D.C.

———. 1984. Productivity of Antarctic waters. A reappraisal. *In* "Marine Phytoplankton and Productivity" (O. Holm-Hansen, L. Bolis & R. Gilles, eds.), pp. 19–34. Springer-Verlag, Berlin.

El-Sayed, S. Z. & S. Taguchi. 1981. Primary production and standing crop of phytoplankton along the ice-edge in the Weddell Sea. *Deep-Sea Res.* **28**: 1017–1032.

El-Sayed, S. Z. & J. T. Turner. 1977. Productivity of the Antarctic and tropical/subtropical regions: A comparative study. *In* "Polar Oceans" (M. Dunbar, ed.), pp. 463–503. Arctic Inst. North Am., Calgary.

El-Sayed, S. Z. & L. H. Weber. 1982. Spatial and temporal variations in phytoplankton biomass and primary productivity in the southwest Atlantic and the Scotia Sea. *Polar Biol.* **1**: 83–90.

Eppley, R. W. 1972. Temperature and the regulation of phytoplankton growth in the sea. *Fish. Bull.* **70**: 1063–1085.

Eppley, R. W. & B. J. Peterson. 1979. Particulate organic matter flux and plankton new production in the deep ocean. *Nature (London)* **282**: 677–680.

Falkowski, P. G. 1983. Light-shade adaptation and vertical mixing of marine phytoplankton: A comparative field study. *J. Mar. Res.* **41**: 215-237.

———. 1984. Kinetics of light intensity adaptation in *Dunaliella tertiolecta*, a marine plankton chlorophyte. *Photosynthetica* **18**: 62-68.

Falkowski, P. G. & T. G. Owens. 1980. Light-shade adaptation: Two strategies in marine phytoplankton. *Plant Physiol.* **66**: 592-595.

Fay, R. R. 1974. Significance of nanoplankton in primary production of the Ross Sea, Antarctica, during the 1972 austral summer. Ph.D. Dissertation, Texas A&M Univ., College Station.

Fiala, M. & L. Oriol. 1984. Vitamine B_{12} et phytoplancton dans l'Océan Antarctique. Distribution et approche expérimentale. *Mar. Biol. (Berlin)* **79**: 325-332.

Fryxell, G. A. & G. A. Kendrick. 1988. Austral spring microalgae across the Weddell Sea ice edge: Spatial relationships found along a northward transect during AMERIEZ 83. *Deep-Sea Res.* **35**: 1-20.

Gallegos, C. L. & T. Platt. 1982. Phytoplankton production and water motion in surface mixed layers. *Deep-Sea Res.* **29**: 65-76.

Gallegos, C. L., T. Platt, W. G. Harrison & B. Irwin. 1983. Photosynthetic parameters of Arctic marine phytolankton: Vertical variations and time scales of adaptation. *Limnol. Oceanogr.* **28**: 698-708.

Garrison, D. L. & K. R. Buck. 1986. Organism losses during ice melting: A serious bias in sea ice community studies. *Polar Biol.* **6**: 237-239.

Garrison, D. L., S. F. Ackley & K. R. Buck. 1983. A physical mechanism for establishing algal populations in frazil ice. *Nature (London)* **306**: 363-365.

Garrison, D. L. & K. R. Buck. 1986. Organism losses during ice melting: A serious bias in sea ice community studies. *Polar Biol.* **6**: 237-239.

Garrison, D. L., C. W. Sullivan, & S. F. Ackley. 1986. Sea ice microbial communities in the Antarctic. *BioScience* **36**: 243-250.

Garrison, D. L., K. R. Buck, & G. A. Fryxell. 1987. Algal assemblages in Antarctic pack ice and in ice-edge plankton. *J. Phycol.* **23**: 564-572.

Geider, R. J. & T. Platt. 1986. A mechanistic model of photoadaptation in microalgae. *Mar. Ecol.: Prog. Ser.* **30**: 85-92.

Gilstad, M. 1987. Effect of photoperiod and irradiance on diatom growth. M. S. Thesis, Univ. of Trondheim, Trondheim, Norway.

Glibert, P. M., D. C. Biggs & J. J. McCarthy. 1982. Utilization of ammonium and nitrate during austral summer in the Scotia Sea. *Deep-Sea Res.* **29**: 837-850.

Goldman, J. C., J. J. McCarthy & D. G. Peavey. 1979. Growth rate influence on the chemical composition of phytoplankton in oceanic waters. *Nature (London)* **279**: 210-215.

Gran, H. H. 1931. On the conditions for the production of plankton in the sea. *J. Cons., Cons. Int. Explor. Mer.* **75**: 37-46.

Hameedi, M. J. 1978. Aspects of water column primary productivity in the Chukchi Sea during summer. *Mar. Biol. (Berlin)* **48**: 37-46.

Hansell, D. & J. J. Goering, 1989. A method for estimating uptake and production rates for urea in seawater using ^{14}C and ^{15}N-urea. *Can. J. Fish. Aquat. Sci.* **46**: 198-202.

Harris, G. P. 1980. Spatial and temporal scales in phytoplankton ecology. Mechanisms, methods, models and management. *Can. J. Fish. Aquat. Sci.* **37**: 877-900.

Harrison, W. G. & T. Platt. 1986. Photosynthesis-irradiance relationships in polar and temperate phytoplankton populations. *Polar Biol.* **5**: 153-164.

Harrison, W. G., T. Platt & B. Irwin. 1982. Primary production and nutrient assimilation by natural phytoplankton populations of the eastern Canadian Arctic. *Can. J. Fish. Aquat. Sci.* **39**: 335-345.

Harrison, W. G., E. J. H. Head, R. J. Conover, A. R. Longhurst & D. D. Sameoto. 1985. The distribution and metabolism of urea in the eastern Canadian Arctic. *Deep-Sea Res.* **32**: 23–42.
Hart, T. J. 1934. On the phytoplankton of the south-west Atlantic and Bellingshausen Sea. *'Discovery' Rep.* **8**: 1–268.
Hasle, G. R. 1969. An analysis of the phytoplankton of the Pacific Southern Ocean: Abundance, composition, and distribution during the Brategg expedition. *Hvalradets Skr.* **52**: 1–168.
Hegseth, E. N. & E. Sakshaug. 1983. Seasonal variation in light- and temperature-dependent growth of marine plankton diatoms in in situ dialysis cultures in the Trondheimsfjord, Norway (63°N). *J. Exp. Mar. Biol. Ecol.* **67**: 199–220.
Heimdal, B. R. 1983. Phytoplankton and nutrients in the waters north-west of Spitsbergen in the autumn of 1979. *J. Plankton Res.* **5**: 901–918.
Hewes, C. D. & O. Holm-Hansen. 1983. A method for recovering nanoplankton from filters for identification with the microscope: The filter–transfer–freeze (FTF) technique. *Limnol. Oceanogr.* **28**: 389–394.
Hewes, C. D., O. Holm-Hansen & E. Sakshaug. 1985. Alternate carbon pathways at lower trophic levels in the Antarctic food web. *In* "Antarctic Nutrient Cycles and Food Webs" (W. R. Siegfried, P. R. Condy & R. M. Laws, eds.), pp. 277–283. Springer-Verlag, Berlin.
Hewes, C. D., E. Sakshaug, F. M. H. Reid & O. Holm-Hansen. 1990. Microbial autotrophic and heterotrophic eucaryotes in Antarctic waters: Relationships between microbial carbon and chlorophyll, adenosine triphosphate, and particulate organic matter. *Mar. Ecol.: Prog. Ser.* (in press).
Heywood, R. B. & J. Priddle. 1987. Retention of phytoplankton by an eddy. *Cont. Shelf Res.* **7**: 937–955.
Heywood, R. B., I. Everson & J. Priddle. 1985. The absence of krill from the south Georgia zone, winter 1983. *Deep-Sea Res.* **32**: 369–378.
Holm-Hansen, O. 1970. ATP in algal cells, as influenced by environmental conditions. *Plant Cell Physiol.* **11**: 689–700.
Holm-Hansen, O. & M. Huntley. 1984. Feeding requirements of krill in relation to food sources. *J. Crustacean Biol.* **4**: 156–173.
Holm-Hansen, O., S. Z. El-Sayed, G. A. Franceschini & R. L. Cuhel. 1977. Primary production and the factors controlling phytoplankton growth in the Southern Ocean. *In* "Adaptations Within Antarctic Ecosystems" (G. A. Llano, ed.), pp. 11–50. Gulf Publ. Co., Houston, Texas.
Horner, R. A. 1976. Sea ice organisms. *Oceanogr. Mar. Biol.* **14**: 167–182.
———. 1985. Ecology of sea ice microalgae. *In* "Sea Ice Biota" (R. Horner, ed.), pp. 83–103. CRC Press, Boca Raton, Florida.
Horner, R. A. & V. Alexander. 1972. Algal populations in Arctic sea ice: An investigation of heterotrophy. *Limnol. Oceanogr.* **17**: 454–458.
Horner, R. A. & G. C. Schrader. 1982. Relative contributions of ice algae, phytoplankton and benthic microalgae to primary production in nearshore regions of the Beaufort Sea. *Arctic* **35**: 485–503.
Iverson, R. L., T. E. Whitledge & J. J. Goering. 1979. Chlorophyll and nitrate fine structure in the southeastern Bering Sea shelf break front. *Nature (London)* **281**: 664–666.
Jacques, G. 1983. Some ecophysiological aspects of the Antarctic phytoplankton. *Polar Biol.* **2**: 27–33.
Jennings, J. C., Jr., L. I. Gordon & D. M. Nelson. 1984. Nutrient depletion indicates high primary productivity in the Weddell Sea. *Nature (London)* **308**: 51–54.

Jerlov, N. G. 1968. "Optical Oceanography." Elsevier, New York.
Johnson, T. O. & W. O. Smith, Jr. 1986. Sinking of phytoplankton assemblages in the Weddell Sea marginal ice zone. *Mar. Ecol.: Prog. Ser.* **33**: 131–137.
Karl, D. M., B. D. Tilbrook & G. Tien. 1990. Seasonal coupling of organic matter production and particle flux in the Bransfield Strait, Antarctica. *Deep-Sea Res.* (in press).
Kiefer, D. A. & B. G. Mitchell. 1983. A simple, steady state description of phytoplankton growth based on absorption cross section and quantum efficiency. *Limnol. Oceanogr.* **28**: 770–776.
Kirk, J. T. 1975. A theoretical analysis of the contribution of algal cells to the attenuation of light within natural waters. *New Phytol.* **75**, 11–20.
———. 1983. "Light and Photosynthesis in Aquatic Ecosystems." Cambridge Univ. Press, Cambridge, 401 pp.
Koike, I., O. Holm-Hansen & D. C. Biggs. 1986. Inorganic nitrogen metabolism by Antarctic phytoplankton with special reference to ammonium cycling. *Mar. Ecol.: Prog. Ser.* **30**: 105–116.
Kottmeier, S. T. & C. W. Sullivan. 1987. Late winter primary production and bacterial production in sea ice and seawater west of the Antarctic Peninsula. *Mar. Ecol.: Prog. Ser.* **36**: 287–298.
Krebs, W. N. 1983. Ecology of neritic marine diatoms, Arthur Harbor, Antarctica. *Micropaleontology* **29**: 267–297.
Kremer, J. N. & S. Nixon. 1978. "A Coastal Marine Ecosystem. Simulation and Analysis." Springer-Verlag, Berlin.
Kristiansen, S. & B. A. Lund. 1989. Nitrogen cycling in the Barents Sea. I. Uptake of nitrogen in the water column. *Deep-Sea Res.* **36**: 255–268.
Laws, E. A., P. K. Bienfang, D. A. Ziemann & L. D. Conquest. 1988. Phytoplankton population dynamics and the fate of production during the spring bloom in Auke Bay, Alaska. *Limnol. Oceanogr.* **33**: 57–65.
Ledford-Hoffman, P. A., D. J. DeMaster & C. A. Nittrouer. 1986. Biogenic-silica accumulation in the Ross Sea and the importance of Antarctic continental-shelf deposits in the marine silica budget. *Geochim. Cosmochim. Acta* **50**: 2099–2110.
Lewis, M. R. & J. C. Smith. 1983. A small volume, short-incubation-time method for measurement of photosynthesis as a function of incident irradiance. *Mar. Ecol.: Prog. Ser.* **13**: 99–102.
Lewis, M. R., J. J. Cullen & T. Platt. 1984. Relationships between vertical mixing and photoadaptation of phytoplankton: Similarity criteria. *Mar. Ecol.: Prog. Ser.* **15**: 141–149.
Lewis, M. R., W. G. Harrison, N. S. Oakey, D. Hebert & T. Platt. 1986. Vertical nitrate fluxes in the oligotrophic ocean. *Science* **234**: 870–873.
Li, W. K. W. 1985. Photosynthetic response to temperature of marine phytoplankton along a latitudinal gradient (16°N to 74°N). *Deep-Sea Res.* **32**: 1381–1391.
Li, W. K. W., J. C. Smith & T. Platt. 1984. Temperature response of photosynthetic capacity and carboxylase activity in Arctic marine phytoplankton. *Mar. Ecol.: Prog. Ser.* **17**: 237–243.
Marshall, P. T. 1957. Primary production in the Arctic. *J. Cons., Cons. Int. Explor. Mer* **23**: 173–177.
Martin, J. H. & S. E. Fitzwater. 1988. Iron deficiency limits phytoplankton growth in the north-east Pacific subarctic. *Nature (London)* **331**: 341–343.
Martin, J. H. & R. M. Gordon. 1988. Northeast Pacific iron distribution in relation to phytoplankton production. *Deep-Sea Res.* **35**: 177–196.
Maske, H. & H. Haardt. 1987. Quantitative in vivo absorption spectra of phytoplankton:

Detrital absorption and comparison with fluorescence excitation spectra. *Limnol. Oceanogr.* **32**: 620–633.

Matheke, G. E. M. & R. Horner. 1974. Primary productivity of benthic microalgae in the Chukchi Sea near Barrow, Alaska, *J. Fish. Res. Board Can.* **31**: 1779–1786.

Maykut, G. A. & T. C. Grenfell. 1975. The spectral distribution of light beneath first-year sea ice in the Arctic Ocean. *Limnol. Oceanogr.* **20**: 554–563.

Maynard, N. G. & D. K. Clark. 1987. Satellite color observations of spring blooming in Bering Sea shelf waters during the ice edge retreat in 1980. *J. Geophys. Res.* **92**: 7127–7140.

McCarthy, J. J. 1981. The kinetics of nutrient utilization. *Can. Bull. Fish. Aquat. Sci.* **210**: 211–233.

McRoy, C. P. & J. J. Goering. 1976. Annual budget of primary production in the Bering Sea. *Mar. Sci. Commun.* **2**: 255–267.

Mitchell, B. G. & D. A. Kiefer. 1988. Chlorophyll *a* specific absorption and fluorescence excitation spectra for light limited phytoplankton. *Deep-Sea Res.* **35**: 639–664.

Müller-Karger, F. & V. Alexander. 1987. Nitrogen dynamics in a marginal sea-ice zone. *Cont. Shelf Res.* **7**: 805–823.

Nelson, D. M. & L. I. Gordon, 1982. Production and pelagic dissolution of biogenic silica in the Southern Ocean. *Geochim. Cosmochim. Acta* **46**: 491–501.

Nelson, D. M. & W. O. Smith, Jr. 1986. Phytoplankton bloom dynamics of the western Ross Sea ice edge. II. Mesoscale cycling of nitrogen and silicon. *Deep-Sea Res.* **33**: 1389–1412.

Nelson, D. M., W. O. Smith, Jr., L. I. Gordon & B. A. Huber. 1987. Spring distributions of density, nutrients, and phytoplankton biomass in the ice edge zone of the Weddell–Scotia Sea. *J. Geophys. Res.* **92**: 7181–7190.

Nelson, D. M., W. O. Smith, Jr., R. D. Muench, L. I. Gordon, C. W. Sullivan & D. M. Husby. 1989. Particulate matter and nutrient distributions in the ice-edge zone of the Weddell Sea: Relationship to hydrography during late summer. *Deep-Sea Res.* **36**: 191–209.

Neori, A. & O. Holm-Hansen. 1982. Effect of temperature on rate of photosynthesis in Antarctic phytoplankton. *Polar Biol.* **1**: 33–38.

Niebauer, H. J. 1984. On the effect of El Niño in Alaskan waters. *Bull. Am. Meteorol. Soc.* **65**: 472–473.

———. 1988. Effects of El Niño–southern oscillation and north Pacific weather patterns on interannual variability in subarctic Bering Sea. *J. Geophys. Res.* **93**: 5051–5068.

Niebauer, H. J. & V. Alexander. 1985. Ocenaographic frontal structure and biological production at an ice edge. *Cont. Shelf Res.* **4**: 367–388.

Niebauer, H. J. & W. O. Smith, Jr. 1989. A numerical model of mesoscale physical–biological interactions in the Fram Strait marginal ice zone. *J. Geophys. Res.* **94**: 16151–16175.

Olson, R. J. 1980. Nitrate and ammonium uptake in Antarctic waters. *Limnol. Oceanogr.* **25**: 1064–1074.

Paasche, E. 1980. Silicon content of five marine plankton diatom species measured with a rapid filter method. *Limnol. Oceanogr.* **25**: 474–480.

Paasche, E. & S. R. Erga. 1988. Phosphorous and nitrogen limitation of phytoplankton in the inner Oslofjord (Norway). *Sarsia* **73**: 229–243.

Palmisano, A. C. & C. W. Sullivan. 1983. Sea ice microbial communities (SIMCO). 1. Distribution, abundance and primary production of ice microalgae in McMurdo Sound, Antarctica in 1980. *Polar Biol.* **2**: 171–177.

Palmisano, A. C., J. B. Soo Hoo, D. C. White, G. A. Smith, G. A. Stanton & L. Burckle. 1985. Shade adapted benthic diatoms beneath Anarctic sea ice. *J. Phycol.* **21**: 664–667.

Palmisano, A. C., J. B. SooHoo, R. L. Moe & C. W. Sullivan. 1986a. Sea ice microbial

communities. VII. Changes in under-ice spectral irradiance during the development of Antarctic sea ice microalgal communities. *Mar. Ecol.: Prog. Ser.* **35**: 165–173.
Palmisano, A. C., J. B. SooHoo, S. L. SooHoo, S. T. Kottmeier, L. L. Craft & C. W. Sullivan. 1986b. Photoadaptation in *Phaeocystis pouchetii* advected beneath annual sea ice in McMurdo Sound, Antarctica. *J. Plankton Res.* **8**: 891–906.
Parkinson, C. L., J. C. Comiso, H. J. Zwally, D. J. Cavalieri, P. Gloersen & W. J. Campbell. 1987. "Arctic Sea Ice. 1973–1976: Satellite Passive-Microwave Observations," NASA Spec. Publ. 489. Natl. Aeron. Space Admin., Washington, D.C.
Parson, T. R., M. Takahashi & B. Hargrave. 1984. "Biological Oceanographic Processes." Pergamon, New York.
Platt, T. & D. V. Subba Rao. 1975. Primary production of marine microphytes. *In* "Photosynthesis and Productivity in Different Environments" (J. P. Cooper, ed.), pp. 249–280. Cambridge Univ. Press, Cambridge.
Platt, T., C. L. Gallegos & W. G. Harrison. 1980. Photoinhibition of photosynthesis in natural assemblages of marine phytoplankton. *J. Mar. Res.* **38**: 687–701.
Platt, T., W. G. Harrison, B. Irwin, E. P. Horne & C. L. Gallegos. 1982. Photosynthesis and photoadaptation of marine phytoplankton in the Arctic. *Deep-Sea Res.* **29**: 1159–1170.
Post, A. F., Z. Dubinsky, K. Wyman & P. G. Falkowski. 1985. Physiological responses of a marine planktonic diatom to transitions in growth irradiance. *Mar. Ecol.: Prog. Ser.* **25**: 141–149.
Preisendorfer, R. W. 1957. Exact reflectance under a cardioidal luminance distribution. *Q. J. R. Meteorol. Soc.* **83**: 540.
Prezelin, B. B. 1981. Light reactions in photosynthesis. *Can. Bull. Fish. Aquat. Sci.* **210**: 1–43.
Probyn, T. A. & S. J. Painting. 1985. Nitrogen uptake by size-fractionated phytoplankton populations in Antarctic surface waters. *Limnol. Oceanogr.* **30**: 1327–1331.
Rey, F. & H. Loeng. 1985. The influence of ice and hydrographic conditions on the development of phytoplankton in the Barents Sea. *In* "Marine Biology of Polar Regions and Effects of Stress on Marine Organisms" (J. S. Gray & M. E. Christiansen, eds.), pp. 49–64. Wiley, Chichester.
Rey, F. & H. R. Skjoldal. 1987. Consumption of silicic acid below the euphotic zone by sedimenting diatom blooms in the Barents Sea. *Mar. Ecol.: Prog. Ser.* **36**: 307–312.
Rey, F., H. R. Skjoldal & D. Slagstad. 1987. Primary production in relation to climatic changes in the Barents Sea. *In* "The Effect of Oceanographic Conditions on Distribution and Population Dynamics of Commercial Fish Stocks in the Barents Sea" (H. Loeng, ed.), Proc. 3rd Sov. Norw. Symp., pp. 29–46. Inst. Mar. Res., Bergen.
Riley, G. A. 1963. Theory of food-chain relations in the ocean. *In* "The Sea" (M. N. Hill, ed.), pp. 438–463. Wiley, New York.
Rivkin, R. B. & M. Putt. 1987a. Diel periodicity of photosynthesis in polar phytoplankton: Influence on primary production. *Science* **238**: 1285–1288.
―――. 1987b. Photosynthesis and cell division by Antarctic microalgae: Comparison of benthic, planktonic and ice algae. *J. Phycol.* **23**: 223–229.
―――. 1987c. Heterotrophy and photoheterotrophy by Antarctic microalgae: Light dependent incorporation of amino acids and glucose. *J. Phycol.* **23**: 442–452.
Rönner, U., F. Sörensson & O. Holm-Hansen, 1983. Nitrogen assimilation by phytoplankton in the Scotia sea. *Polar Biol.* **2**: 137–147.
Sakshaug, E. 1989. The physiological ecology of polar phytoplankton. *In* "Proceedings of the Sixth Conference of the Comité Arctique Internationale 13–15 May 1985" (L. Rye, & V. Alexander, eds.). pp. 61–89. Brill, Leiden.
Sakshaug, E. & K. Andresen. 1986. Effect of light regime upon growth rate and chemical

composition of a clone of *Skeletonema costatum* from the Trondheimsfjord, Norway. *J. Plankton Res.* **8:** 619-637.

Sakshaug, E. & O. Holm-Hansen. 1984. Factors governing pelagic production in polar oceans. *In* "Marine Phytoplankton and Productivity" (O. Holm-Hansen, L. Bolis & R. Gilles, eds.), pp. 1-18. Springer-Verlag, Berlin.

―――. 1986. Photoadaptation in Antarctic phytoplankton: Variations in growth rate, chemical composition and P versus I curves. *J. Plankton Res.* **8:** 459-473.

Sakshaug, E. & Y. Olsen. 1986. Nutrient status of phytoplankton blooms in Norwegian waters and algal strategies for nutrient competition. *Can. J. Fish. Aquat. Sci.* **43:** 389-396.

Sakshaug, E., K. Andresen, S. Myklestad & Y. Olsen. 1983. Nutrient status of phytoplankton communities in Norwegian waters (marine, brackish and fresh) as revealed by their chemical composition. *J. Plankton Res.* **5:** 175-195.

Sakshaug, E., E. Grandeli, M. Elbrächter & H. Kayser. 1984. Chemical composition and alkaline phosphatase activity of nutrient-saturated and P-deficient cells of four marine dinoflagellates. *J. Exp. Mar. Biol. Ecol.* **77:** 241-254.

Sakshaug, E., S. Demers & C. M. Yentsch. 1987. *Thalassiosira oceanica* and *T. pseudonana:* Two photoadaptational strategies. *Mar. Ecol.: Prog. Ser.* **41:** 275-282.

Sakshaug, E., K. Andresen, & D. A. Kiefer. 1989. A steady-state description of growth and light absorption in the marine planktonic diatom *Skeletonema costatum*. *Limnol. Oceanogr.* **34:** 198-205.

Sambrotto, R. N., J. J. Goering & C. P. McRoy. 1984. Large yearly production of phytoplankton in the western Bering Strait. *Science* **225:** 1147-1150.

Schandelmeier, L. & V. Alexander. 1981. An analysis of the influence of ice on spring phytoplankton population structure in the southeast Bering Sea. *Limnol. Oceanogr.* **26:** 935-943.

Smetacek, V. S. 1985. Role of sinking in diatom life-history cycles: Ecological, evolutionary and geological significance. *Mar. Biol. (Berlin)* **84:** 239-251.

Smith, R. H., P. Clement, G. F. Cota & W. K. W. Li. 1987. Intracellular photosynthate allocation and the control of Arctic ice algal production. *J. Phycol.* **23:** 124-132.

Smith, R. E. H., J. Anning, P. Clement & G. Cota. 1988. The abundance and production of ice algae in Resolute Passage, Canadian Arctic. *Mar. Ecol.: Prog. Ser.* **48:** 251-263.

Smith, S. L., W. O. Smith, Jr., L. A. Codispoti & D. L. Wilson. 1985. Biological observations in the marginal ice zone of the East Greenland Sea. *J. Mar. Res.* **43:** 693-717.

Smith, W. O., Jr. 1987. Phytoplankton dynamics in marginal ice zones. *Oceanogr. Mar. Biol.* **25:** 11-38.

Smith, W. O. Jr. & G. Kattner. 1989. Inorganic nitrogen uptake by phytoplankton in the marginal ice zone of the Fram Strait. *Rapp. P.-V. Reun. Cons. Int. Explor. Mer* **188:** 90-97.

Smith, W. O., Jr. & D. M. Nelson. 1985. Phytoplankton bloom produced by a receding ice edge in the Ross Sea: Spatial coherence with the density field. *Science* **227:** 163-166.

―――. 1986. Importance of ice edge phytoplankton production in the Southern Ocean. *BioScience* **36:** 251-257.

―――. 1990. Phytoplankton growth and new production in the Weddell Sea marginal ice zone during austral spring and autumn. *Limnol. Oceanogr.* **35** (in press).

Smith, W. O., Jr. M. E. M. Baumann, D. L. Wilson & L. Aletsee. 1987. Phytoplankton biomass and productivity in the marginal ice zone of the Fram Strait during summer 1984. *J. Geophys. Res.* **92:** 6777-6786.

Smith, W. O., Jr., N. K. Keene & J. C. Comiso. 1988. Interannual variability in estimated primary productivity of the Antarctic marginal ice zone. *In* "Antarctic Ocean and Resources Variability" (D. Sahrhage, ed.), pp. 131-139. Springer-Verlag, Berlin.

SooHoo, J. B., A. C. Palmisano, S. T. Kottmeier, M. P. Lizotte, S. L. SooHoo & C. W. Sullivan. 1987. Spectral light absorption and quantum yield of photosynthesis in sea ice microalgae and a bloom of *Phaeocystis pouchetii* from McMurdo Sound, Antarctica. *Mar. Ecol.: Prog. Ser.* **39**: 175–189.

Sorokin, J. I. 1971. On the role of bacteria in the productivity of tropical oceanic waters. *Int. Rev. Gesamten Hydrobiol.* **56**: 1–48.

Spies, A. 1987. Growth rates of Antarctic marine phytoplankton in the Weddell Sea. *Mar. Ecol.: Prog. Ser.* **41**: 267–274.

Subba Rao, D. V. & T. Platt. 1984. Primary production of Arctic waters. *Polar Biol.* **3**: 191–201.

Sullivan, C. W., A. C. Palmisano & J. B. SooHoo. 1984. Influence of sea ice and sea ice biota on downwelling irradiance and spectral composition of light in McMurdo Sound. *Proc. SPIE — Int. Soc. Opt. Eng.* **489**: 159–165.

Sullivan, C. W., C. R. McClain, J. C. Comiso & W. O. Smith, Jr. 1988. Phytoplankton standing crops within an Antarctic ice edge assessed by satellite remote sensing. *J. Geophys. Res.* **93**: 12,487-12,498.

Sverdrup, H. U. 1953. On conditions for the vernal blooming of phytoplankton. *J. Cons., Cons. Int. Explor. Mer* **18**: 287–295.

Tilzer, M. M. & Z. Dubinsky. 1987. Effects of temperature and daylength on the mass balance of Antarctic phytoplankton. *Polar Biol.* **7**: 35–42.

Tilzer, M. M., M. Elbrächter, W. W. Gieskes & B. Beese. 1986. Light–temperature interactions in the control of photosynthesis in Antarctic phytoplankton. *Polar Biol.* **5**: 105–111.

von Bodungen, B., V. S. Smetacek, M. M. Tilzer & B. Zeitzschel. 1986. Primary production and sedimentation during spring in the Antarctic Peninsula region. *Deep-Sea Res.* **33**: 177–194.

Wefer, G., G. Fischer, D. Fütterer & R. Gersonde. 1988. Seasonal particle flux in the Bransfield Strait, Antarctica. *Deep-Sea Res.* **35**: 891–898.

Weiler, C. S. & D. M. Karl. 1979. Diel changes in phased-dividing cultures of *Ceratium furca* (Dinophyceae): Nucleotide triphosphates, adenylate energy charge, cell carbon, and patterns of vertical migration. *J. Phycol.* **15**: 384–391.

Whitaker, T. M. 1982. Primary production of phytoplankton off Signy Island, South Orkneys, the Antarctic. *Proc. R. Soc. London, Ser. B* **214**: 169–189.

Wilson, D. L., W. O. Smith, Jr. & D. M. Nelson. 1986. Phytoplankton bloom dynamics of the western Ross Sea ice edge. I. Primary productivity and species-specific production. *Deep-Sea Res.* **33**: 1375–1387.

Zwally, H. J., J. C. Comiso, C. L. Parkinson, W. J. Campbell, F. D. Carsey & P. Gloersen. 1983. "Antarctic Sea Ice. 1973–1976: Satellite Passive-Microwave Observations." NASA Spec. Publ. 459. Natl. Aeron. Space Admin., Washington, D.C.

10 Polar Zooplankton

Sharon L. Smith
Oceanographic Sciences Division
Department of Applied Science
Brookhaven National Laboratory
Upton, New York

Sigrid B. Schnack-Schiel
Alfred-Wegener-Institut für Polar-und Meeresforschung
2850 Bremerhaven,
Federal Republic of Germany

I. Introduction 527
II. The Arctic 528
   A. The Central Arctic Ocean 528
   B. The Marginal Seas 531
   C. The Bering Sea and Subarctic North Pacific Ocean 535
   D. Seasonal Cycles and Specialized Adaptations of the Major Planktonic Copepods and Euphausiids 538
III. The Antarctic 554
   A. The Southern Ocean 554
   B. The Major Seas South of 50°S 557
   C. Seasonal Cycles and Special Adaptations of Euphausiids 566
   D. Seasonal Cycles and Special Adaptations of Other Zooplankton 572
IV. Concluding Remarks 579
   References 581

I. Introduction

Life cycles of zooplankton in polar latitudes are governed not only by low temperatures, which prevail throughout much of the year, but also by extremes in solar radiation and associated cycles in pelagic primary production. Throughout the polar environment, light is virtually absent for four or more months per year and nearly continually present another four or more months, the intervening periods being ones of rapid change from one extreme to the other. The adaptations that have evolved in pelagic zooplankton in response to the extreme conditions of the polar habitat are part of the subject of this chapter. Comprehensive, quantitative, mechanistic understanding of how pelagic zooplankton have responded to their polar habitats is still unrealized, though in recent years a variety of factors have combined to focus international attention on the polar environment, especially its oceans and seas, once again.

In this chapter we also attempt to identify other factors in both polar environments that are of great importance in shaping the communities of organisms found and to emphasize differences between the two poles that may give insight through comparison. Effects of circulation, seasonal ice cover, freshwater input, and bathymetry are some of the other factors which have great influence on community structure and population growth of pelagic zooplankton in the marine environments of the Arctic and Antarctic.

II. The Arctic

A. The Central Arctic Ocean

The primary connection between the North Atlantic and the Arctic Ocean at both the surface and middepth is the passage between northern Greenland and Spitsbergen (Coachman and Aagaard, 1974), known as Fram Strait (Fig. 10.1). In terms of volume, the North Atlantic Ocean has the most pervasive influence because it forms both the Atlantic water found approximately between 200 and 900 m and the bottom water between 900 and 2000 m (see Chapter 4). Potentially, therefore, planktonic communities of the North Atlantic could be distributed more widely in the Arctic Ocean than their Pacific Ocean counterparts because at middepth the inflow via Fram Strait is carried throughout the Canadian and Eurasian basins (Coachman and Aagaard, 1974). This is not the case, however, and taxa typically found in the North Atlantic (e.g., *Calanus finmarchicus, Metridia lucens*) occur only occasionally in the Arctic Ocean (Minoda, 1967). Taxa from the Pacific Ocean remain essentially confined to the Eurasian basin of the Arctic Ocean (Grice, 1962). Among copepods, *Calanus hyperboreus, Calanus glacialis,* and *Metridia longa* are most frequently encountered throughout the Arctic Ocean (Grice, 1962).

Early studies of the pelagic fauna of the Arctic Ocean used the Barents Sea and Spitsbergen area for access (Sars, 1900; Nansen, 1902) and thus the species reported are occasionally those also described for coastal Norway (e.g., *Euchaeta norvegica, Calanus finmarchicus, Thysanoessa longicaudata*). The *Nautilus* expedition to the ice edge north of Spitsbergen (Hardy, 1936) also found pelagic communities whose species were mixtures of Arctic and boreal taxa. With the addition of ice islands as study platforms in the

10 Polar Zooplankton

Figure 10.1 Basins of the Arctic Ocean and its marginal seas.

1950s and 1960s, it was possible to penetrate to the central portion of the Arctic basin and to gather data on seasonal variations in pelagic communities. These long-term studies revealed that copepods predominated in the upper 500 m of the water column, with *C. hyperboreus, Microcalanus pygmaeus,* and *Oithona similis* present during summer but only *M. pygmaeus* and *O. similis* present during winter (Minoda, 1967). Biomass was highest in July and was accounted for almost entirely by development of the *C. hyperboreus* population, which spawned under the ice and showed evidence of ontogenetic migration (Minoda, 1967). Other copepod taxa showed relatively constant abundance throughout the year (e.g., *C. glacialis* and *O. similis*) or two peaks in abundance (e.g., *Metridia longa;* Minoda, 1967).

Studies conducted from ice island Alpha, drifting in the Canadian basin from July through February, revealed that the biomass of zooplankton in the upper 200 m of the water column consisted primarily of *C. glacialis, C. hyperboreus,* and *M. longa* (Johnson, 1963). *Calanus glacialis* reached maximum abundance between 50 and 100 m, *C. hyperboreus* between 50 and 200 m, and *M. longa* between 25 and 50 m. Numbers of species increased with increasing depth to 1500 m, and only the carnivore *Pareuchaeta glacialis* was noted to be reproductively active during February (Johnson, 1963).

The life cycle of *C. hyperboreus* was investigated in detail from ice island T-3 in the Canadian basin (Dawson, 1978). The cycle seems to require 3 years, with year one encompassing growth from egg to copepodid stages II and III, year two to copepodid stages IV and V, and year three to adulthood and spawning. Copepodids were found below 300 m in winter (November–May), and they migrated to the surface (0–100 m) in summer. Gravid females were near the surface prior to any phytoplankton bloom, and females slowly descended from 100 to 300 m in spring. Males were always few and deeper (below 400 m) than the females (Dawson, 1978).

Comparisons of zooplankton communities sampled from ice islands drifting in the Eurasian and Canadian basins were made by T. L. Hopkins (1969). The highest biomass in the Eurasian basin was approximately 3 mg dry wt m^{-3}, but because this was measured in January and no summer data were reported, it must be considered a lower limit. The highest biomass in the Canadian basin was approximately 1 mg dry wt m^{-3}, measured in July and August. These data suggest that there may be substantial differences in total biomass of the two basins. In both basins, however, *Calanus* species accounted for 50% of the biomass in the 0–1500-m layer (T. L. Hopkins, 1969).

Later (1975–1977) studies using ice islands in the Canadian basin have shown that the vertical distribution of biomass has seasonal variations that are most readily explained by ontogenetic migration. That is, in the May–July period, biomass in the 50–100-m stratum increased tenfold while biomass in the 200–1000-m stratum decreased two to fourfold (Kosobokova, 1982). The biomass comprised mostly (64–99%) copepods, with *C. hyperboreus* dominant. This species, along with *C. glacialis,* dominated the upper 500 m in summer and the upper 1000 m in spring and autumn. The numerical dominants, however, were *Oithona similis, Microcalanus pygmaeus,* and *Oncaea borealis* (Kosobokova, 1982). There was a slight tendency for *C. glacialis* females and stage V copepodids to undertake diel vertical migrations in summer, but *C. hyperboreus* did not (Kosobokova, 1978). Similar conclusions were reached on the basis of data from the Canadian Arctic in August: diel migrations were practically absent but ontogenetic migrations occurred and were the dominant factor controlling vertical distributions of herbivores (Longhurst et al., 1984).

Copepod taxa restricted to deeper water of the Canadian basin of the Arctic Ocean have also received attention. In the case of *Spinocalanus* species, two (*S. longicornis, S. antarcticus*) have overlapping distributions from 100 to 400 m and two others (*S. horridus, S. elongatus*) overlap from 500 to 2500 m (Damkaer, 1975). The two most abundant species of Aetideidae, *Aetideopsis rostrata* and *Gaidius previspinus*, also showed overlapping distributions between 200 and 750 m (Markhaseva, 1984). But in the case of *Lucicutia* species, vertical distributions of three major taxa were essentially nonoverlapping (Vidal, 1971). *Lucicutia anomola* was collected only deeper than 3000 m, whereas *L. pseudopolaris* and *L. polaris* partitioned the 1000 – 3000-m stratum. It is possible that vertical separation lessens competition among the three species for a limited food supply (Vidal, 1971).

B. The Marginal Seas

The physiography of the Arctic Ocean is such that broad, shallow continental shelves coincide with the outflow of major rivers such as the Ob, Yenisei, Pechora, and Lena, and these two factors combine to shape the planktonic communities of the Kara, Laptev, White, and East Siberian seas. These shallow, coastal, marginal seas of the Arctic basin, which occupy approximately 70% of its area (Herman, 1974), contain a pelagic fauna that combines the ubiquitous community of the upper layer of the deeper, central basins and neritic taxa that are constrained by temperature and salinity characteristics of the nearshore environment. In the Chukchi Sea meroplanktonic larvae of benthic invertebrates were much more abundant than they were in the Beaufort Sea, but typically widespread and Arctic species of copepod (e.g., *C. hyperboreus, M. longa*) were collected in both seas (Johnson, 1958). Numerically dominant species of copepod found in both the Chukchi and Beaufort seas were *C. glacialis* (called *C. finmarchicus*), *C. hyperboreus, M. longa, Pseudocalanus minutus, M. pygmaeus, O. similis*, and *O. borealis*. Euphausiids were collected rarely but, when present, *Thysanoessa raschi* was most abundant. Chaetognaths and medusae were also generally abundant (Johnson, 1958).

Coastal areas of the Beaufort Sea have a thin (5 – 10-m) upper layer that in summer may be quite warm (ca. 10 – 12°C) and fresh (<10 per mil; Grainger, 1965; Evans and Grainger, 1980). This layer contains medusae (*Sarsia princeps, Euphysa flammea, Halitholus cirratus*) and copepods (*Eurytemora herdmani, Limnocalanus grimaldi, Acartia clausi, Acartia longiremis, Drepanopus bungei*) primarily, with *E. herdmani* and *L. grimaldi* confined to the upper 1 m (Grainger, 1965; Evans and Grainger,

1980). These neritic taxa appear to disperse eastward from the Chukchi Sea, being found continuously from Point Barrow (156°W) to Amundsen Gulf (120°W; Grainger, 1965). The neritic copepods generally have one generation per year and reach their biomass maxima in August (Evans and Grainger, 1980). In the Chukchi and western Beaufort Sea, both *A. clausi* and *E. herdmani* were absent in 1976, a year when temperatures were less than 7°C (English and Horner, 1977), which suggests that the development of neritic populations depends on increased temperatures found in the thin surface layer. Most of the widespread taxa (*C. glacialis, O. similis*) are present in the western Beaufort Sea; in addition, an hypothesized upwelling circulation in the vicinity of Prudhoe Bay may explain the capture of species characteristic of offshore, deeper waters (*C. hyperboreus, M. longa, M. pygmaeus*) in the upper 20 m (English and Horner, 1977). Johnson (1958) also noted the clear separation of nearshore and offshore pelagic communities in the vicinity of Pt. Barrow.

The calanoid copepods of the shallow Kara, Laptev, and East Siberian seas are *D. bungei, L. grimaldi, Pseudocalanus* spp., *Derjuginia tolli, O. similis,* and *Senecella calanoides* (Jaschnov, 1946; Pirozhnikov, 1985). Yearly fluctuations in the distribution of the species depend on salinity and, therefore, on the amount of riverine outflow (Jaschnov, 1946). *Pseudocalanus* spp. were an abundant and common component of the coastal plankton off Pt. Barrow in the Beaufort Sea (Johnson, 1958) and a major constituent (81–97% of numbers) of the winter and spring community in Stefansson Sound, a shallow (10 m) embayment adjoining the Beaufort Sea (Horner and Murphy, 1985). In collections between 40 and 90 m in the Chukchi and Beaufort Seas, *Pseudocalanus* spp. constituted 89% of total copepods collected (Grice, 1962). *Pseudocalanus* are here denoted as several species because a recent revision of the genus describes four different species for this area (Frost, 1989).

Extensive work has been conducted in the White Sea, which adjoins the Barents Sea to the east of Murmansk and is fed by the Vashka, Dvina, and Onega rivers. Recurrent group analysis (Kolosova, 1975, 1978) has defined Arctic and boreal components of the zooplanktonic community of the White Sea. *Calanus glacialis, M. longa, O. borealis, Pseudocalanus elongatus,* and *Oikopleura vanhoffeni* dominate the Arctic portion, while *A. longiremis, Temora longicornis,* and *Evadne nordmani* are members of the boreal portion. In general, the Arctic forms are found below 30 m while the boreal taxa are in the upper 15 m (Kolosova, 1978).

In exploring the distribution of *P. elongatus* [again this is probably not the correct identification of the animal (Frost, 1989)], which constitutes more than 50% of the biomass below 25 m in the White Sea, Prygunkova (1985) found large interannual variations in abundance in summer which could be linked to the warming rate of the surface layer in spring. *Pseudocalanus elongatus* left the surface layer once it warmed and joined other Arctic forms at depth. The *P. elongatus* population was large in years when the surface layer remained cool and phytoplankton biomass was high. In warm years she suggests that *P. elongatus* is forced from the surface before food is available, thereby reducing fecundity, reproductive success, and population growth. The life cycle of *P. elongatus* in the White Sea is such that spawning takes place from March through August, with the August females having been spawned in March (Pertsova, 1981), resulting in two generations per year. *Temora longicornis,* on the other hand, flourished when temperatures were warm, and *A. longiremis* was always abundant.

Studies of feeding of two of the larger calanoids of the White Sea, *M. longa* and *C. glacialis,* showed that males, females, and copepodid stage V (CV) of *M. longa* ingested animal food, primarily copepod nauplii and invertebrate eggs. Females and males were strictly predatory (Perueva, 1984). Feeding activity was greatest at night, with fullest guts observed between 0300 and 0900 h (Perueva, 1983). Maximum gut fullness of females and CVs was greater than that of males and CIVs. The daily ration was estimated to be 3–5% of bodily weight, and the defecation rate was approximately one pellet h^{-1}. *Metridia longa* began diel migrations at the end of summer, which continued into autumn and winter, when reproduction took place (Perueva, 1984). The life cycle was completed in 1 year, and Perueva (1984) argues that the carnivorous diet allowed a relatively shortened life cycle. *Calanus glacialis,* on the other hand, migrated to deeper water when surface temperatures rose in summer, feeding there as an herbivore and probably remaining there in diapause until spring (Perueva, 1984). Copepodid stage IV of *C. glacialis* showed increasing rates of fecal pellet production as food concentration increased, with ingestion in 8 h equaling 13% of bodily weight (Perueva, 1976). A fecal pellet was released each 12 min (Perueva, 1976).

The shallow Barents Sea has essentially no riverine input and is an area where water of North Atlantic origin, modified by Norwegian coastal influences, meets water of Arctic Ocean origin (see Chapter 4). Plankton studies have emphasized the influence of the North Atlantic in that *C. finmarchicus* is frequently dominant (Nesmelova, 1968; Degtereva and Nesterova, 1975; Fleminger and Hulsemann, 1977). By the end of April, nauplii and early copepodids dominated in the southern parts of the Barents Sea, much earlier than they do in the open Norwegian Sea, and by the end of July the population of *C. finmarchicus* was primarily copepodid stage V (Bliznichenko *et al.,* 1976). Detailed studies of major components of the zooplankton in the Barents Sea are surprisingly few and relatively recent, focusing on two dominant euphausiids, *Thysanoessa raschi* and *T. inermis* (Soboleva, 1972, 1975; Zelikman *et al.*, 1979), and two copepods, *C. finmarchicus* and *C. glacialis* (Tande and Båmstedt, 1985; Båmstedt and Tande, 1985; Tande *et al.*, 1985). *Thysanoessa raschi* dominated the euphausiid population and was collected in shallower waters (75–200 m) than *T. inermis* (100–250 m; Bliznechenko *et al.*, 1976), similar to the distributions observed in the Bering Sea (Smith *et al.*, 1983). *Thysanoessa raschi* and *T. inermis* were generally distributed in an aggregated manner (Zelikman *et al.*, 1978), with *T. raschi* inhabiting the colder, less saline coastal waters and *T. inermis* inhabiting offshore waters (Timofeev, 1987). Spawning of *T. inermis* (mid-April through early June) occurred before that of *T. raschi* (early May through late June), and it was observed that the abundance of eggs was reasonably constant from year to year even though adult numbers of *T. raschi* increased 30-fold (Zelikman *et al.*, 1980). These euphausiids may undertake diel vertical migrations (Zelikman *et al.*, 1979) but the data are equivocal owing to the problems of sampling aggregated organisms, the escape capabilities of these large crustaceans that allow them to avoid nets, and their predisposition to occur very near the sediment–water interface (Mauchline and Fisher, 1969). In the Bering Sea, lateral distribution of these euphausiids was similar but evidence for diel vertical migrations was lacking (Smith *et al.*, 1983).

Although the main outflow of the Arctic Ocean is into the North Atlantic Ocean through Fram Strait and the Greenland Sea, the passages of the Canadian Archipelago also connect these two oceans. Flow is generally southward in summer through the passages of the archipelago (Muench, 1971, cited in Coachman and Aagaard, 1974) into Davis Strait, where, on the eastern side, warmer water from the North Atlantic enters around the southern tip of Greenland and flows northward (Pavshtiks, 1969; Huntley *et al.*, 1983). The zooplankton of the passages of the archipelago and contiguous fjords are therefore derived primarily from the Arctic Ocean, as evidenced by the widespread abundance of species such as *C. glacialis, C. hyperboreus, M. longa,* and *O. similis* (Grainger, 1965; Jaschnov, 1970; Buchanan and Sekerak, 1982). *Calanus glacialis,* which is a surface-dwelling species in the Canadian Archipelago at least in summer, has been used to indicate unmixed Arctic water, and its abundance relative to that of *C. finmarchicus* showed clearly the proportional influence of Arctic and Atlantic waters in the Labrador Sea, Davis Strait, and Baffin Bay (Grainger, 1961, 1963). Studies late in the summer have revealed that most of the zooplankton of the Canadian Archipelago were not actively migrating on a diel basis (Longhurst *et al.*, 1984; Sameoto, 1987), that in August some zooplankton had already begun to descend to overwintering depths (Longhurst *et al.*, 1984), that many of the copepod species that dominated the community were most abundant in the upper 50 m (Buchanan and Sekerak, 1982), and that the dominant copepods, three species of *Calanus*, had their maximal abundance located in or above the subsurface chlorophyll maximum regardless of the time of day at which sampling was conducted (Herman, 1983). Seasonal periods of reproduction by copepods in Davis Strait varied depending on when ice-free conditions were realized, being earlier in the southeast (Pavshtiks, 1969; Huntley *et al.*, 1983).

C. The Bering Sea and Subarctic North Pacific Ocean

The only connection between the Pacific Ocean and the Arctic Ocean is via the shallow, narrow Bering Strait connecting the Bering and Chukchi seas. Both coastal plankton from the eastern Bering Sea and oceanic plankton from the central Bering Sea and North Pacific enter the Chukchi Sea (Johnson, 1958; Springer, 1988). The northward-flowing slope water of the Bering Sea (Anadyr water) impinges on the continental shelf north of St. Lawrence Island, where its nutrients stimulate primary production (Sambrotto *et al.*, 1984) and its oceanic zooplankton, consisting of large copepods and euphausiids, supports a diverse and abundant seabird community (Springer and Roseneau, 1985; Springer *et al.*, 1987). Once in the Arctic Ocean, the current splits, with some plankton carried around the north coast of Alaska and into the Beafort Sea and some carried around the coast of Siberia and into the East Siberian Sea. Planktonic species associated with the North Pacific (*Neocalanus plumchrus, N. cristatus, Eucalanus bungii, Metridia pacifica,* and *Calanus marshallae,* which is often mistakenly called *C. glacialis*) have been found as far east as Cape Halkett (151°W; English and Horner, 1977) and Martin Point (146°W; Johnson, 1958), but apparently not much beyond Cape Lisburne (166°W; Springer, 1988) or Point Barrow (156°W; Johnson, 1958). Surface circulation in the Canadian basin is largely a gyre (Coachman and Aagaard, 1974) and therefore tends to retain whatever plankton are introduced via Bering Strait.

Bering Strait is fed in part by the southeastern Bering Sea, which is a complex system of three oceanographic fronts and two interfrontal zones (slope/outer shelf and middle shelf) which persist throughout most of the year and have distinct hydrographic properties (Kinder and Coachman, 1978; Coachman, 1986). The circulation in most of the southeastern Bering Sea is tidally driven, with lateral diffusion extending from the coast to the shelf break. Net flow at the shelf break and on the outer shelf is $1-5$ cm s^{-1} paralleling the bathymetry (Coachman, 1986).

In spring and summer the zooplankton of the southeastern Bering Sea consists of an outer shelf and slope group (*N. plumchrus, N. cristatus, E. bungii, Pseudocalanus* spp., and *M. pacifica*) and a middle shelf group (*T. raschi, C. marshallae, Pseudocalanus* spp., *Oithona* spp., and *Acartia* spp.;

Motoda and Minoda, 1974; Cooney and Coyle, 1982; Smith and Vidal, 1984, 1986). The biomass of the outer shelf and slope community was always dominated by *N. plumchrus,* with the proportion of total biomass in the upper 120 m varying from 33% at the beginning of April (total biomass = 0.6 g C m^{-2}) to 94% at the end of May (total biomass = 5.6 g C m^{-2}; Vidal and Smith, 1986). Either *N. cristatus* or *M. pacifica* was subdominant. Total biomass was maximized in May. In the middle shelf area, biomass in early spring was dominated by euphausiids (primarily *T. inermis* and *T. raschi*), but euphausiid biomass remained fairly constant (0.5 g C m^{-2}) from late March through early June (Vidal and Smith, 1986). Biomass of the copepod *C. marshallae* increased throughout this period, and by early June it dominated total biomass (58% of the 1.39 g C m^{-2}; Vidal and Smith, 1986). The dominant copepods in these two habitats overlap only in the region of the 115-m isobath (Smith and Vidal, 1984). Over the slope and outer shelf, growth of the zooplanktonic community peaked in May and was due to growth of large copepods, whereas over the middle shelf growth had still not reached a peak by early June (Vidal and Smith, 1986). Small copepods contributed little to either biomass or growth in either habitat (Vidal and Smith, 1986).

The spring bloom of phytoplankton, which is generally a 3- to 4-week event in May, resulted in increased abundance of most zooplanktonic taxa over the middle shelf, whereas over the outer shelf and slope, most taxa were not significantly more abundant during the bloom (Smith and Vidal, 1984, 1986). This is because major biological differences separate the species inhabiting the slope/outer shelf and middle shelf areas. The two dominant species of the outer shelf and oceanic habitat, *N. plumchrus* and *N. cristatus,* have monocyclic life cycles with ontogenetic vertical migrations, in which growth and development of the copepodid stages take place at the surface but reproduction occurs in deep water during winter (Heinrich, 1962, 1968; Smith and Vidal, 1984). These taxa produced offspring as a single cohort per year over the outer shelf and slope (Smith and Vidal, 1984) which grew at rates of 14% dry body wt day^{-1} for early copepodids of *N. plumchrus* and 8% day^{-1} for early stages of *N. cristatus* (Vidal and Smith, 1986). Late copepodids of both species grew at approximately 4% dry body wt day^{-1} (Vidal and Smith, 1986). Diel vertical migrations were absent. Other dominant species of the outer shelf and slope either reproduced continuously (*M. pacifica*) or produced more than one cohort (*E. bungii;* Smith and Vidal, 1984). Both these taxa grew at approximately 11–13% dry wt day^{-1} (Vidal and Smith, 1986). *Metridia pacifica* females and stage V copepodids undertook diel vertical migrations from late April through mid-July (Smith *et al.,* 1983). Juvenile stages of *E. bungii* also migrated on a diel basis (Minoda, 1971).

Of the dominant species of copepod on the middle shelf, *Acartia* spp.

showed continuous reproduction while *C. marshallae* produced two cohorts (Smith and Vidal, 1986). *Calanus marshallae* grew at 14–15% dry wt day^{-1} (Vidal and Smith, 1986). The most abundant euphausiid of the southeastern Bering Sea is *T. raschi*, which is a neritic and euryhaline species of wide distribution (Fukuchi, 1977; Hopkins *et al.*, 1984). It is probably a filter feeder (Nemoto, 1967) and breeds after the onset of the spring bloom of phytoplankton (Minoda and Marumo, 1974). Although the biomass of euphausiids, mainly *T. raschi*, remained fairly constant over the middle shelf in March–June, numbers decreased from 500 to 100 m^{-2} (Vidal and Smith, 1986). Growth of calyptopis stages was 14% dry wt day^{-1}, and adolescents in the first year class grew at 3% dry wt day^{-1} (Vidal and Smith, 1986).

The major copepods of the outer shelf and slope of the Bering Sea (Cooney and Coyle, 1982; Smith and Vidal, 1984; Vidal and Smith, 1986) also occur throughout the subarctic North Pacific (Heinrich, 1962, 1968; Minoda, 1971; Motoda and Minoda, 1974; Miller *et al.*, 1984; Parsons and Lalli, 1988), where *N. cristatus* spawned continuously below 250 m with a peak in November and *E. bungii* produced two cohorts in the upper mixed layer (early May and early July; Miller *et al.*, 1984). The November peak in spawning of *N. cristatus* showed up as a peak in copepodid stage V abundance in the upper layer in May (Miller *et al.*, 1984), similar to observations over the outer shelf and slope of the Bering Sea (Smith and Vidal, 1984). Diapause in these two copepods began between July and October (Miller *et al.*, 1984). The strategy of *N. plumchrus* has just been clarified (Miller and Clemons, 1988) because what has been called *N. plumchrus* is, in fact, two species (Miller, 1988). *Metridia pacifica* spawned throughout the year with peaks in March, July, and September (Batchelder, 1985), consistent with the spring and summer results from the Bering Sea (Smith and Vidal, 1986). In summer, late copepodids and adult females migrated daily between 0 and more than 250 m (Batchelder, 1985).

Maximum ingestion rate for *N. plumchrus* copepodid stage V in the slope region of the southeastern Bering Sea during the spring bloom was 6.1 µg C ind^{-1} h^{-1} (ind = individual), and for copepodid stage V of *N. cristatus* maximum ingestion was 18.3 µg C ind^{-1} h^{-1} (Dagg and Wyman, 1983). Much lower maximum rates were found when individuals were confined to bottles for 16–22 h (Dagg *et al.*, 1982), but these lower rates were quite adequate for the estimated growth and respiration needs of the organisms. Measured respiration rates were approximately 5% bodily carbon daily and ammonium excretion rates were approximately 5% bodily nitrogen daily (Dagg *et al.*, 1982). Other analyses suggest that *N. plumchrus* and *N. cristatus* respire 2% of their bodily carbon daily, with rates for *M. pacifica* being 5% bodily carbon daily and for the euphausiids *T. raschi* and *T. inermis*, 3% bodily carbon daily (Vidal and Whitledge, 1982).

In contrast with the Bering Sea, where much of the annual growth of grazing zooplankton takes place during a pronounced and predictable spring bloom, the subarctic North Pacific shows no seasonal bloom, yet it has seasonally varying populations of *N. plumchrus* and *N. cristatus* in its zooplankton community (Miller *et al.*, 1984). The feeding of these species in the North Pacific was shown to be omnivorous, with *N. cristatus* more predatory than *N. plumchrus* (Frost *et al.*, 1983; Greene and Landry, 1988). Model results suggested that microzooplankton may play a key role as grazers of small phytoplankton cells and prey for the larger *Neocalanus* species in this system (Frost, 1987). The *Neocalanus* species themselves, however, were also capable of ingesting phytoplankton of a wide range of sizes and at quite low cell concentrations (Frost *et al.*, 1983).

D. Seasonal Cycles and Specialized Adaptations of the Major Planktonic Copepods and Euphausiids

The biology of the major copepods of the Arctic Ocean has been studied in detail in the fjords of Norway, the Norwegian Sea, the Greenland Sea, and the Canadian Archipelago. Despite the severe constraints placed on the pelagic fauna at high latitudes, life cycles of the various species show significant variations (Conover, 1988). Because the question of how animals have adjusted to long winters with little food or light is critical to an understanding of polar food webs, and despite the difficulties of conducting experimental and basic observational programs in polar areas, our concepts have become fairly sophisticated in recent years. Large calanoid copepods have been the organisms most often studied because of previous work on some of the species in other locations, their dominance of the biomass during spring and summer, and the relative ease of collection and isolation.

1. *Calanus glacialis* Jaschnov

The Barents Sea and the Greenland Sea are areas of sympatry of the closely related species *C. finmarchicus* and *C. glacialis*. In the earliest investigations of both areas, small and large forms of *C. finmarchicus* were discussed (Ussing, 1938; Digby, 1954; Wiborg, 1954); although the naming of the larger form *C. glacialis* (Jaschnov, 1955) did not end the confusion (Matthews, 1967), there is now agreement that these are two species (Frost, 1971; Fleminger and Hulsemann, 1977) and that their presence together indicates a mixed or subarctic water mass (Grainger, 1961). When *C. finmarchicus* and *C. glacialis* females are collected alive, their swimming patterns and pigmentation are sufficiently different that distinguishing them is not difficult. Using live samples isolated for egg production experiments (Smith, 1990), we have found that female *C. glacialis* can be considerably smaller

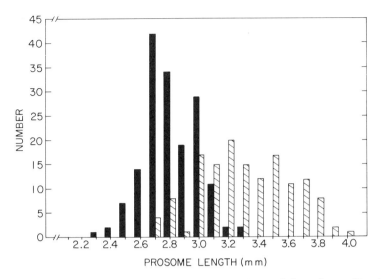

Figure 10.2 Sizes (prosome length in millimeters) of adult female *Calanus finmarchicus* (solid bars) and adult female *Calanus glacialis* (hatched bars) from the Fram Strait area of the Greenland Sea.

than is generally thought in the Greenland Sea, where *C. glacialis* and *C. finmarchicus* are sympatric (Fig. 10.2).

Calanus glacialis, a species endemic to the Arctic Ocean (Grainger, 1963; Pavshtiks, 1971) but found as far south as Nova Scotia (Runge *et al.*, 1985), has a life cycle with some equivocal characteristics. In Foxe Basin (west of Baffin Island), for example, the *C. glacialis* population has some individuals apparently with a 1-year life cycle, overwintering as copepodid stages V and VI, and some showing a 2-year cycle, overwintering as copepodid stage III (Grainger, 1963). Based on collections in the Barents Sea during May–August, *C. glacialis* is thought to spawn in May and June and to have a 2-year life cycle (Tande *et al.*, 1985). However, in the Godthaab fjord of western Greenland, the life cycle was completed in 1 year, with fertilization in December, spawning primarily in March, copepodid stage I (CI) abundant in April or later, copepodid stage III (CIII) abundant in May or later, and copepodid stage V (CV) abundant in September/October (Maclellan, 1967). The timing of the appearance of copepodid stage I varied from early June in the eastern Labrador Sea to late September in western Baffin Bay (Grainger, 1963), consistent with circulation and general springtime development of plankton communities in this area (Huntley *et al.*, 1983).

Maclellan (1967) observed that *C. glacialis* females were capable of prolonged spawning, an observation subsequently documented experimentally by Hirche and Bohrer (1987). This leads to the presence of different copepodid stages in the water during the summer months and confusion concerning the length of the life cycle. Maclellan (1967) implies that in the Godthaab fjord the individuals spawned early in the year, which reach copepodid stage V by September/October, were responsible for much of the spawning in the subsequent year, since she found only a few early copepodids present in February when ovaries were ripening and spawning was beginning. Whether *C. glacialis* overwinters as copepodid stage III remains to be verified. A 1-year life cycle is also possible for *C. glacialis* in the Greenland Sea (Smith, 1990) and off Nova Scotia, based on molting rates observed in the field (Runge *et al.*, 1985) and rearing experiments conducted in the laboratory (Corkett *et al.*, 1986). The rearing experiments suggest a growth rate for early copepodids of approximately 12% body wt day^{-1} (Corkett *et al.*, 1986).

Understanding the life cycle of *C. glacialis* is further complicated by experimental evidence that females are long-lived, laying eggs for 10 months when fed, and that egg laying is food dependent (Hirche and Bohrer, 1987). These observations led Hirche and Bohrer (1987) to suggest that the life cycle of *C. glacialis* resembled that of *C. finmarchicus*. In Hudson Bay, egg laying was also dependent on food concentration, with ice algae providing nutrition for egg production prior to pelagic algal blooms (Runge and Ingram, 1988). Observations began on March 31, however, and therefore could have missed an earlier spawning or the main reproductive event described by Maclellan (1967). Similarly, Hirche and Bohrer's field observations were made in June and July, well after any potential early spawning and within a subsequent period of prolonged reproduction. Active spawning by *C. glacialis* was observed in the Greenland Sea in late March prior to any phytoplankton growth, when measured ingestion rates would have balanced respiration only (Smith, 1990). This early egg laying appeared to be supported by stored lipids (Smith, 1990) and would result in a 1-year life cycle.

During the spring bloom in the Labrador Sea, females of *C. glacialis* did not ingest cells smaller than 20 μm, but all cells larger than 20 μm were ingested in proportion to their abundance, with no evidence of any saturation of the functional response (Huntley, 1981). Not surprisingly, when chlorophyll *a* concentrations reached approximately 7 mg m^{-3}, ingestion rates of female *C. glacialis* showed saturation, as did those of copepodid stage V (Tande and Båmstedt, 1985). Copepodid stage IV showed saturation at about 2 mg m^{-3}.

Maximum ingestion rates of copepodid stage V of *C. glacialis* were on the order of 2 μg C ind^{-1} h^{-1} or 12% of body wt day^{-1} (Tande and Båmstedt, 1985; Head, 1986; Head's data converted from pigment concentrations based on a ratio of C to chlorophyll of 50). Maximum ingestion rate of copepodid stage IV was 0.3 μg C ind^{-1} h^{-1} (4% body wt day^{-1}) and of adult females was 11.0 µg C ind^{-1} h^{-1} (54% body wt day^{-1}; Tande and Båmstedt, 1985). Rates of gut clearance for *C. glacialis* did not vary with stage of development, ranging between 1.4 and 1.7% min^{-1} for copepodid stage IV through adult females (Tande and Båmstedt, 1985), whereas *C. finmarchicus* adult females showed clearance rates approximately twice those of copepodid stage V (3.2 and 1.7% min $^{-1}$, respectively; Tande and Båmstedt, 1985). Examination of the gut contents of copepodid stage IV of *C. glacialis* revealed that tintinnids and nauplii as well as phytoplankton were ingested, which suggests that it can grow successfully in habitats other than diatom blooms (Perueva, 1977a). Studies of diel feeding rhythms (Perueva, 1977b; Båmstedt, 1984; Head *et al.*, 1985) give the general impression that feeding intensity was highest in the 0100–1000-h period. Digestive enzyme activity and gut fullness were not correlated (Båmstedt, 1984; Head *et al.*, 1985).

Weight-specific respiration rate in *C. glacialis* at ambient temperature (−1.7°C) increased with decreasing size, from 0.6 µl O$_2$ (mg dry wt)$^{-1}$ h^{-1} in adult females to 0.95 µl O$_2$ (mg dry wt)$^{-1}$ h^{-1} in copepodid stage III (Båmstedt and Tande, 1985). Copepodid stage V and adult females had essentially the same respiration rate. Weight-specific ammonium excretion rates followed the same pattern, with females releasing 4.7 nmol NH$_4^+$ (mg dry wt)$^{-1}$ h^{-1} and copepodid stage IV releasing 9.5 nmol NH$_4^+$ (mg dry wt)$^{-1}$ h^{-1}. Although the study was conducted during the spring diatom bloom, the O/N ratios suggested that protein was the substrate catabolized, possibly in conjunction with the synthesis of wax esters (Båmstedt and Tande, 1985). Hirche (1987) found somewhat lower respiration rates for *C. glacialis* at 0°C that would correspond to daily ingestion rates of 2–3% bodily carbon.

2. *Calanus hyperboreus* (Krøyer)

Calanus hyperboreus, also endemic to the Arctic Ocean (Grainger, 1965) and probably its most abundant species of *Calanus* (Grainger, 1963), has a better-understood life cycle owing to its widespread distribution and its large size. A detailed 23-month study of its vertical distribution in the Arctic Ocean revealed what seems to be a 3-year life cycle (Dawson, 1978). Copepodids migrated between the surface (summer) and depths greater than 200–300 m (winter), while males remained deeper than 400 m and females descended from 100 to 300 m in spring and early summer (Dawson, 1978). Gravid females were present in the spring in both years, suggesting that spawning occurred prior to any phytoplankton bloom. In the fjords of southwestern Norway (58°N; Matthews *et al.*, 1978), fjords of northwestern Norway (65°N; Wiborg, 1954), and eastern Greenland (Ussing, 1938; Digby, 1954), *C. hyperboreus* has an annual life cycle in which spawning occurs in January and February in the fjords of Norway and somewhat later off eastern

Greenland. Active egg laying by *C. hyperboreus* was observed in late March in the Greenland Sea (Smith, 1990).

Laboratory studies of the development rates of *C. hyperboreus* have shown that 250 days are required for the egg to reach adulthood at 2-4°C (Conover, 1965) and 114 days to reach copepodid stage III at 4-6°C (Conover, 1967). Total developmental duration from time taken for embryonic development at 4°C was estimated to be 106 days (Corkett *et al.*, 1986). In the field, however, this animal was reproducing at zero and subzero temperatures, with embryonic development times measured in the field at -2°C being approximately 8 days (Smith, 1990). This equals a total developmental time of 201-350 days. The developmental times for *C. hyperboreus* at 0°C based on the ratios of Corkett *et al.*, (1986) seem too short, since they do not consider the possibility of prolonged diapause known to occur in copepodid stage V (Conover, 1965). Based on the rate of embryonic development measured at -2°C in the field, the life span measured at 2-4°C in the laboratory, and field observations of population development, a 1-year life cycle seems to be a reasonable assumption in areas outside the Arctic Ocean.

Attributes of the life cycle of *C. hyperboreus* have been investigated fairly extensively in the laboratory. The molt to adulthood took place primarily in November, December, and January, with none taking place in May and June (Conover, 1965). Copepodid stage V was achieved by the end of May (Conover, 1965). Egg laying depleted lipid reserves (Conover, 1967; Head and Harris, 1985; Smith, 1990) to the extent that spent females may weigh 10% of what they did immediately following a phytoplankton bloom (Conover, 1967). Spent females in the Canadian Archipelago have lost 80% of their soluble protein and 60% of their lipid compared with their gravid counterparts, and they respire at a 30% lower rate (35 μl O_2 ind^{-1} day^{-1}; Head and Harris, 1985). Lipid-rich nauplii do not feed prior to the nauplius V stage (Conover, 1962). Egg laying began within 1-9 weeks following molting to adulthood (Conover, 1967). Measurements of egg production in the Greenland Sea prior to the spring bloom of phytoplankton and at -2°C showed a maximum clutch size of 57 eggs (female)$^{-1}$ (Smith, 1990). Laboratory studies have also shown that ambient food supplies are not required for either gonad maturation (Conover, 1962) or egg laying, since starved females maintained at a relatively constant temperature laid their eggs in November and December as well (Conover, 1967).

Respiration rates of copepodid stages IV and V and adult females of *C. hyperboreus* in the Gulf of Maine increased throughout March, reaching their annual maxima (18-22 μl O_2 ind^{-1} day^{-1} for CV and 24-35 μl O_2 ind^{-1} day^{-1} for adult females) in April and May (Conover, 1960, 1962; Conover and Corner, 1968) (Fig. 10.3). At other times of the year respiration rates of copepodid stage V were reduced, and the combination of reduced

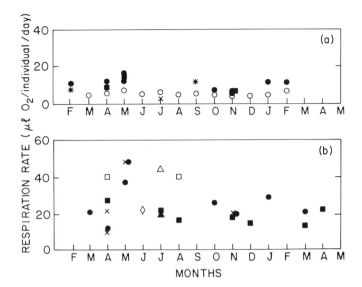

Figure 10.3 (a) Respiration rate of *Metridia longa* adult females. (●) Conover and Corner, 1968; (*) Båmstedt et al., 1985; (○) Haq, 1967; (■) Conover, 1968; (×) Conover and Cota, 1985. (b) Respiration rate of *Calanus hyperboreus* adult females. (●) Conover and Corner, 1968; (×) Conover, 1968; (■) Conover, 1962; (□) Head and Harris, 1985; (◇) Smith, 1988; (△) Kalland, 1972; (▲) Conover and Cota, 1985.

respiration and utilization of lipid as the metabolic substrate allow prolonged survival of this stage in the absence of food (Conover, 1962). Gravid females had respiration rates (20 μl O_2 ind^{-1} day^{-1}) that were 25% higher than those of nongravid females (Conover, 1962).

Other measurements at higher latitudes (e.g., Canadian Archipelago: Conover and Cota, 1985; Greenland Sea: Smith, 1988), primarily in summer, were in agreement with the April and August observations from the Gulf of Maine. However, in the Arctic Ocean (Kalland, 1972) and Jones Sound (Head and Harris, 1985), respiration rates of female *C. hyperboreus* were found to be approximately 40 μl O_2 ind^{-1} day^{-1} (Fig. 10.3). Reasons for the higher rates were not specified, although Head and Harris (1985) consider the higher rates typical for the high Arctic and calculate that even with such high respiration rates, females could survive in an overwintering state for 125 days. This contention contradicts evidence from the Gulf of Maine, where respiration rates of stage V copepodids in November and December (6 μl O_2 ind^{-1} day^{-1}) during overwintering were less than half the rates of the actively growing and feeding individuals captured in April (18 μl O_2 ind^{-1} day^{-1}; Conover and Corner, 1968; Conover, 1968). In *C. finmarchicus*, as another example, overwintering respiration rates of females and stage V copepodids were only approximately half those of actively feeding animals (Hirche, 1983).

Rates of nitrogen excretion, apart from their usefulness in evaluating the extent to which a pelagic system is dependent on regenerated nitrogen for its overall productivity (Dugdale and Goering, 1967), can be used as an additional indicator of metabolic status and potential metabolic substrates (Conover, 1968). In the Greenland Sea, normal rates of ammonium excretion for females (5 μg N ind^{-1} day^{-1}) and stage V copepodids (3 μg N ind^{-1} day^{-1}) of *C. hyperboreus*, combined with rates of respiration and low rates of ingestion, suggested that the animals were metabolizing protein, possibly their own tissues, since their body weight decreased during the study (Smith, 1988). In the Canadian Archipelago, ammonium excretion rates of *C. hyperboreus* were lower [2 μg N (female)$^{-1}$ d^{-1}] than those in the Greenland Sea (Conover and Cota, 1985; Smith, 1988), and excretion of urea was measured for most stages of *C. hyperboreus* and averaged 9% of ammonium excretion (Conover and Cota, 1985). Spent females in the Canadian Archipelago released nitrogen in spring at a very low rate (0.1 μg N ind^{-1} day^{-1}) that was half that of gravid females (Head and Harris, 1985). Stage V copepodids at depth released virtually no nitrogen, suggesting the onset of overwintering (Head and Harris, 1985).

Calanus hyperboreus is primarily herbivorous, ingesting a variety of diatom cells offered it in the laboratory (Conover, 1960) and showing growth when fed *Thalassiosira fluviatilis, T. nordenskioldii, Rhizosolenia setigera,* and *Ditylum brightwellii* (Conover, 1964). Assimilation efficiency of copepodid stage V varied with the species of diatom offered (Conover, 1964). In the Greenland Sea when phytoplankton biomass was low, adult female and copepodid stage V of *C. hyperboreus* obtained most of their ingested carbon from the few diatoms present, although they did ingest some nanoplankton (Smith, 1988). In all cases ingestion was estimated to be less than 1% bodily carbon day^{-1} (Smith, 1988). In the Labrador Sea, however, during the spring diatom bloom, *C. hyperboreus* did not ingest cells smaller than 10 μm and ingestion was proportional to concentration (Huntley, 1981).

Activity of digestive enzymes was not correlated with gut fullness in *C. hyperboreus* in the Canadian Archipelago (Head *et al.*, 1985), but digestive enzymes in the Greenland Sea showed two trends. In copepodid stage V, animals captured below 200 m had much lower activity than animals captured in the upper 100 m, and activities were highest at the surface when temperatures were less than 0°C and the animals were under ice cover (Hirche, 1989).

The nutritional value of *Phaeocystis pouchetii*, often a very abundant component of the phytoplankton community, to high-latitude copepods is a continual question. The flagellate form of this species seems to be consumed

10 Polar Zooplankton

Table 10.1 Lipid Content of *Calanus Thyperboreus, Calanus glacialis, Calanus finmarchicus, Metridia longa, Thysanoessa inermis,* and *thysanoessa raschi*

	Dry weight (μg ind^{-1})	Lipid weight (μg ind^{-1})	Percent
Calanus hyperboreus			
♀	(4,067)	2,875^a (maximum)	71
♀ (?)	1,770	980^b	55
♀	3,168	1,077^c (mean)	34
♀	3,300	2,100^d (maximum)	64
♀	3,600	1,908^e (maximum)	53
♀	1,893	502^f (mean)	26
CV	(2,736)	1,745^a (maximum)	64
CV	1,920	703^c (mean)	38
CV	2,700	1,539^e (maximum)	57
CIV	575	218^c (mean)	38
CIV	450	194^e (maximum)	43
CIII	122	30^c (mean)	25
Calanus glacialis			
♀	440	220^g	50
♀	714	400^h	56
♀	533	187^f (mean)	35
Calanus finmarchicus			
♀	275	32^i	12
♀	344	136^e	40
♀	340	103^f (mean)	30
♀	429	99^c (mean)	23
CV	418	130^c (mean)	31
CV	350	40^i	11
CV	275	82^f (mean)	30
CV	510	250^e	49
Metridia longa			
♀	176	60^g	34
♀	340	105^e (maximum)	31
Thysanoessa inermis			
I	4,670	708^j	15
II	10,667	1,222^j	11
Thysanoessa raschi			
I	4,556	389^j	9
II	7,111	807^j	12

^{a} Head and Harris (1985).
^{b} Conover (1964).
^{c} Smith (1988).
^{d} Lee (1974).
^{e} Conover and Corner (1968).
^{f} Smith (1989).
^{g} Lee *et al.* (1971).
^{h} Lee (1975).
^{i} Marshall and Orr (1972).
^{j} Falk-Petersen *et al.* (1982).

in small amounts (Huntley et al., 1987) if at all (Smith, 1988) by C. hyperboreus. The colonial form, however, was ingested by copepodid stage IV in amounts that should support respiration and growth (Huntley et al., 1987). The lipid content of C. hyperboreus can be considerable (Table 10.1) and varies seasonally in females in the Arctic Ocean from 64% of dry body weight during and immediately after the phytoplankton bloom (August) to 29% in June (Lee, 1974). Wax esters account for 82–91% of the lipid (Lee, 1974). With starvation in the laboratory, presumably simulating the overwintering period, females lost 5.25 µg day^{-1} dry body weight (0.3% day^{-1}) and 4.6 µg day^{-1} of lipid (0.5% day^{-1}; Conover, 1964). The same daily rate of lipid loss was reported by Lee (1974) for females starved for 90 days. In the Canadian Archipelago during the presumed transition into overwintering, stage V copepodids that were deep (>200 m) had 32% more lipid (1720 µg ind^{-1}) and 55% more soluble protein (876 µg ind^{-1}) than surface-caught individuals (Head and Harris, 1985). All indications are that these considerable lipid reserves are the metabolic substrate used by C. hyperboreus during the long overwintering period it experiences at high latitudes.

The RNA concentration of C. hyperboreus peaks in March, coincident with its reproduction (Båmstedt, 1983). The genome size of female C. hyperboreus is the same as that of C. glacialis and twice that of C. finmarchicus, while all three species have essentially the same number of nuclei (McLaren et al., 1988).

The density of various zooplanktonic taxa has been determined, and C. hyperboreus is less dense than seawater (1.022–1.025 g cm^{-3}) for most of the year, May through January (Køgeler et al., 1987). Just before spawning in February, however, C. hyperboreus was more dense (1.036 g cm^{-3}) than seawater (Køgeler et al., 1987).

3. Calanus finmarchicus (Gunnerus)

Understanding of the biology of C. finmarchicus in its Arctic and subarctic habitats has grown slowly, owing partly to its widespread distribution north of 40°N (Fleminger and Hulsemann, 1977), which makes the Arctic edge of its distribution less attractive for study, and to the fact that C. glacialis, the Arctic congener of C. finmarchicus, was not distinguished prior to 1955 (Grainger, 1961; Jaschnov, 1972). Calanus finmarchicus has long been known to be a major prey item for fish of subarctic and boreal parts of the North Atlantic (e.g., herring and capelin) and is so abundant off Norway that commercial fisheries for the copepod itself have developed (Wiborg, 1976).

Female C. finmarchicus showed the highest weight-specific ingestion rates of the three Calanus species studied in the Labrador Sea, with ingestion rate increasing as particle concentration increased (Huntley, 1981). Of the three species, only C. finmarchicus females ingested particles in the 5–10-µm size range (Huntley, 1981). In Fram Strait, however, female *C. finmarchicus* did not ingest nanoplankton or *P. pouchetii* (Smith, 1988). Although a very distinct daily grazing rhythmn has been observed in copepodid stage V of *C. finmarchicus* in the St. Lawrence estuary (Simard *et al.*, 1985), the feeding of this stage in higher latitudes showed no such rhythmn (Smith, 1988). Even in summer, stage V copepodids in Fram Strait were not feeding and in the Barents Sea showed gut evacuation rates that were half those of adult females (Tande and Båmstedt, 1985). Ingestion rates of female *C. finmarchicus* in Fram Strait in summer were less than 2% bodily carbon day^{-1} (Smith, 1988), and in the Barents Sea the daily ingestion rates for females varied between 3 and 32% bodily carbon (Tande and Båmstedt, 1985). Similarly, ingestion rates of copepodid stage V in Fram Strait were very low (less than 1% bodily carbon daily; Smith, 1988) compared with those of the Barents Sea (1–17% bodily carbon daily; Tande and Båmstedt, 1985).

The activities of digestive enzymes of copepodid stage V of *C. finmarchicus* in Fram Strait in June and July were high in the upper 100 m compared with the 200–500-m layer and were higher in the marginal ice zone than in the underice habitat (Hirche, 1989). In the Balsfjorden, activities of digestive enzymes of copepodid stage V were never as high as those of adult females and were extremely low from August through February, indicating dormancy during overwintering (Tande and Slagstad, 1982). Similarly, off the coast of Norway in November, *C. finmarchicus* (stage CV presumably) had empty guts and very reduced digestive enzyme activity (Båmstedt and Ervik, 1984).

The range in respiration rates of *C. finmarchicus* captured in summer in Fram Strait was 11–26 μl O_2 (mg dry wt)$^{-1}$ day^{-1} for copepodid stage V and 20–43 μl O_2 (mg dry wt)$^{-1}$ day^{-1} for females at temperatures between 0° and 3°C (Smith, 1988), somewhat higher than those measured during a study of the effects of temperature on respiration and activity of *C. finmarchicus* from the same area at approximately the same time of year [12 μl O_2 (mg dry wt)$^{-1}$ day^{-1} for adult females at 3°C; Hirche, 1987].

A seasonal study off the west coast of Sweden (in which stages of *C. finmarchicus* were not identified) showed highest ammonium and urea excretion rates in March and April, followed by a steady decline into October (Båmstedt, 1985). In the Fram Strait in summer, daily ammonium excretion rates were 0.6–5.6 μg N (mg dry wt)$^{-1}$ day^{-1} for copepodid stage V and 0.5–3.8 μg N (mg dry wt)$^{-1}$ day^{-1} for adult females (Smith, 1988).

Prior to egg laying and during gonadal development in Fram Strait, the total lipid content of *C. finmarchicus* females is low compared with that of *C. hyperboreus* and *C. glacialis* and similar to that of *M. longa* (Table 10.1). The *C. finmarchicus* females collected in Fram Strait in March and early April (Smith, 1990) were not laying eggs and their ovaries were not well enough developed to do so. Thus, their relatively low proportion of lipid does not reflect losses due to egg laying. The high percentage of wax ester in the lipid of *C. finmarchicus* suggests that it is sequestered for use in egg production (Sargent and Whittle, 1981). The lipid content and wax ester proportion of the lipid increased exponentially with developmental stage in *C. finmarchicus* captured in the North Sea (Kattner and Krause, 1987). The density of *C. finmarchicus* increased at the end of February in the Tromsø area at the time of reproduction, possibly because the relatively light wax esters were transformed into denser lipids constituting the eggs (Køgeler et al., 1987).

The life cycle of *C. finmarchicus* in subarctic habitats has been described as an annual one (e.g., Sømme, 1934; Lie, 1965; Maclellan, 1967; Davis, 1976), although there may be more than one period of reproduction (Wiborg, 1954; Matthews et al., 1978). The cycle off Norway begins with overwintering individuals located deeper than 200 m rising to the upper 50 m before mid-March (Sømme, 1934). This is seen as a peak in the abundance of adult females in March and April virtually everywhere along the coast of Norway (Wiborg, 1954). Peak abundances of nauplii were evident in April off Norway (Tande et al., 1985) and west Greenland (Maclellan, 1967). By July the population was primarily copepodid stages IV and V, with the proportion of copepodid stage V increasing through February (Wiborg, 1954; Lie, 1965).

The life cycle of *C. finmarchicus* in the waters of the Barents Sea originating in the North Atlantic takes 1 year, with spawning occurring in April and May (Tande et al., 1985). Female *C. finmarchicus* collected north of the Polar Front, however, were immature, suggesting that reproductive development ceases once *C. finmarchicus* leaves water from the North Atlantic and enters water from the Arctic Ocean. Thus, the *C. finmarchicus* north of the Polar Front are probably nonreproductive expatriates (Tande et al., 1985). When *C. finmarchicus* were growing in water of North Atlantic origin in the Greenland Sea in summer, all copepodid stages were present, indicating active reproduction and development, but when they were collected from water of Arctic Ocean origin, only late-stage copepodids and females were found (Fig. 10.4). Thus, expatriate *C. finmarchicus* not only are nonreproductive but also seem to cease development at copepodid stage V, judging by the large proportion of individuals in that stage.

The overwintering strategy of *C. finmarchicus* is of interest because several studies have shown that rates of egg production in this species are dependent on food supply (Marshall and Orr, 1972; Runge, 1985). Such a dependence implies that the spring phytoplankton bloom must occur prior to egg laying. However, it has been suggested that egg development occurs in females after the upward ontogenetic migration in February/March but before the spring bloom in subarctic habitats (Sømme, 1934). Furthermore,

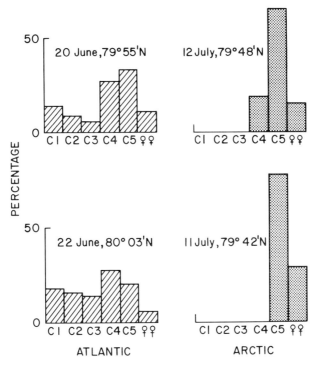

Figure 10.4 Age structure of *Calanus finmarchicus* captured in the Atlantic and Arctic portions of the Fram Strait area of the Greenland Sea.

near Tromsø in northern Norway, spawning itself occurred prior to the spring bloom (Davis, 1976). In the Fram Strait in April, however, *C. finmarchicus* females captured in the upper 100 m were not yet ripe and laid no eggs (Smith, 1990).

Gonadal development has been observed in overwintering *C. finmarchicus* (copepodid stage V) in Balsfjorden, northern Norway, beginning in January, coincident with the period (December–March) when the overwintering stock was losing weight (Tande, 1982). Weight loss by adult females in the January–April period was 37% (from 204 μg dry wt ind^{-1} to 128 μg; Hopkins *et al.*, 1984). Depletion of the oil sac was correlated with the appearance of females with semiripe ovaries, suggesting that in January through March lipid reserves from the previous summer were used for ovarian development (Tande, 1982). Sexual differentiation in copepodid stage V occurred 2 months before spawning in Balsfjorden (Tande and Hopkins, 1981) and in starved individuals maintained in the laboratory (Grigg and Bardwell, 1982). However, final development of the ovaries and the rate of egg release were dependent on external food supply (Tande and Hopkins, 1981; Tande, 1982). Overwintering stage V copepodids show several adaptations that would enhance prolonged survival in the absence of food, such as reduced digestive enzyme activity (Hallberg and Hirche, 1980; Tande and Slagstad, 1982) accompanied by poorly developed secretory epithelium in the gut (Hallberg and Hirche, 1980), low rates of ammonium excretion (Tande, 1982), and low rates of respiration (Hirche, 1983). When deep-living, overwintering individuals were collected, their guts were empty, they had large oil sacs, and their swimming behavior was sluggish (Hirche, 1983). These observations suggest that the overwintering period is one of lipid metabolism. Overwintering strategies and the environmental and physiological cues involved are research areas in need of urgent attention. It seems likely that hormonal controls are involved (Grigg and Bardwell, 1982; Hirche, 1983), but direct evidence is nonexistent.

4. *Metridia longa* (Lubbock)

Metridia longa is an abundant but less well known member of Arctic and subarctic plankton communities. It is characterized by turquoise bioluminescence and extensive diel vertical migrations (Vinogradov, 1970). Its life cycle and strategy are distinct from those of the *Calanus* species with which it shares space and food. Off northern Norway, the life cycle is an annual one with spawning in April and May (Tande and Grønvik, 1983). The proportion of copepodid stage V with ovaries in an advanced state increased from September to December, with ovaries of adult females developing rapidly in February and March (Tande and Grønvik, 1983). Ripe females were collected in April and May, and the peak in the proportion of copepodid stage I in the population was observed in June (Grønvik and Hopkins, 1984). Maturation of females in December differed from year to year, suggesting that development during overwintering has interannual variation and that the winter environment may influence timing and magnitude of reproduction in this species (Tande and Grønvik, 1983). Off northern Norway, the *M. longa* population was primarily copepodid stage V by September, with overwintering being undertaken by adults (Grønvik and Hopkins, 1984).

Based on the time elapsing between peaks in various stages observed in Balsfjorden (Grønvik and Hopkins, 1984) and dry weights of various stages of *M. longa* measured in the Greenland Sea (S. L. Smith, unpublished), growth from copepodid stage I to copepodid stage IV proceeded at a rate of 8% of dry body wt day^{-1}, and growth between copepodid IV and V slowed to 2–4% dry body wt day^{-1}. This is considerably slower than the growth estimated for *M. pacifica* in the Bering Sea in spring and early summer (Vidal and Smith, 1986). The dry weight of females from Norway declined during winter from 150 μg ind^{-1} in October to 90 μg ind^{-1} in March (Grønvik and Hopkins, 1984). Stage V copepodids similarly lost weight in winter (from 98 μg ind^{-1} in October to 73 μg ind^{-1} in January and February; Grønvik and Hopkins, 1984). The decrease in body weight suggests that even though *M. longa* can feed as a carnivore (Haq, 1967), it is unable to meet its metabolic needs in winter.

The noontime distribution of *M. longa* in the water column of Balsfjorden varied with stage, copepodid stage I remaining in the upper 50 m and copepodid IV through adults remaining deeper than 110 m (Grønvik and Hopkins, 1984). In the Greenland Sea in July, there was no detectable diel vertical migration by any stage of *M. longa;* copepodid stages I through III were scarce and captured only below 150 m (Smith, 1988). Females and stage V copepodids were consistently deeper than 50 m (Smith, 1988). Furthermore, *M. longa* was most abundant at the ice edge, similar to *C. finmarchicus* (Smith, 1988), suggesting the possibility that increased food supplies, especially microzooplankton (Bolms and Lenz, 1989), are linked to increased reproduction. Adult females dominated the age structure of *M. longa* in the Greenland Sea in July (Smith *et al.*, 1986), whereas off northern Norway the population of *M. longa* was dominated by copepodid stages II and III in July (Grønvik and Hopkins, 1984). Thus, the annual life cycle described for northern Norway did not apply in the Greenland Sea.

Adult females dominated all the collections deeper than 100 m in the Greenland Sea in March and April (Smith *et al.*, 1988). Subadult stages were collected only deeper than 250 m and were restricted to copepodid stages IV and V (Smith *et al.*, 1988). Off northern Norway adult males and females also predominated at this time and were restricted to depths greater than 100 m (Grønvik and Hopkins, 1984). There was no evidence of seasonal vertical migration in this species. Thus, in late winter, the age structure of *M. longa* in Balsfjorden is similar to that in the Greenland Sea. In the Arctic Ocean and Canadian Archipelago, reproduction occurred over a long period of time (Grainger, 1965), an observation which is consistent with summer observations of age structure in the Greenland Sea. These inconsistencies in age structure point to variations in the life cycle of *M. longa* that are of unknown cause.

Measurements of respiration rates of *M. longa* in the Canadian Archipelago in July showed low rates (3 μl O_2 ind^{-1} day^{-1}; Conover and Cota, 1985) (Fig. 10.3), but off northern Norway in September, respiration rates were high (11 μl O_2 ind^{-1} day^{-1}; Båmstedt *et al.*, 1985) (Fig. 10.3). The dynamic range in respiration rate on an annual basis is much smaller for *M. longa* than for *C. hyperboreus* (Fig. 10.3), perhaps because of *M. longa's* tendency to swim continuously regardless of season (Conover and Corner, 1968). At all times of the year, weight-specific respiration rates of *M. longa* exceeded those of *C. finmarchicus* (Conover and Corner, 1968). The highest rates of excretion of ammonium by *M. longa* (probably adults) were measured off the west coast of Sweden [5.4 µg N (mg dry wt)$^{-1}$ day^{-1}] in April, followed by a rapid decline to roughly half the April rate in June, with low rates maintained throughout summer and autumn (Båmstedt, 1985; data converted assuming dry weight was 50% protein). In the Canadian Archipelago in summer, excretion rates were approximately 1 µg NH_4^+-N (mg dry wt)$^{-1}$ day^{-1} (Conover and Cota, 1985). Excretion of urea is equivocal; in the Canadian Archipelago none was detected (Conover and Cota, 1985), whereas off Sweden the highest rates were seen in May [1.7 µg N (mg dry wt)$^{-1}$ day^{-1}] and the lowest rates in June [0.3 µg N (mg dry wt)$^{-1}$ day^{-1}; Båmstedt, 1985].

The nutrition and lipid reserves of *M. longa* differ from those of the *Calanus* species, primarily because *M. longa* is distinctly omnivorous (Haq, 1967). Feeding studies off northern Norway using mixed stages of *M. longa* showed that highest ingestion rates occurred near sunrise and sunset in September, with the animals undertaking diel vertical migrations (Båmstedt *et al.*, 1985). Because these animals were primarily (94%) in copepodid stage V, the maximum ingestion rate measured (13 µg wet wt h^{-1}; Båmstedt *et al.*, 1985) converts to approximately 33% dry body wt day^{-1} based on dry weights for CVs determined in the Greenland Sea (S. L. Smith, unpublished). In January there was no feeding or diel migration by the stage V copepodids, males, or females (Båmstedt *et al.*, 1985). In the fjords of southwestern Norway in November, mixed stages of *M. longa* had lower lipid content, higher digestive enzyme activity, and higher gut-fullness than did *C. finmarchicus,* suggesting that *M. longa* continues to feed much later in the season and feeds omnivorously (Båmstedt and Ervik, 1984). In the Canadian Archipelago, the maximum ingestion rate for *M. longa* in summer was 9 µg C (mg dry wt)$^{-1}$ day^{-1}, although stages and sex were not identified (Conover and Cota, 1985).

The extent to which *M. longa* feeds omnivorously, the duration of its seasonal activity in the water column, and the degree to which it stores lipid are some of the questions currently under debate. Analysis of patterns of dry weight in which lowest dry weights of adult females (90 µg dry wt ind^{-1}) and males (40 µg dry wt ind^{-1}) occurred in March following peaks in the preceding July and August led Hopkins *et al.* (1984) to conclude that abundant phytoplankton was essential for growth. The almost monotonic increase in weight of copepodid stage V between July and November, corresponding to growth of approximately 1% dry body wt day^{-1}, however, took place when primary productivity was declining from 450 mg C m^{-2} day^{-1} to zero (Hopkins *et al.*, 1984). The analysis of Hopkins *et al.*, (1984) failed to include microzooplankton, on which *M. longa* can feed (Haq, 1967) and which may be abundant during periods of high concentrations of phytoplankton owing to the reproduction of more strictly herbivorous species. Naupliar stages of most crustacean plankton are microzooplankton.

The lipid content of *M. longa* off northern Norway varied seasonally, being greatest in October (84% of dry body wt) and least in March (27% dry body wt; Falk-Petersen *et al.*, 1987). At all times wax esters were the dominant component of total lipid (Hopkins *et al.*, 1985; Falk-Petersen *et al.*, 1987). In the Arctic Ocean in October, *M. longa* (probably adult females) was 57% lipid by weight, and 76% of that was wax esters (Lee, 1975). The composition of the fatty alcohols of wax esters changed seasonally in *M. longa* off northern Norway from 14:0 and 16:0 fatty alcohols in October and January to 22:1 and 20:1 fatty alcohols in March then back to 16:0 in June (Falk-Petersen *et al.*, 1987). The change during winter suggests either selective retention of the 22:1 and 20:1 fatty alcohols during lipid metabolism or ingestion of zooplanktonic prey containing 22:1 and 20:1 fatty alcohols, since 22:1 and 20:1 fatty alcohols are synthesized *de novo* primarily by calanoid copepods (Sargent and Whittle, 1981; Falk-Petersen *et al.*, 1987).

The evidence thus suggests that *M. longa* reproduces later in the year, feeds for a larger proportion of the year, feeds more omnivorously and carnivorously, is active for a longer period of the year, has less seasonal variability in its respiration rate, and is less able to synthesize lipid than are Arctic and subarctic species of *Calanus* which co-occurred with *M. longa*.

5. Pseudocalanus spp.

Two species of *Pseudocalanus* occur in the Arctic, *P. minutus* and *P. acuspes* (Frost, 1989). Since a revision of this widespread and abundant genus has just been published (Frost, 1989), an earlier review of the biology of the genus (Corkett and McLaren, 1978), a study of its Arctic ecology (Conover *et al.*, 1986), and this taxonomic revision provide the most complete summary on this genus in the Arctic.

6. Thysanoessa raschi (M. Sars) and *T. inermis* (Krøyer)

In the Balsfjorden of northern Norway, both *T. inermis* and *T. raschi* spawned as 2-year-olds in April and May, but by September the juvenile *T. inermis* were clearly larger than *T. raschi* (Hopkins *et al.*, 1984). *Thysanoessa inermis* maintained its size differential over *T. raschi* even as 3-year-olds, but during the growth periods of these two species, roughly March through October, the specific daily growth rates of *T. raschi* in terms of dry weight, protein, or lipid exceeded those of *T. inermis* (Hopkins *et al.*, 1984). *Thysanoessa inermis* accumulated more lipid than did *T. raschi* during spring (Table 10.1). Apparently, during overwintering of the first year, both species lose little weight, but in the second overwintering period the relatively lipid-rich *T. inermis* loses 47% of its lipid and 37% of its dry weight and *T. raschi* loses 15% of its lipid and 44% of its dry weight (Hopkins *et al.*, 1984). The rather small lipid reserves in overwintering *T. raschi* suggest that it may feed omnivorously during that period, but evidence is inconclusive (Hopkins *et al.*, 1985). Feeding, growth, and egg production by these species have been studied in the Bering Sea during March–July (S. L. Smith and D. Ninivaggi, unpublished).

III. The Antarctic

Investigation of Antarctic zooplankton dates back to the *Challenger* expedition of 1872–1876. Intensive biological studies followed at the beginning of this century, with the first continuous scientific program being conducted from the research vessel *Discovery* in 1926–1939, followed by *Scoresby* in 1950–1951. These extensive collections contributed valuable insights into distributional patterns as well as life cycles of many zooplankton (e.g., Ommaney, 1936; Bargmann, 1945; Marr, 1962; Andrews, 1966; Foxton, 1966; Mackintosh, 1972).

Intensive studies of composition, standing crops, and distributional patterns combined with environmental factors were first carried out by Mackintosh (1934) for the Atlantic sector, by Hardy and Gunther (1935) for the South Georgia area, and by Baker (1954) and Foxton (1956) in the Circumpolar Current. Mackintosh (1934) distinguished different zones and defined boundaries of distribution based on the association of zooplankton species with certain water masses and identified four groups: warm-water species, cold-water species, neritic species, and widespread species. In the mid-1960s interest was focused on the Southern Ocean because of apparently large concentrations of the Antarctic krill, *Euphausia superba*.

A. The Southern Ocean

The Southern Ocean consists of the southern parts of the Atlantic, Indian, and Pacific oceans (Hempel, 1985). Most of the circumpolar ocean is influenced by westerly winds, which cause easterly surface currents known as the West Wind Drift (Fig. 10.5). Its northern boundary is the Subtropical Convergence, corresponding approximately to 40°S. The Antarctic Convergence or Polar Front is embedded in the West Wind Drift and divides the system at 50 to 60°S into a subantarctic and an Antarctic zone (see Chapter 4). The Polar Front influences the distribution of many epipelagic organisms and forms a northern biogeographic barrier. Large parts of the West Wind Drift are continuously ice free.

10 Polar Zooplankton

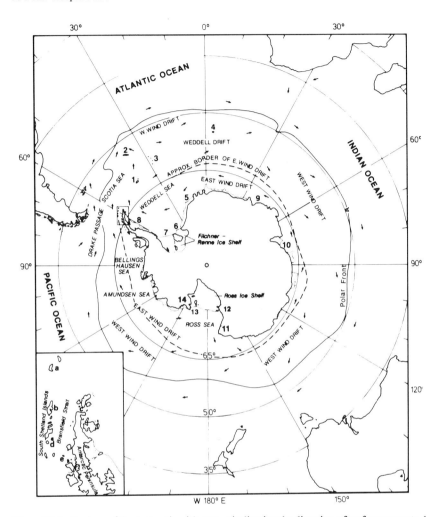

Figure 10.5 Regions of the Antarctic with arrows indicating the direction of surface currents. 1, South Orkney Islands; 2, South Georgia; 3, South Sandwich Islands; 4, Bouvet Island; 5, Riiser–Larsen Ice Shelf; 6, Vahsel Bay; 7, Gould Bay; 8, Larsen Ice Shelf; 9, Lutzow-Holm Bay; 10, Prydz Bay; 11, Victoria Land; 12, McMurdo Sound; 13, Bay of Whales; 14, Edward VII Land. Inset of Antarctic archipelago: a, Elephant Island, b, King George Island; c, Antarctic Sound; d, Deception Island; e, Palmer Archipelago; f, Gerlache Strait. After Mackintosh (1934).

Easterly winds, however, predominate close to the Antarctic continent, driving the surface water in westerly directions. The Antarctic Divergence is the boundary between the eastward and the westward currents and is accompanied by upwelling (see Chapter 4). The East Wind Drift is partly broken up in large gyres and eddies (e.g., the Weddell Sea gyre). A front results where cold water from the Weddell Sea meets warmer water from the Scotia Sea (the Weddell–Scotia Sea confluence). The East Wind Drift and its associated gyres and eddies are ice covered in winter and spring. A belt of ice and shelf water adjacent to the Antarctic continent and ice shelves is separated from the East Wind Drift by a front, the continental convergence. This zone is continuously covered by ice with open leads and polynyas and is broadest in the Weddell, Bellingshausen, and Amundsen seas.

Three different latitudinal zones for zooplankton communities (northern, intermediate, and southern zones) coincide with the distribution of water masses and with the patterns of ice cover (Voronina, 1971; Hempel, 1985). The northern zone occupies the ice-free part of the oceanic West Wind Drift and is relatively poor in biomass and production (Maruyama *et al.*, 1982; Hempel, 1985). Within the West Wind Drift higher zooplankton biomass occurs near the Antarctic Convergence and decreases toward the south. (Foxton, 1956; T. L. Hopkins, 1971; Yamada and Kawamura, 1986). Copepods, salps, small euphausiids, chaetognaths, and amphipods dominate the zooplankton, with copepods constituting more than 60% of the biomass in most parts and seasons (Hopkins, 1971; Seno *et al.*, 1963; Maruyama *et al.*, 1982). *Calanoides acutus, Calanus propinquus, Rhincalanus gigas,* and *Metridia gerlachei* form the bulk of the copepod biomass. Among chaetognaths, *Sagitta gazellae* and *Eukrohnia hamata* are common, and swarms of *Salpa thompsoni, Themisto gaudichaudii,* and *Thysanoessa* spp. are frequently encountered. Krill are mostly absent except near South Georgia (Hempel, 1985).

The intermediate zone is the seasonal pack-ice zone, which is ice covered in winter/spring and ice free in summer/autumn. This zone includes most of the East Wind Drift, the northern branch of the Weddell Gyre, and the Antarctic Peninsula. Large parts of the East Wind Drift are relatively shallow, and this region is the most productive one in the Southern Ocean, with the highest annual primary production (Hempel, 1985). *Euphausia superba* is the dominant species, and its summer distribution coincides with that of the sea ice cover in winter (Marr, 1962; Mackintosh, 1972). The main feature is the coincidence of large concentrations of *E. superba* in areas where water masses of different origins mix or where sharp changes in bottom topography occur. Highest concentrations of krill were found over the shelf and slope in Bransfield Strait (Stein and Rakusa-Suszczewski, 1984) and in the vicinity of fronts and eddies such as the zones of the Antarctic Divergence and the Weddell–Scotia Confluence (Marr, 1962; Macaulay *et al.*, 1984; Nast *et al.*, 1988).

The southern zone is the permanent pack ice zone and includes cold water along the continental shelf, especially in the shallow parts of the inner Weddell and Ross seas (Hempel, 1985). Pelagic phytoplankton production is limited to a short but intense season, but there is a longer period of ice algal production (Hempel, 1985). Zooplankton abundance and biomass are low. The small neritic euphausiid species *Euphausia crystallorophias* (Thomas and Green, 1988) and juveniles of the pelagic silverfish *Pleurogramma antarcticum* are typical species of this zone (Boysen-Ennen and Piatkowski, 1988).

B. The Major Seas South of 50°S

1. The Scotia Sea, the Bransfield Strait, and the waters adjacent to the Antarctic Peninsula

The Scotia Sea is bordered to the south by the Weddell Sea, to the west by the Drake Passage, and on all other sides by the Scotia Arc (Fig. 10.5). Cold Antarctic Surface Water constitutes a shallow, upper layer of 100–250 m which flows northward with a strong easterly component. Beneath the surface water is a thick layer of warmer water, the Circumpolar Deep Water, which is directed southward. The major oceanographic feature in the Scotia Sea is the Weddell–Scotia Confluence separating the cold water of the Weddell Sea from the warmer water of the Antarctic Circumpolar Current. This boundary has a major effect on zooplankton distributions.

The waters west of the Antarctic Peninsula are influenced by a northeasterly current from the southeast Pacific basin and the Bellingshausen Sea, approaching the coast of the Antarctic Pensinsula in the Antarctic Circumpolar Current (Fig. 10.5). In the vicinity of the South Shetland Islands, greater parts of the Bellingshausen Sea water pass north of the islands and flow in a northeasterly direction along the southern Drake Passage. Part of the Bellingshausen Sea water enters Bransfield Strait, a passage between the South Shetland Islands and the Antarctic Peninsula, between the westernmost islands of the South Shetland group, forming a strong northeasterly flow close to the southern shores of the islands. Cold Weddell Sea water enters the Bransfield Strait at the northeastern end and flows northward across the strait and southwestward along the shelf off the Antarctic Peninsula.

The southern part of the Scotia Sea, the seas west of the Antarctic Peninsula, and the Bransfield Strait are covered by sea ice part of the year and belong to the seasonal pack ice zone (Hempel, 1985). The open waters of the northern Scotia Sea are relatively poor in phytoplankton. A pronounced seasonality occurs in phytoplankton biomass and production in the Bransfield Strait, with a dense bloom in austral spring that progressively decreases with time. Generally speaking, the shallow coastal zones are relatively productive on a yearly basis (see Chapter 9).

Euphausiids and copepods dominate the zooplankton assemblages (e.g., Jażdżewski *et al.*, 1982; Montù and Oliveira, 1986). However, they can be outnumbered by salps, mainly *S. thompsoni* (Witek *et al.*, 1985; Montù and Oliveira, 1986). Salps, however, were not found at all or were very scarce at stations influenced by water from the Weddell Sea (Piatkowski, 1985, 1987).

Subantarctic and Antarctic species are typical components of the zooplankton community around South Georgia (Hardy and Gunther, 1935; Atkinson and Peck, 1988). The Weddell–Scotia Confluence is the southern limit of subantarctic species such as *Calanus simillimis* (Vladimirskaya, 1978), but it does not significantly separate two distinct herbivorous copepod communities (Marin, 1987) and is not a border for the occurrence of krill (Rakusa-Suszczewski, 1984). The zooplankton population in the Antarctic Circumpolar Current consists of more advanced developmental stages of copepods and euphausiids than occur in the Weddell Sea (Makarov, 1979; Brinton, 1985; Marin, 1987).

When *E. superba* is absent, copepods constitute 40–96% by numbers of the total zooplankton population of the Scotia Sea and Bransfield Strait (Mujica and Torres, 1982; Mujica and Asencio, 1985). Cyclopoid copepods (*Oithona* spp. and *Oncaea* spp.) often dominate the copepod assemblage and account for 40–80% of total numbers (T. L. Hopkins, 1985b; Schnack *et al.*, 1985a). Among the calanoid species, *M. gerlachei* is the most numerous, occurring throughout the area (Jażdżewski *et al.*, 1982; Zmijewska, 1985). *Metridia gerlachei* and the small calanoid species *Microcalanus pygmaeus* made up 70% of the total calanoids in Croker Passage between Gerlache Strait and Bransfield Strait (T. L. Hopkins, 1985b) and up to 90% in northern Bransfield Strait and south of Elephant Island (Schnack *et al.*, 1985a). Other abundant calanoid species are *C. acutus, C. propinquus, R. gigas, Ctenocalanus* sp., *Clausocalanus* sp., *Stephus longipes,* and *Euchaeta antarctica,* all with large spatial and temporal variations.

Among the euphausiids *E. superba* dominates (Weigmann-Haass and Haass, 1980), with major concentrations in areas where water masses of different origins mix (Witek *et al.*, 1981; Nast *et al.*, 1988). Gravid and spawning adults occur mainly along the continental slope and in offshore waters west of the Antarctic Peninsula and in the Drake Passage, while juveniles and subadults dominate in the shelf areas along the Antarctic Peninsula and at the edges of pack ice (Witek *et al.*, 1981; Nast, 1982; Quetin and Ross, 1984). Adult *E. superba* occur in high concentrations around South Georgia (e.g., Everson, 1982). Highest concentrations of *E. superba* adults usually are associated with low densities of larvae and vice versa (Brinton, 1985).

In January–March 1981, very high concentrations of krill larvae were found (Kittel and Jażdżewski, 1982; Brinton and Townsend, 1984); the greatest concentration was found north of 60°S (e.g., north of the Weddell–

Scotia Confluence), where the water depth exceeded 2000 m and surface water temperatures were above 1 °C (Rakusa-Suszczewski, 1984). This area was characterized by low larval concentrations in other years, when krill larvae occurred mainly in the eastern part of the Bransfield Strait, in the Antarctic Sound, and in the vicinity of Elephant Island (Hempel and Hempel, 1977/1978; Hempel et al., 1979; Schnack et al., 1985a).

Thysanoessa macrura is the second dominant euphausiid species (Kittel and Stepnik, 1983), occasionally being more abundant than *E. superba* (Pires, 1986; Mujica and Asencio, 1985; Piatkowski, 1987). *Euphausia crystallorophias* is a neritic species which occurs over the shelf of the Bransfield Strait and west of the Antarctic Peninsula (John, 1936; Weigmann-Haass and Haass, 1980; Piatkowski, 1987). Swarms of *E. crystallorophias* occur in Admiralty Bay, King George Island (Jackowska, 1980; Stepnik, 1982), and the caldera of Deception Island (Piatkowski, 1985; Everson, 1987). *Euphausia frigida* and *E. triacantha* are found only in areas with waters of the West Wind Drift which have not been influenced by the Weddell Sea (Hempel and Marschoff, 1980; Kittel et al., 1985). The shelf and slope waters around the tip of the Antarctic Peninsula and in the Bransfield Strait are regarded as important spawning areas for krill (Marr, 1962; Hempel et al., 1979; Witek and Kittel, 1985). Krill larvae drift passively from the peninsula region into the Scotia Sea (Brinton and Townsend, 1984).

Zooplankton in assemblages can be distinguished by differences in species composition and abundance and are associated with different water masses (Jażdżewski et al., 1982). Radiolarians, the polychaete *Tomopteris carpenteri,* and the crustaceans *Rhincalanus gigas* and *Themisto gaudichaudii* are dominant members in the southern Drake Passage and Scotia Sea influenced by the West Wind Drift (Jażdżewski et al., 1982). Many species are abundant in the shelf waters west and north of the Palmer Archipelago influenced by the Bellingshausen Sea (Witek et al., 1985), but small amounts of zooplankton are found in waters of Bransfield Strait (Witek et al., 1985), where *Oncaea curvata* and *Pelagobia longicirrata* are typical inhabitants (Jażdżewski et al., 1982). Areas influenced by water from the Weddell Sea are characterized by the occurrence of rare copepod species such as *Paralabidocera antarctica* and by the scarcity of *R. gigas, E. frigida, T. gaudichaudii, Viblila antarctica,* and *S. thompsoni* (Witek et al., 1985; Schnack et al., 1985a; Zmijewska, 1985; Piatkowski, 1987).

2. The Weddell Sea

The Weddell Sea forms the southern part of the Atlantic Ocean and is bounded by the Antarctic Peninsula, the Antarctic continent (including the Larsen Ice Shelf and the Filchner–Rønne Ice Shelf), the Scotia Ridge, and an artificially straight line from the South Sandwich Islands to Cape Norvegia (Fig. 10.5). The continental shelf occupies one-quarter of the area with narrow regions (90 km wide) in the east and wider regions (400–500 km) in the south off the Filchner–Rønne Ice Shelf. The East Wind Drift, characterized by a cold surface layer and a deep core of warmer water, moves into the Weddell Sea and circulates in a clockwise manner. A divergence zone exists near Halley Bay at the northeastern boundary of the Filchner depression, where part of the coastal current of the East Wind Drift follows the contours of the continental slope to the west and part propagates along the coastline to

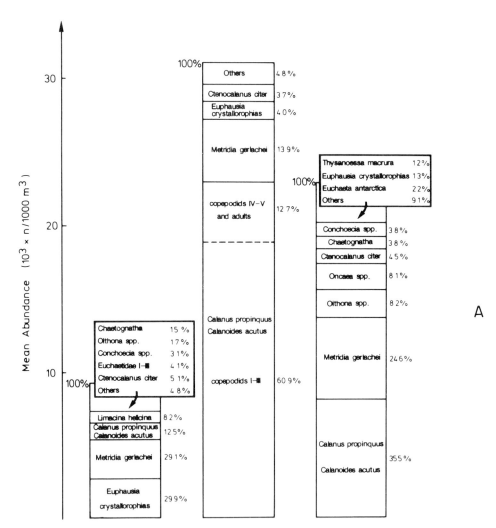

Figure 10.6 Zooplankton communities of the Weddell Sea. (A) Mean abundances and percentage composition of zooplanktonic species. (B) Geographic extension as given by cluster analysis. After Boysen-Ennen and Piatkowski (1988). (*Figure continues.*)

the southwest toward the Filchner–Rønne Ice Shelf. Cold shelf water (about −1.8°C), transported in the coastal current in the vicinity of the ice shelves, is separated from warmer water (about 0°C) of the Weddell Gyre by a coastal convergence at the continental slope.

The Weddell Sea is part of the permanent ice zone (Hempel, 1985), with large portions along the Antarctic Peninsula covered with sea ice year-round (Fig. 10.5). The ice cover along the eastern coast breaks up irregularly, forming shore leads and polynyas. The ice-free region expands during summer along the eastern coast and sometimes reaches the Filchner–Rønne Ice Shelf. The northern Weddell Sea is ice free for several months, whereas the southern part is ice free for only a few weeks. During summer, the coastal polynya has relatively high phytoplankton stocks and productivity (El-Sayed and Taguchi, 1981; von Bodungen et al., 1988).

Different patterns in the distribution of zooplankton and in community structure (e.g., Boysen-Ennen and Piatkowski, 1988; Hubold et al., 1988) occur during summer in the Weddell Sea. Zooplankton abundances decrease along the coastal current of the shelf from the northeastern to the southwestern portion of the Weddell Sea (Hubold et al., 1988). Moreover, there are differences in the composition of developmental stages, with younger stages occurring in the south, consistent with delayed development of summer conditions in general in the southernmost part of the Weddell Sea.

Cluster analyses have defined three distinct communities of epipelagic zooplankton in the eastern Weddell Sea during summer (Fig. 10.6). These are southern shelf, northeastern shelf, and oceanic. Separation into commu-

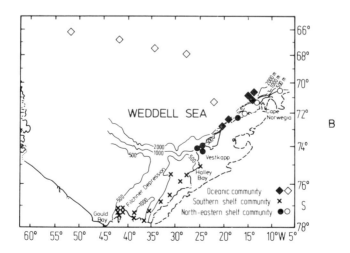

Figure 10.6 (*Continued*)

nities is not caused by complete absence or presence of any particular species. Communities are characterized by differing numbers of species, relative abundances, and trophic diversities and are in agreement with the topography and distribution of water masses (Boysen-Ennen and Piatkowski, 1988).

The epipelagic inhabitants of the deep oceanic part of the central Weddell Sea and the area located off the coast over deep water in the northeastern Weddell Sea (Oceanic Community) are typical members of the Warm Deep Water of the East Wind Drift, which does not occur in the shelf region. The coastal convergence separating the Warm Deep Water from the cold shelf water can be considered the distributional border for these oceanic species. These oceanic taxa include gelatinous species such as the hydromedusae *Calycopsis borchgrewinki;* the scyphomedusae *Atolla wyvillei;* the siphonophore *Vogtia serrata;* the salp *S. thompsoni;* the pteropod *Clio pyramidata;* the copepods *R. gigas, Euchirella rostromagna, Gaidius* sp., *Scaphocalanus vervoorti, Racovitzanus antarcticus, Heterorhabdus* sp., *Haloptilus ocellatus,* and *H. oxycephalus;* the euphausiid *T. macrura;* the hyperiid amphipod *Primno macropa;* and the chaetognath *Sagitta marri* (Ottestad, 1932; Vervoort, 1965; Siegel, 1982; Boysen-Ennen and Piatkowski, 1988; Hubold *et al.,* 1988). Within the oceanic community no single species predominates, but more species occur in large numbers. Omnivorous and carnivorous species account for about two-thirds of the total abundance. The circumpolar copepod species *M. gerlachei, Ctenocalanus* sp., *Oithona* spp., and *Oncaea* spp. dominate numerically.

The divergence zone near Halley Bay divides shelf zooplankton communities in the eastern Weddell Sea into southern and northeastern components. The Northeastern Shelf Community is located on the narrow shelf of the northeastern Weddell Sea, where a coastal convergence divides the cold shelf water from warmer water of the East Wind Drift. Thus, this community consists of neritic and oceanic species. Total abundance is high, mainly copepodid stages of *C. acutus* and *C. propinquus.* Five species, the copepods *C. acutus, C. propinquus, M. gerlachei,* and *Ctenocalanus* sp. and larvae of the euphausiid *E. crystallorophias,* constitute 95% of the total abundance. Filter-feeding species prevail and diversity is low. Euphausiids (*E. superba, E. crystallorophias*) and chaetognaths (*Eukrohnia hamata, Sagitta gazellae*) are the major constituents of the macrozooplankton. Among the euphausiids, *E. superba* dominates and occurs in large numbers on the shelf and in deep waters of the slope region, whereas *E. crystallorophias* is a typical shelf species occurring almost exclusively in relatively shallow waters (Siegel, 1982; Boysen-Ennen and Piatkowski, 1988).

The distribution of larvae of *E. crystallorophias,* however, is not restricted to the shelf region; they also occur over greater depths (Fevolden, 1980; Hempel and Hempel, 1982). Larvae of *E. superba* occur, like the adults, over the shelf and slope along the northeastern Weddell Sea, but not before February and then only as calyptopis stages (Hempel and Hempel, 1982; Hubold *et al.,* 1988). Two euphausiid species, *E. superba* and *T. macrura,* occur mainly in the northeastern and oceanic parts of the Weddell Sea, respectively. Their southernmost occurrence coincides with the divergence near Halley Bay. Only single specimens occur south of the divergence, and they have been rarely recorded over the southern shelf off the Filchner–Rønne Ice Shelf (Fevolden, 1980; Boysen-Ennen and Piatkowski, 1988). The third euphausiid species in the Weddell Sea, *E. crystallorophias,* is a key species of Antarctic shelf areas (Boysen-Ennen and Piatkowski, 1988). Its distribution extends from the northeastern to the southwestern part of the Weddell Sea close to the Antarctic Peninsula (Hempel and Hempel, 1982). It is the prevailing member of the Southern Shelf Community, which is located on the extensive southern shelf where cold water constitutes the epipelagic zone. The community is characterized by many neritic species including larval stages of benthic organisms (e.g., echinospira forms of lamellariid larvae; Boysen-Ennen and Piatkowski, 1988) and of the pelagic nototheniid fish *Pleurogramma antarcticum* (Boysen-Ennen and Piatkowski, 1988; Hubold *et al.,* 1988). Filter-feeding and omnivorous species predominate, although total abundance is low. *Euphausia crystallorophias* and the copepod *M. gerlachei* dominate by numbers. Other typical inhabitants of this Antarctic shelf area are the amphipods *Eusirus propeperdentatris* and *Epimeriella macronyx,* the pteropod *Limacina helicina,* and the copepods *Aetideopsis* spp. (Boysen-Ennen and Piatkowski, 1988).

Cluster analysis has shown similarities between the community off Vestkapp at the end of January 1985 and the southern shelf community along the ice shelf from Vahsel Bay to Gould Bay (400 nautical miles apart) 10 days later, which can be explained by the downstream transport of organisms in the surface layers of the coastal current (Hubold *et al.,* 1988). Horizontal transport by the coastal current can also explain the temporal variability observed in zooplankton at Vestkapp between the end of January 1985 and the middle of February 1985 (Hubold *et al.,* 1988). During this time, major shifts in zooplankton composition and abundance of certain species took place, especially in the abundance of larvae of *E. superba.* In January larvae were nearly absent, but 3 weeks later they occurred in high abundance over the shelf and slope. *Euphausia superba* larvae are probably transported horizontally by the coastal current from the northeast, where relatively high concentrations of adults and larvae normally occur (Hempel and Hempel, 1982; Siegel, 1982).

The abundance of large zooplankton (> 500 μm), except *E. superba,* decreases from January to February in the Weddell Sea. Copepodid stages IV and V and adults of *M. gerlachei, C. acutus,* and *C. propinquus* are significantly less abundant in February than in January within the upper 200 m (Hubold et al., 1988). Smaller copepodid stages (I–III) increased through the middle of February in the surface layers, whereas older stages and adults became more abundant at greater depths (>200 m), indicating the onset of the annual downward ontogenetic migration. Copepodid stages I–III and IV–adult of *C. acutus* and *C. propinquus* made up 9 and 33%, respectively, of the population in January 1980, but in February 1980 these stages constituted 78 and 2%, respectively, of the copepod population in the upper 100 m (Kaczmaruk, 1983). Within the entire zooplanktonic community, copepodids I–III of *C. acutus* and *C. propinquus* accounted for 61% of total numbers, and stages IV, V, and adult accounted for 13% of total numbers in the upper 300 m by the end of February 1983 (Boysen-Ennen and Piatkowski, 1988).

In addition to decreases in abundance of zooplankton, a shift toward older developmental stages was observed in *E. crystallorophias* and *T. macrura* (Hempel and Hempel, 1982). They also found higher abundances of larval *E. crystallorophias* and *T. macrura* in February than 6 weeks earlier. This fluctuation in the occurrence of larvae of these two euphausiid species, the high concentration of appendicularians in January and February 1985 but their absence in February and March 1983, and the dominance of krill larvae in February 1980 and 1985 but not in 1983 all suggest the important influence of advective processes (Hempel and Hempel, 1982; Boysen-Ennen and Piatkowski, 1988; Hubold et al., 1988).

In winter and early spring, phytoplankton are scarce in the water column, particularly beneath pack ice, but blooms of phytoplankton occur at the ice edge, in sea ice, and in small layers of platelet ice under the annual pack ice (see Chapter 9). Paralleling the phytoplankton distributions, surface zooplankton (>500 μm) in October–November 1986 were one-tenth of the January–February 1985 population at Vestkapp. Copepods (*M. gerlachei, C. acutus, C. propinquus*), ostracods, and chaetognaths formed the bulk of zooplankton larger than 500 μm in the upper 200 m (Hubold and Hempel, 1987). *Euphausia superba,* absent from net samples in the pack ice, was frequently observed on overturned ice floes (Hubold and Hempel, 1987).

3. The Ross Sea

The Ross Sea is bordered by Edward VII Land, the Ross Ice Shelf, and Victoria Land (Fig. 10.5). Along the western boundary, glaciers of Victoria Land flow into the Ross Sea. There is a wide, deep (ca. 600 m) continental shelf, and the shelf break is at 800 m. A strong cyclonic gyre characterizes the summer surface circulation on the continental shelf, which means that the surface current flows westward along the ice shelf and then north along the coast of Victoria Land. Three general water types can be defined (see Chapter

4): the Upper Water, the Circumpolar Deep Water (Antarctic Circumpolar Water), and the Shelf Water. The Ross Sea is entirely covered by sea ice for at least 9 months of the year, although coastal polynyas are frequently observed during winter and spring.

Summer phytoplankton blooms occur along the barrier edge of the Ross Ice Shelf and at the edge of receding pack ice (see Chapter 9). Zooplankton biomass is low (T. L. Hopkins, 1987) and an inverse relationship between chlorophyll *a* and zooplanktonic biomass was found in the southwestern part of the Ross Sea in summer (Biggs *et al.*, 1985).

Investigations at McMurdo Sound show increased zooplankton biomass with increasing depth. High zooplankton biomass below 200 m near the McMurdo Ice Shelf edge was probably the result of tides transporting enriched open-sea water from the north into the sound (Hicks, 1974). Ross Sea zooplankton is dominated by copepods, followed by pteropods, euphausiids, ostracods, polychaetes, and radiolarians (T. L. Hopkins, 1987; Foster, 1987).

Copepod biomass was higher near the Ross Ice Shelf and at the shelf-slope front of the Ross Sea than at offshore stations in the southern Ross Sea or in its extreme southwest corner (Biggs, 1982). *Oithona similis, Ctenocalanus* sp., *C. acutus, M. gerlachei*, and *O. curvata* form the bulk of the Ross Sea copepods (T. L. Hopkins, 1987). Bradford (1971) described a distinct copepod fauna at the outer and inner positions of the Ross Sea. The outer Ross Sea (northern and northeastern part) has many faunal members of the Antarctic Circumpolar Water, with large numbers of deep, warm-water species (e.g., *R. gigas, Metridia curticauda, Onchocalanus wolfendeni, Heterorhabdus austrinus, H. farrani, Haloptilus ocellatus*, and *Racovitzanus antarcticus*) which were found only seaward of the 500-m isobath. The inner portion in the southwest corner of the Ross Sea is not directly influenced by Antarctic Circumpolar Water, which is reflected in the scarcity of *R. gigas* and the occurrence of neritic species such as *Aetideopsis antarctica* and *Paralabidocera antarctica* (Bradford, 1971). The latter species seems to be endemic to coastal waters and is often encountered just beneath the undersurface of fast ice (Fukuchi and Sasaki, 1981). The neritic nature of the southwestern corner is also seen in the presence of numerous invertebrate larvae: polychaetes, barnacles, nemerteans, ascidians, decapod crustaceans, and echinospira larvae of a lamellarian gastropod (Foster, 1987). In McMurdo Sound small copepods such as *Ctenocalanus* sp., *O. similis,* and *O. curvata* are most common, particularly in the upper 100 m. With increasing depth the large copepods *M. gerlachei* and *C. acutus* also become abundant components of the community (Foster, 1987).

Among the euphausiids, *E. crystallorophias* is the dominant species over the whole Ross Sea shelf area, but it is absent north of the shelf-slope front, where *T. macrura* occurs. *Euphausia superba* is rare in the Ross Sea, with a few single specimens found near the shelf-slope front (Biggs, 1982). Hyperiid amphipods were found almost exclusively north of the shelf-slope front, whereas gammeridean amphipods (*Orchomenella, Epimeriella*) dominated over the shelf (Biggs, 1982). A few hyperiids, mainly *Hyperiella dilatata*, were encountered throughout the Ross Sea (Biggs, 1982; Foster, 1987). The pteropods *Clione* sp. and *L. helicina* are common, the latter being most dominant in the southwest part (T. L. Hopkins, 1987) and absent north of the shelf-slope front (Biggs, 1982).

Larval stages of the nototheniid fish *Pleurogramma antarcticum* have been found along the ice shelf with high concentrations in the Bay of Whales (Biggs, 1982). The siphonophore *Pyrostephos vanhoeffeni* was ubiquitous in the southwest area close to the ice shelf (Biggs, 1982; Foster, 1987). This species, along with calanoid (*Xanthocalanus, Tharybis*), cyclopoid (*Oithona*), and harpacticoid (*Tisbe, Longipedia*) copepods, has been found in the water column underneath the Ross Ice Shelf (Totton and Bargmann, 1965; Azam *et al.*, 1979; Bradford and Wells, 1983).

In summary, the zooplankton assemblage in the outer northeastern Ross Sea is similar to the oceanic community of the northern Weddell Sea, whereas the zooplankton assemblage of the inner Ross Sea corresponds to the shelf communities of the southeastern Weddell Sea (Boysen-Ennen and Piatkowski, 1988).

C. Seasonal Cycles and Special Adaptations of Euphausiids

A major feature of the Southern Ocean is the large (up to about 60 mm) euphausiid species *E. superba*. Krill biomass has been estimated to be much higher than the total annual global harvest of fish and shellfish (El-Sayed, 1985); thus, krill are not only of scientific but also of commercial value. Large seasonal and interannual variations in krill biomass are known (e.g., Priddle *et al.*, 1988). For example, 1983–1984 was a season of unusually low krill biomass (Heywood *et al.*, 1985; Brinton *et al.*, 1986).

Krill are typical members of the East Wind Drift. Their distribution is circumpolar and patchy (Everson, 1977; D. G. M. Miller, 1986). Major concentrations have been found in the meanders and gyres of the East Wind Drift, i.e., in the Bellingshausen Sea, north of the Ross Sea, near the Kerguelen–Gaussberg Ridge, and in the Scotia Sea. Low abundances occur in regions located between these highly concentrated populations. It has also been suggested that stocks are self-maintaining (Mackintosh, 1973). Early studies of genetic variations suggested at least two discrete, genetically different krill populations in the area of the Antarctic Peninsula (Ayala *et al.*, 1975; Fevolden and Ayala, 1981). However, later electrophoretic analyses of krill from different regions (Bellingshausen Sea, Bransfield Strait, Scotia Sea, Weddell Sea, Prydz Bay) gave no evidence of distinct genetic stocks. Hence, it is concluded that a single, genetically homogeneous population occurs in the Antarctic (Schneppenheim and MacDonald, 1984; MacDonald et al., 1986). Using morphometric parameters, Siegel (1986) did not detect significant separations of krill populations in samples from the northern Bellingshausen Sea, Drake Passage, Bransfield Strait, and the Weddell Sea. He described the southern Drake Passage and Bransfield Strait as thoroughfares for *E. superba* between the Bellingshausen Sea and the Scotia Sea. Whereas krill larvae drift with the water currents, adults are known to be strong swimmers with short-term horizontal movements of 6–30 cm s^{-1} (Kils, 1981a; Hamner, 1984). Krill are able to swim against most currents and can migrate over long distances (Kanda et al., 1982; Kils, 1983).

Krill can form dense swarms that move rapidly, primarily in a horizontal direction (Kanda et al., 1982; Hamner, 1984), from a few meters to > 500 m horizontally (Witek et al., 1981; Hamner, 1984). Most swarms, however, have a length of < 100 m (Klindt and Zwack, 1984). A very large swarm ("superswarm") was encountered in March 1981 north of Elephant Island. Based on acoustic observations, estimated total biomass in the superswarm was 2.1×10^6 tons distributed over 450 km^2 (Macaulay et al., 1984). The superswarm was not homogeneous in population structure but contained mixed size groups. The eastern part consisted mostly of males > 50 mm in size, whereas the western part had small females and males, with females dominating (Brinton and Antezana, 1984).

Krill occur in the top 200 m of the water column but are found primarily between 20 and 100 m (e.g., Everson, 1983). Krill undergo diel vertical migrations. They rise to the surface at night and sink and concentrate into dense swarms in the daytime, although reports of more irregular behavior also exist (Arimoto et al., 1979; Everson and Ward, 1980; Everson and Murphy, 1987). The vertical extent of diel migration and its period depend on the size of the krill; large krill ascend and descend over greater distances than small individuals. Accordingly, aggregations of large krill can be detected above as well as below small krill (Everson and Ward, 1980; Nast, 1978/1979).

The major determinants of vertical migration of krill swarms are irradiance and food levels (Everson, 1977; Kalinowski and Witek, 1980). Active feeding may occur during swarming (Antezana et al., 1982; Antezana and Ray, 1984). However, a sharp decline in filtration rate with increasing krill density has been observed (Morris et al., 1983). Krill feed at all times of the day but at higher rates during darkness, an observation which is consistent with reports of highest alimentary tract fillings of krill in the upper layers at night (Kalinowski and Witek, 1980). Within a rapidly swimming swarm, *E.*

superba is orientated parallel to the water surface and does not feed. When food is encountered, krill turn and roll upside down; the animals have no particular orientation during intensive feeding (Hamner, 1984).

Great variability exists among krill swarms. Accordingly, swarms close together can have as many differences as swarms far apart (Watkins et al., 1986; Buchholz, 1988). The biological variability among krill swarms includes body length, maturity stage, sex ratio, molt stage, and feeding state (e.g., Brinton and Antezana, 1984; Morris and Ricketts, 1984; Buchholz, 1985; Watkins, 1986). At least part of the variability among swarms may be generated by sorting mechanisms, such as size, sex (Marr, 1962), molt stage (Buchholz, 1985), differential swimming speeds (Mauchline, 1980), and sinking rates (Kils, 1981b). Within a swarm, krill are unevenly distributed with respect to these parameters.

Molt frequency and growth are dependent on size, food availability, and temperature. When food is abundant, intermolt periods are much shorter than when phytoplankton are scarce (Morris and Priddle, 1984; Buchholz, 1988). Intermolt periods ranged from 8 to 15 days in summer and 50 to 60 days in winter. This is in agreement with values from experimental work (e.g., Poleck and Denys, 1982; Morris and Keck, 1984). The interaction of temperature and feeding and their effect on growth are not clearly differentiated (Buchholz, 1988), although a seasonal growth pattern is clear from field data (Siegel, 1986).

Euphausiids are long-lived animals. Initially it was thought that *E. superba* had a life span of 2 years (Marr, 1962; Mackintosh, 1972), but now a range of 5 to 8 years seems more probable (Ettershank, 1984, 1985; Rosenberg et al., 1986; Ikeda and Thomas, 1987a; McClatchie, 1988). During winter, slow growth rates (Everson, 1977; McClatchie, 1988) and shrinking body size (Ikeda and Dixon, 1982) have been observed.

Most Antarctic euphausiids spawn twice during their lifetime. However, *E. superba* is suspected to spawn at least three times, whereas *E. frigida* spawns only once (Siegel, 1987). The more northerly euphausiid species reach maturity earlier than the southerly ones, and they have a shorter life span (Fig. 10.7; Siegel, 1987). The periods of reproduction among the euphausiid species differ; *T. macrura* spawns earliest. Gravid females were found in August–October in Admiralty Bay, King George Island (Stepnik, 1982) and calyptopis stages were observed in October (Makarov, 1979). At the same time, *E. frigida* and *E. triacantha* begin to spawn (Makarov, 1977, 1979). *Euphausia crystallorophias* follows with gravid females in December (Baker, 1959; Stepnik, 1982). Spawning by *E. superba* takes place latest in the season, between November and April (see Mackintosh, 1972), with greatest intensity in January–March, although there is some year-to-year variation (e.g., Mackintosh, 1972; Witek et al., 1980; Brinton et al., 1986; Hosie et al., 1988).

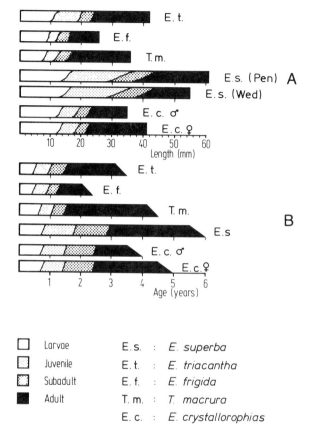

Figure 10.7 Life cycles of Antarctic Euphausiacea. (A) Minimum and maximum sizes of developmental stages for the different species. (B) Occurrence of developmental stages at age of the species. (Pen = Antarctic Peninsula; Wed = Weddell Sea.) After Siegel (1987).

According to Marr (1962), eggs are released near the surface either over the continental shelf or in oceanic waters and sink to depth (Fig. 10.8). The second naupliar stage migrates upward (Hempel and Hempel, 1985), and during the ascent the nauplii molt through the metanaupliar stage. The first calyptopis stage reaches the surface. Hempel and Hempel (1985) found 1-day-old eggs in water depths greater than 1000 m and supposed at least some spawning had occurred in deeper waters. This "developmental ascent" (Marr, 1962) is characteristic of many Antarctic euphausiids (Makarov, 1978).

Krill larvae develop to late furcilia and juveniles during austral fall (Marr, 1962) or winter (Siegel, 1987) in their first year of life. These stages are frequently encountered in the vicinity of pack ice or underneath the ice

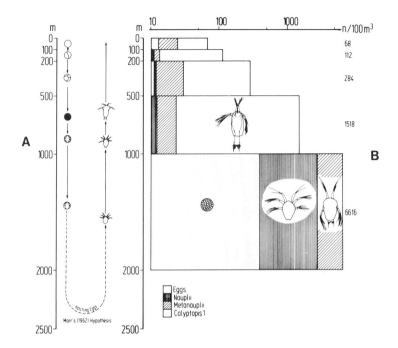

Figure 10.8 The developmental ascent of krill. (A) Hypothesis of Marr (1962). (B) Total abundance and relative abundance of eggs and early larvae at a station west of the South Shetland Islands. After Hempel and Hempel (1985).

during autumn and winter, where they feed (D. G. M. Miller, 1986; Daly and Macaulay, 1988). During spring, the krill population consists mainly of juveniles and subadults. With retreat of the pack ice, concentrations of all developmental stages increase. Highest concentrations occur during the spawning season, when krill are found primarily near the continental slope or in oceanic waters, often in dense swarms. Krill abundance declines in late summer. In autumn, adult krill migrate from the oceanic waters toward neritic areas, and an extensive seasonal horizontal migration occurs (Mackintosh, 1972; Siegel, 1988). In winter, swarming is minimal and krill are scarce in the water column (e.g., Marr, 1962; Heywood et al., 1985; Siegel, 1988).

Euphausia superba is able to feed on a wide variety of particles including nanoflagellates (McClatchie and Boyd, 1983; Hamner, 1988) as well as large organisms such as *Artemia* nauplii (Ikeda and Dixon, 1984), copepods (Price et al., 1988), its own furcilia larvae (Boyd et al., 1984), and larger individuals (Ikeda, 1984; Ishii et al., 1985). Studies of digestive enzymes, however, characterized *E. superba* as omnivorous with affinities for phytoplankton (Mayzaud et al., 1987).

Feeding rates varied with food concentrations (Antezana et al., 1982; Schnack, 1985). Krill ingested the most abundant food item (Schnack, 1985) or the largest particles (Ikeda and Dixon, 1984; Ishii, 1986). However, when phytoplankton became scarce, krill showed reduced clearance rates and a diet that was less specialized (Kato et al., 1982).

Species composition of phytoplankton may shift toward nanoflagellates in the field due to preferential feeding by krill on larger cells (Quetin and Ross, 1985). Phytoplankton biomass is often negatively correlated with that of krill (Uribe, 1982; Nast and Gieskes, 1986). Hence, heavy grazing pressure may cause the decline of phytoplankton blooms (Uribe, 1982). However, the spring bloom of phytoplankton occurs prior to the appearance of high krill concentrations as well as the distinct decline of that bloom (Witek et al., 1982). The buildup of phytoplankton blooms during spring is apparently not controlled by krill and intense grazing may be only of local significance within dense krill swarms (Holm-Hansen and Huntley, 1984; von Bodungen, 1986). Consumption by krill swarms was calculated to be between 58 and 81% of daily primary production, whereas the mean *E. superba* population consumed about 3% of daily primary production (Holm-Hansen and Huntley, 1984; D. G. M. Miller et al., 1985). On the other hand, higher phytoplankton biomass was positively correlated with individual krill swarms of high biomass but negatively correlated with mean integrated krill abundance (Weber and El-Sayed, 1987), suggesting greater grazing pressure when krill were dispersed.

The occurrence of *E. superba* in the pack ice in association with ice floes has been found in many parts of the Antarctic (e.g., Fraser, 1936; Daly and Macaulay, 1988; Stretch et al., 1988). Krill feed on the underside of the ice by raking the attached algae with their thoracic endopodites (O'Brien, 1988; Stretch et al., 1988), as described from experimental work with ice algae frozen into blocks as food (Hamner et al., 1983). Melting ice releases ice algae into the water column, which induces foraging behavior in krill. Krill increase their turning and swimming rates with rapid opening and closing of the thoracic appendages when the ice algae are released. This foraging behavior is modified by the rate of ice melting (and therefore by the rate of release of ice algae) and by the spatial patchiness of algae in the ice (Stretch et al., 1988).

The underside of the sea ice provides a feeding ground for adult krill preparing to reproduce and a nursery ground for krill larvae, and the uneven undersurface of sea ice provides shelter from predators and plays an important role in the survival of overwintering zooplankton, especially krill (Stretch et al., 1988). In the shallow waters of Lutzow-Holm Bay, krill were observed to overwinter in a layer close to the sea bottom, feeding on detrital material on the seabed (Kawaguchi et al., 1986).

The growth rate of *E. superba* reflects the amount of food present in the water column, in both the intermolt period and the linear size increase at molt (Buchholz, 1983; Ikeda and Thomas, 1987b). Their ability to adapt to the long Antarctic winter is enhanced by their ability to utilize low phytoplankton concentrations, feed on nanoflagellates and detritus, and scrape ice algae from ice (e.g., Stretch *et al.*, 1988). It has also been shown experimentally that krill can be carnivorous (Price *et al.*, 1988) and cannibalistic (Ikeda, 1984). This diverse feeding behavior, together with reduced metabolic rates (Kawaguchi *et al.*, 1986) and high resorption rate of protein and chitin from the exoskeleton to be shed at molt (Buchholz, 1988), allows *E. superba* to be optimally adapted to the conditions of the Southern Ocean.

D. Seasonal Cycles and Special Adaptations of Other Zooplankton

There is little seasonal variation in the total standing crop (displacement volume) of zooplankton of the subantarctic and Antarctic in the top 1000 m in open waters (Foxton, 1956, 1964), although for the Pacific sector a slightly lower mean value (2.45 g dry wt m^{-2}) was recorded in the winter relative to summer (2.96 g m^{-2}; T. L. Hopkins, 1971). Larger seasonal differences were observed in restricted geographic areas. For example, total zooplanktonic biomass in winter was 68% of the summer value in the upper 1000 m around South Georgia (Atkinson and Peck, 1988), and a distinct seasonal change in biomass occurred in the water column (660 m) under fast ice in Lutzow–Holm Bay (Fukuchi *et al.*, 1985). Highest biomass occurred in August (15.2 mg wet wt m^{-3}, compared with 1.5 and 8.7 mg wet wt m^{-3} in May and October). The increase in winter was due to large numbers of siphonophores.

There is a pronounced seasonal vertical migration pattern, with biomass maxima concentrated in surface layers in summer but in deeper layers in winter. The ontogenetic migrations of major copepod species have a strong influence on the zooplankton standing stock in the surface layer during summer (Mackintosh, 1937; Voronina, 1972a; Atkinson and Peck, 1988). Between 4 and 33% of total biomass is concentrated in winter in the top 250 m of the water column (T. L. Hopkins, 1971), while below 200 m, 35–55% of the biomass is situated in the 500–1000-m layer (Vladimirskaya, 1975). A different vertical distribution pattern in summer and winter is known for the chaetognaths *Sagitta gazellae* (David, 1955, 1965) and *Eukrohnia hamata* (Mackintosh, 1937; David, 1958) and the salp *S. thompsoni* (Foxton, 1964). The upward and downward migrations in summer and winter maintain the animals within their geographic range. They drift northeasterly toward the Antarctic Convergence with the Antarctic Surface Water in summer, whereas in winter they are transported south with Warm Deep Water toward the Antarctic continent (Mackintosh, 1937).

Foxton (1956) described two seasonal surface maxima in zooplankton concentration, one in November–December and the second one in February–March. The first peak was attributed to the overwintering population migrating to upper water layers, and the second peak was due to a new generation (Vinogradov and Naumov, 1964; Voronina, 1968). Additional evidence comes from Admiralty Bay (King George Island in the South Shetland group), the Weddell Sea, and Prydz Bay, where biomass near the surface was higher in December and February than in January and older copepodid stages were captured in December than in February (Chojnacki and Weglénska, 1984; Kaczmaruk, 1983; Yamada and Kawamura, 1986). The reproductive cycle of the mainly herbivorous species is strongly influenced by the distinct seasonal cycle in primary production.

The common copepod species, *C. acutus, C. propinquus,* and *R. gigas,* show annual vertical migration patterns. They rise to the surface sequentially and with subsequent temporal lags among their spawning periods (Voronina, 1972b). When the pack ice recedes in spring, stage IV and V copepodids of *C. acutus* first migrate into the surface layers and lay eggs during summer months. After an intensive period of feeding and growth, *C. acutus* descends as copepodids IV and V to overwintering depths in late autumn (Andrews, 1966). The overwintering population of *C. propinquus* migrates to the surface next, followed by *R. gigas* (Voronina, 1972b). The onset of descent into deeper overwintering layers is again led by *C. acutus.* In more southerly habitats the individuals' rise to the surface is delayed, and the duration of their occurrence in the epipelagic environment is shorter.

Whether *R. gigas* has one or two spawning periods per year is controversial. Bradford (1971) suggested two spawning periods, one in summer and one in late fall, derived from the occurrence of nauplii and young copepodid stages. Voronina (1968), however, suggested only one spawning during the year with timing depending on the latitude (e.g., later spawning farther south). As Everson (1984) pointed out, this is also possible for *C. acutus* and can result from a prolonged spawning time rather than two spawning periods in both species.

The tunicate *S. thompsoni* occurs in two forms in the Antarctic: solitary and aggregate. Solitary forms dominate the population during winter in the deeper layers. Aggregate forms are produced asexually by budding in response to phytoplankton blooms (Foxton, 1964). A quick response to phytoplankton development together with rapid multiplication can give rise to enormous concentrations. For example, Nast (1986) counted 889 individuals $(1000 \ m^3)^{-1}$ north of Elephant Island in March 1985. In summer, aggregate and solitary forms occur in surface waters. At the end of summer, the aggregate forms (which are present only in summer) sexually reproduce solitary forms. *Salpa thompsoni* grows throughout the year but considerably slower during winter (Foxton, 1966).

Most carnivorous species of zooplankton have a reproductive cycle independent of the season and grow throughout the year. One exception is the hyperiid amphipod *T. gaudichaudii,* which has an annual cycle with juveniles released primarily in spring and summer. Their main growth occurs in summer, with the bulk of specimens becoming sexually mature in August. *Themisto gaudichaudii* has a 1-year life cycle (Kane, 1966). Females of the calanoid copepod *Euchaeta* spp. are found with attached spermatophores and egg sacs during winter under the sea ice in McMurdo Sound (Littlepage, 1964; Bradford, 1981). Studies in the South Georgia area revealed reproduction by *E. antarctica* during winter but also during summer (Ward and Robins, 1987). According to Ferrari and Dojiri (1987), mating by *E. antarctica* was at a maximum in late autumn and early winter. Young chaetognaths remain in upper layers for several months. With increasing maturity, they descend into deeper waters to spawn (David, 1958; Hagen, 1985). Chaetognaths are carnivorous and grow continuously (David, 1955).

During the short period of phytoplankton productivity, food is consumed and stored largely as depot lipids, which are used for survival during overwintering and reproduction. It has been shown that zooplankton in polar regions contain much higher percentages of total lipids than related species in temperate and tropical zones (Lee *et al.,* 1972). Because carnivores and omnivores can feed throughout the year, they do not require large amounts of depot lipids. Relatively low concentrations of depot lipid are found in gelatinous species such as *Calycopsis borchgrevinki, Diphyes antarctica, Pyrostephos vanhoeffeni;* in the polychaetes *Tomopteris carpenteri* and *Vanadis antarctica;* and in the chaetognaths *Sagitta gazellae* and *S. marri* (Clarke, 1984a; Hagen, 1988). However, amphipods, mysids, and euchaetid copepods can store relatively large amounts of depot lipids, which is contrary to their omnivorous or carnivorous feeding habits (Reinhardt and Van Vleet, 1986; Hagen, 1988).

The lipid concentration of *Euchaeta antarctica* is high throughout the year and does not vary greatly except during the reproductive period. Between June and August, females have larger concentrations of lipid, which are correlated with the production of eggs rich in lipids (Fig. 10.9; Littlepage, 1964). The herbivorous copepods *Calanoides acutus* and *Calanus propinquus* differ in their seasonal cycle, with *C. acutus* migrating to the surface layer earlier in spring and returning earlier to deeper overwintering layers in autumn. In January–February 1985, for example, copepodid stage V and female *C. acutus* had already accumulated high concentrations of lipids (Table 10.2), indicating the probable reduction of their metabolism during winter, and most *C. acutus* occurred in deeper layers. *Calanus propinquus,* however, was still feeding near the surface in January and reached maximal lipid concentrations similar to those of *C. acutus* only in February (Hagen,

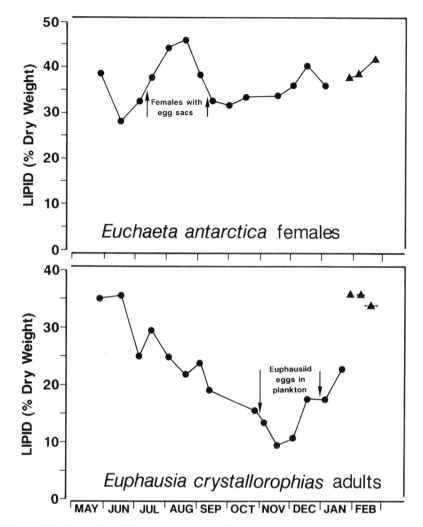

Figure 10.9 Seasonal variation in total lipid of the copepod *Euchaeta antarctica* and the euphausiid *Euphausia crystallorophias*. Data from McMurdo Sound, Ross Sea (●, after Littlepage, extract from "Biologie Antarctique," under the direction of R. Carrick, Hermann, Paris, 1964) and from the Vestkapp, Weddell Sea (▲, after Hagen, 1988).

1988). *Euphausia crystallorophias* accumulates large amounts of lipids in summer when food is available, and the stored lipids decrease steadily during winter (Fig. 10.9; Littlepage, 1964).

The presence of triacylglycerols, the main reserve in carnivorous organisms, suggests feeding year-round (Sargent *et al.*, 1981). Notable exceptions are the medusae *Atolla wyvillei*, ctenophores, and the euchaetiid copepods,

Table 10.2 Lipid Content and Major Lipid Classes of Three Copepod Species and Two Euphausiid Species from the Weddell Sea in January and February 1985[a]

	Total lipid (% dry wt)		Major lipid classes (% of total lipid)					
			Wax esters and sterol esters		Triacylglycerols		Phosphatidylcholine	
	Jan.	Feb.	Jan.	Feb.	Jan.	Feb.	Jan.	Feb.
Calanoides acutus V	47.0	42.4	92.5	92.4	3.3	2.9	2.0	2.7
Calanoides acutus ♀	46.9	—	92.0	—	4.3	—	2.3	—
Calanus propinquus V	25.6	47.1	4.1	1.1	84.7	93.8	3.6	2.3
Calanus propinquus ♀	25.3	42.8	3.0	1.9	83.3	92.8	6.0	2.5
Euchaeta antarctica ♀	38.0	39.4	90.8	92.3	3.1	3.6	2.7	1.8
Euphausia superba, subadult/adult	24.9	27.9	6.3	7.5	43.5	51.1	35.8	31.0
Euphausia crystallorophias, subadult/adult	35.6	33.7	49.1	47.1	4.9	5.7	35.1	35.2

[a] After Hagen (1988).

which are high in wax esters (Clarke, 1984a; Reinhardt and Van Vleet, 1986). *Calanus propinquus* and *E. superba* store mainly triacylglycerols and not wax esters, which in *C. propinquus* can be more than 90% of total lipids (Hagen, 1988). The high concentrations of triacylglycerols found in *C. propinquus* and *E. superba* may indicate that these species are feeding throughout the year. A third major lipid class in euphausiids is phospholipids (Clarke, 1984b; Hagen, 1988), mainly phosphatidylcholine (Table 10.2), which also serves as an energy reserve (Hagen, 1988).

The traditional concept of the Antarctic food web is that it is simple and linear, including only large phytoplankton, krill, and consumers of krill (e.g., Tranter, 1982). However, this short food chain is only part of a much more complex interaction. In some regions of the Southern Ocean, for example, krill are nearly absent and copepods, salps, or small euphausiids are the most important components of zooplankton. In addition, the biomass of microheterotrophs can be significant (ca. 10–25% of the phytoplankton biomass; von Brökel, 1981; von Bodungen *et al.*, 1988), and their roles as grazers as well as prey may be of major importance in the Antarctic food web (Hewes *et al.*, 1985; Marchant, 1985). T. L. Hopkins (1985a, 1987) studied the food web in two different, relatively small areas. McMurdo Sound, near the permanent ice shelf, has a low standing stock of zooplankton with a total biomass of 3.5 g dry wt m^{-2}, whereas Croker Passage (Gerlache Strait) is an area with a high standing stock (total biomass of 58 g dry wt m^{-2}) dominated by *E. superba*.

T. L. Hopkins (1987) calculated that roughly 2% of the phytoplankton biomass in McMurdo Sound could be consumed by the total net-caught zooplankton on a daily basis, and Schnack *et al.*, (1985b) found that daily consumption by copepods was about 1% of daily primary production during austral spring in eastern Bransfield Strait. Both environments had low zooplankton stocks but large phytoplankton blooms. Mass sedimentation, not grazing, may cause the decline of some phytoplankton blooms (von Bodungen, 1986). However, in the open Drake Passage, where low phytoplankton stocks but high copepod biomass were encountered (Schnack *et al.*, 1985b), copepod grazing was 55% of daily primary production.

A shift in diet demonstrates a change in the trophic role of a species during ontogeny, and this is most evident in copepods and euphausiids (Fig. 10.10). The lipid and amino acid composition of fecal pellets of *E. superba* suggests that small krill (about 22 mm in length) are feeding mostly on diatoms, whereas large krill (about 51 mm in length) are feeding on a mixture of phytoplankton, choanoflagellates, and nanozooplankton (Tanoue, 1985a,b). A shift in dietary composition toward larger food items with growth is also known for *Pleurogramma antarcticum,* with cyclopoid copepods being the main dietary component in postlarvae but calanoid copepods

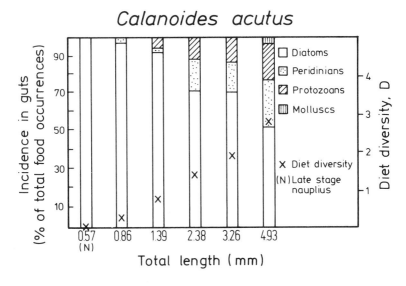

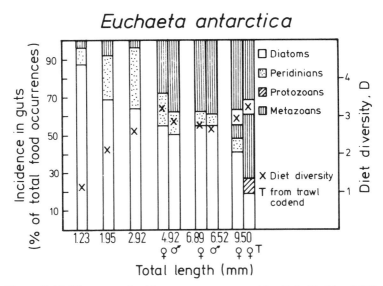

Figure 10.10 Dietary trends with respect to ontogeny. After T. L. Hopkins (1987).

forming most of the diet of juveniles (Kellermann, 1987). Debris of *E. superba* was a major food item in fall in Croker Passage. Krill debris was found throughout the food web (e.g., in most small-particle grazers such as the copepods *Scolecithricella* spp., ostracods, and *E. superba* itself) as well as in mysids and amphipods (T. L. Hopkins, 1987).

IV. Concluding Remarks

There are many major differences between the Arctic and the Antarctic as habitats for pelagic zooplankton. Among the most outstanding and obvious are the pathways by which recruitment of pelagic organisms can take place and the seasonal extent and movement of ice cover on an annual basis. In the Antarctic the Circumpolar Current in principle can carry pelagic organisms completely around the continent and can introduce organisms entering the polar area from any particular subpolar sector into all parts of the Antarctic polar environment. In the Arctic, on the other hand, access to the polar environment is primarily via two restricted passages, the Bering and Fram straits. The communities of organisms entering via these passages remain quite separate owing to the circulation and bathymetry within the Arctic Ocean basin. Similarly, organisms endemic to coastal regions of the Antarctic could be entrained and widely dispersed by the Circumpolar Current, while in the Arctic Ocean there are only three outlets by which organisms can gain access to subpolar seas and be dispersed more widely (i.e., Bering Strait, Fram Strait, and the Canadian Archipelago).

The development, dissipation, and movement of sea ice create important differences between the two poles. In the Antarctic most of the ice cover develops and recedes annually (Zwally *et al.*, 1983). The implication is that an advancing or retreating ice edge passes a particular position of the pelagic environment only once per year. Any impact that an ice edge has on pelagic fauna and processes happens once per year for a given location. Because ice algae and associated communities using the undersurface of ice can be major nutritional sources for pelagic zooplankton and nekton, the presence of ice and its duration in a particular location will be important factors shaping the pelagic food web in the area. In the Arctic, ice development and retreat in the Bering Sea are similar to those seen in the Antarctic (Niebauer, 1983), with the result that much of the primary production associated with retreating ice in spring is not utilized by the pelagic food web in this shallow sea but may support a large benthic community. In the Greenland Sea, however, ice is continually being advected from the Arctic Ocean, creating ice edge conditions year-round in Fram Strait (Smith *et al.*, 1985). There is relatively little seasonality associated with the ice edge itself in Fram Strait, making the area unusual among ice-impacted pelagic environments.

The distinct seasonality in primary production in both polar environments is probably the most important factor influencing life cycles and adaptation (Clarke, 1983). In both the Arctic and Antarctic there exist pelagic organisms equipped with a capacity to store lipid when food is plentiful and overwinter on this store. The ability to synthesize and/or store lipids is a critical aspect of life cycles of zooplankton at high latitudes, since these depot lipids provide a respiratory substrate during starvation that accompanies overwintering (Lee et al., 1971, 1972) and the materials for egg production (Lee et al., 1972; Sargent et al., 1981) and naupliar development (Lee, 1974; Sargent et al., 1977). In copepods the depot lipid is primarily wax esters (Lee, 1975; Sargent and Whittle, 1981), which are synthesized de novo by the animals from their phytoplankton diet. Synthesis of wax esters is a mechanism whereby large lipid reserves can be accumulated in very short periods of time (Clarke, 1983). Wax esters can be transferred through the food web and incorporated directly into sperm whale blubber, myctophids, and bathypelagic fish, for example, or they can be converted into triacylglycerols in the guts of fish, many of which contain hydrolyzing enzymes (Sargent and Whittle, 1981). Other polar plankton (e.g., certain euphausiids and amphipods) store triglycerides instead of wax esters (Lee, 1975). Considering the great importance of depot lipids in allowing high-latitude plankton to exploit brief periods of phytoplankton growth and to maximize subsequent reproductive effort, our studies of this phenomenon in high-latitude plankton to date have been superficial, and most of the physiology of lipid synthesis, storage, mobilization, and utilization by polar zooplankton is unknown.

In both polar environments there are many neritic taxa whose life history strategies are virtually uninvestigated. The extent to which these organisms store lipids, produce resting eggs, or have other adaptations for surviving long, dark, foodless winters requires exploration, as do the strategies of smaller, widespread taxa such as *Oithona* spp. and *Oncaea* spp. We have made much progress on the hypotheses of Heinrich (1962), but there is still much more to understand about the ways in which ontogenetic migrations are controlled and accomplished and their quantitative significance for particular populations.

We have mentioned microzooplankton several times, and although we are beginning to appreciate their importance in many food webs, it is possible that microzooplankton are a critical component of the polar food web. Except in the subarctic North Pacific, most of our knowledge of microzooplankton in polar food webs is cursory. Quantifying the role of microzooplankton in the nutrition of other polar zooplankton is critical and timely, in view of the increasing number of studies suggesting widespread occurrence of microzooplankton in polar seas and the fact that juvenile stages of many polar species are part of the microzooplankton. The most striking and obvious difference between the Arctic and Antarctic, of course, is the presence of the large and abundant *Euphausia superba* in the Antarctic and the complete absence of a similar organism in the Arctic. This may signify the importance of a deep-water habitat in the evolution of this species, because deep water is present over most of the polar and subpolar Antarctic but absent over much of the polar and subpolar Arctic where primary production occurs.

Acknowledgments

The assistance of P. V. Z. Lane, E. Schwarting, and B. Beck in the research was invaluable. The careful work of E. Schwarting and B. Beck in separating *C. glacialis* and *C. finmarchicus* has shed new light on the Fram Strait habitat. P. Aud and M. Apelskog cheerfully undertook gathering obscure references, and E. T. Premuzic translated Russian publications. Editorial assistance of S. French and C. Lamberti was most helpful. Throughout the chapter, conversions between carbon and dry weight were performed according to Cushing *et al.* (1958) and between wet and dry weight according to Mullin (1969). Grants from the National Science Foundation, Office of Naval Research, and Department of Energy supported this work. We sincerely thank E. Boysen-Ennen, F. Buchholz, and V. Siegel for critical review of the Antarctic part of the manuscript. This is Contribution No. 167 of the Alfred-Wegener-Institut für Polar- und Meeresforschung.

References

Andrews, K. J. H. 1966. The distribution and life history of *Calanoides acutus* (Giesbrecht). *Discovery Rep.* **34**: 117–162.
Antezana, T. & K. Ray. 1984. Active feeding of *Euphausia superba* in a swarm north of Elephant Island. *J. Crustacean Biol.* **4**: 142–155.
Antezana, T., K. Ray & C. Melo. 1982. Trophic behavior of *Euphausia superba* Dana in laboratory conditions. *Polar Biol.* **1**: 77–82.
Arimoto, T., K. Matuda, E. Hamada & K. Kanda. 1979. Diel vertical migration of krill swarm in the Antarctic ocean. *Trans. Tokyo Univ. Fish.* **3**: 93–97.
Atkinson, A. & J. M. Peck. 1988. Summer–winter differences in copepod distribution around South Georgia. *Polar Biol.* **8**: 463–473.
Ayala, F. J., J. W. Valentine & G. S. Zumwalt. 1975. An electrophoretic study of the Antarctic zooplankter *Euphausia superba*. *Limnol. Oceanogr.* **20**: 635–640.
Azam, F., J. B. Beers, L. L. Campbell, A. F. Carlucci, O. Holm-Hansen, F. M. H. Reid, & D. M. Karl, 1979. Occurrence and metabolic activity of organisms under the Ross Ice Shelf, Antarctica, at station J9. *Science* **203**: 451–453.
Baker, A. de C. 1954. The circumpolar continuity of Antarctic plankton species. *Discovery Rep.* **27**: 201–218.
Baker, A. de C. 1959. The distribution and life history of *Euphausia triacantha* Holt and Tattersall. *Discovery Rep.* **29**: 309–340.

Båmstedt, U. 1983. RNA concentration in zooplankton: Seasonal variation in boreal species. *Mar. Ecol.: Prog. Ser.* **11**: 291–297.

———. 1984. Diel variations in the nutritional physiology of *Calanus glacialis* from lat. 78°N in the summer. *Mar. Biol. (Berlin)* **79**: 275–267.

———. 1985. Seasonal excretion rates of macrozooplankton from the Swedish west coast. *Limnol. Oceanogr.* **30**: 607–617.

Båmstedt, U. & A. Ervik. 1984. Local variations in size and activity among *Calanus finmarchicus* and *Metridia longa* (Copepoda: Calanoida) over-wintering on the west coast of Norway. *J. Plankton Res.* **6**: 843–857.

Båmstedt, U. & K. S. Tande. 1985. Respiration and excretion rates of *Calanus glacialis* in Arctic waters of the Barents Sea. *Mar. Biol. (Berlin)* **87**: 259–266.

Båmstedt, U., K. S. Tande & H. Nicolajsen. 1985. Ecological investigations on the zooplankton community of Balsfjorden, northern Norway: Physiological adaptations in *Metridia longa* (Copepoda) to the overwintering period. *In* "Marine Biology of Polar Regions and Effects of Stress on Marine Organisms" (J. S. Gray & M. E. Christiansen, eds.), pp. 313–327. Wiley, Chichester.

Bargmann, H. E. 1945. The development and life history of adolescent and adult krill (*Euphausia superba*). *Discovery Rep.* **23**: 103–178.

Batchelder, H. P. 1985. Seasonal abundance, vertical distribution, and life history of *Metridia pacifica* (Copepoda: Calanoida) in the oceanic subarctic Pacific. *Deep-Sea Res.* **32**: 949–964.

Biggs, D. C. 1982. Zooplankton excretion and NH_4^+ cycling in near-surface waters of the Southern Ocean. I. Ross Sea, austral summer 1977–1978. *Polar Biol.* **1**: 55–67.

Biggs, D. C., A. F. Amos & O. Holm-Hansen. 1985. Oceanographic studies of epipelagic ammonium distributions: The Ross Sea NH_4^+ flux experiment. *In* "Antarctic Nutrient Cycles and Food Webs" (W. R. Siegfried, P. R. Condy & R. M. Laws, eds.), pp. 93–103. Springer-Verlag, Berlin.

Bliznichenko, T. E., A. Degtereva, S. Drobysheva, V. Nesterova, E. Rossova & V. Ryzhov. 1976. Results of plankton investigations in the Norwegian and Barents seas in 1976. *Annee Biol.* **33**: 54–56.

Bolms, G. & J. Lenz. 1989. On the significance of microzooplankton in the Fram Strait during MIZEX 1984. *Rapp. P.-V. Réun., Cons. Int. Explor. Mer.* **188**: 171.

Boyd, C. M., M. Heyraud & C. N. Boyd. 1984. Feeding of the Antarctic krill *Euphausia superba*. *J. Crustacean Biol.* **4**: 123–141.

Boysen-Ennen, E. & U. Piatkowski. 1988. Meso- and macrozooplankton communities in the Weddell Sea. *Polar Biol.* **9**: 17–35.

Bradford, J. M. 1971. The fauna of the Ross Sea. 8. Pelagic Copepoda. *Bull.—N. Z. Dep. Sci. Ind. Res.* **206**: 9–31.

———. 1981. Records of *Paraeuchaeta* (Copepoda: Calanoida) from McMurdo Sound, Antarctica, with a description of three hitherto unknown males. *N. Z. J. Mar. Freshwater Res.* **15**: 391–402.

Bradford, J. M. & J. B. J. Wells. 1983. New calanoid and harpacticoid copepods from beneath the Ross Ice Shelf, Antarctica. *Polar Biol.* **2**: 1–15.

Brinton, E. 1985. The oceanographic structure of the eastern Scotia Sea. III. Distributions of euphausiid species and their developmental stages in 1981 in relation to hydrography. *Deep-Sea Res.* **32**: 1153–1180.

Brinton, E. & T. Antezana. 1984. Structure of swarming and dispersed populations of krill (*Euphausia superba*) in Scotia Sea and South Shetland waters during January–March 1981, determined by bongo nets. *J. Crustacean Biol.* **4**: 45–66.

Brinton, E. & A. W. Townsend. 1984. Regional relationships between development and growth

in larvae of Antarctic krill, *Euphausia superba*, from field samples. *J. Crustacean Biol.* **4:** 224–246.
Brinton, E., M. Huntley & A. W. Townsend. 1986. Larvae of *Euphausia superba* in the Scotia Sea and Bransfield Strait in March 1984—development and abundance compared with 1981 larvae. *Polar Biol.* **5:** 221–234.
Buchanan, R. A. & A. D. Sekerak. 1982. Vertical distribution of zooplankton in eastern Lancaster Sound and western Baffin Bay, July–October 1978. *Arctic* **35:** 41–45.
Buchholz, F. 1983. Moulting and moult physiology in krill. *Ber. Polarforsch., Sonderh.* **4:** 81–88.
———. 1985. Moult and growth in euphausiids. *In* "Antarctic Nutrient Cycles and Food Webs" (W. R. Siegfried, P. R. Condy & R. M. Laws, eds.), pp. 339–345. Springer-Verlag, Berlin.
———. 1988. Zur Lebensweise des antarktischen und des nordischen Krills *Euphausia superba* und *Meganyctiphanes norvegica. Ber. Inst. Meereskunde* **177:** 1–249.
Chojnacki, J. & T. Weglénska. 1984. Periodicity of composition, abundance, and vertical distribution of summer zooplankton (1977/1978) in Ezcurra Inlet, Admiralty Bay (King George Island, South Shetland). *J. Plankton Res.* **6:** 997–1017.
Clarke, A. 1983. Life in cold water: THe physiological ecology of polar marine ectotherms. *Oceanogr. Mar. Biol.* **21:** 341–453.
———. 1984a. The lipid content and consumption of some Antarctic macrozooplankton. *Br. Antarct. Surv. Bull.* **63:** 57–70.
———. 1984b. Lipid content and composition of Antarctic krill, *Euphausia superba* Dana. *J. Crustacean Biol.* **4:** 285–294.
Coachman, L. K. 1986. Circulation, water masses and fluxes on the southeastern Bering Sea Shelf. *Cont. Shelf Res.* **5:** 23–108.
Coachman, L. K. & K. Aagaard. 1974. Physical oceanography of Arctic and subarctic seas. *In* "Marine Geology and Oceanography of the Arctic Seas" (Y. Herman, ed.), pp. 1–72. Springer-Verlag, Berlin.
Conover, R. J. 1960. The feeding behavior and respiration of some marine planktonic crustacea. *Biol. Bull. (Woods Hole, Mass.)* **119:** 399–415.
———. 1962. Metabolism and growth in *Calanus hyperboreus* in relation to its life cycle. *Rapp. P.-V. Reun., Cons. Int. Explor. Mer.* **153:** 190–197.
———. 1964. "Food Relations and Nutrition of Zooplankton," Proc. Symp. Exp. Mar., Ecol., Occas. Publ. No. 2, pp. 81–91. Grad. School Oceanogr., Univ. of Rhode Island, Kingston.
———. 1965. Notes on the molting cycle, development of sexual characters and sex ratio in *Calanus hyperboreus. Crustaceana* **8:** 308–320.
———. 1967. Reproductive cycle, early development, and fecundity in laboratory populations of the copepod *Calanus hyperboreus. Crustaceana* **13:** 61–72.
———. 1968. Zooplankton—life in a nutritionally dilute environment. *Am. Zool.* **8:** 107–118.
———. 1988. Comparative life histories in the genera *Calanus* and *Neocalanus* in high latitudes of the northern hemisphere. *Hydrobiologia* **167/168:** 127–142.
Conover, R. J. & E. D. S. Corner. 1968. Respiration and nitrogen excretion by some marine zooplankton in relation to their life cycles. *J. Mar. Biol. Assoc. U. K.* **48:** 49–75.
Conover, R. J. & G. F. Cota. 1985. Balance experiments with Arctic zooplankton. *In* "Marine Biology of Polar Regions and Effects of Stress on Marine Organisms" (J. S. Gray & M. Christiansen, eds.), pp. 217–236. Wiley, Chichester.
Conover, R. J., A. W. Herman, S. J. Prinsenberg & L. R. Harris. 1986. Distribution of and feeding by the copepod *Pseudocalanus* under fast ice during the Arctic spring. *Science* **232:** 1245–1247.

Cooney, R. T. & K. O. Coyle. 1982. Trophic implications of cross-shelf copepod distributions in the southeastern Bering Sea. *Mar. Biol. (Berlin)* **70**: 187–196.
Corkett, C. J. & I. A. McLaren. 1978. The biology of *Pseudocalanus*. *Adv. Mar. Biol.* **15**: 1–231.
Corkett, C. J., I. A. McLaren & J.-H. Sevigny. 1986. The rearing of the marine calanoid copepods *Calanus finmarchicus* (Gunnerus), *C. glacialis* Jaschnov and *C. hyperboreus* Kroyer with comment on the equiproportional rule. *In* "Proceedings of the Second International Conference on Copepoda" (G. Schriever, H. K. Schminke & C.-T. Shih, eds.), pp. 539–546. Syllogeus No. 58, National Museums of Canada, Ottawa.
Cushing, D. H., G. F. Humphrey, K. Banse & T. Lawvastu. 1958. Report of the committee on terms and equivalents. *Rapp. P.-V. Reun., Cons. Int. Explor. Mer* **144**: 15–16.
Dagg, M. J. & K. D. Wyman. 1983. Natural ingestion rates of the copepods *Neocalanus plumchrus* and *N. cristatus* calculated from gut contents. *Mar. Ecol.: Prog. Ser.* **13**: 37–46.
Dagg, M. J., J. Vidal, T. E. Whitledge, R. L. Iverson & J. J. Goering. 1982. The feeding, respiration, and excretion of zooplankton in the Bering Sea during a spring bloom. *Deep-Sea Res.* **29**: 45–63.
Daly, K. L. & M. C. Macaulay. 1988. Abundance and krill in the ice edge zone of the Weddell Sea, austral spring 1983. *Deep-Sea Res.* **35**: 21–41.
Damkaer, D. M. 1975. Calanoid copepods of the genera *Spinocalanus* and *Mimicalanus* from the central Arctic Ocean, with a review of the Spinocalanidae. *NOAA Tech. Rep. NMFS Circ.* **391**: 1–88.
David, P. M. 1955. The distribution of *Sagitta gazellae* Ritter-Zahony. *Discovery Rep.* **27**: 235–278.
———. 1958. The distribution of the Chaetognatha of the Southern Ocean. *Discovery Rep.* **29**: 200–229.
———. 1965. The Chaetognatha of the Southern Ocean. *In* "Biogeography and Ecology in Antarctica" (J. von Mieghem, P. von Oye & J. Schell, eds.), pp. 296–323. Junk Publ., The Hague.
Davis, C. C. 1976. Overwintering strategies of common planktonic copepods in some north Norway fjords and sounds. *Astarte* **9**: 37–42.
Dawson, J. K. 1978. Vertical distribution of *Calanus hyperboreus* in the central Arctic Ocean. *Limnol. Oceanogr.* **23**: 950–957.
Degtereva, A. A. & V. Nesterova. 1975. The plankton development off the northwestern coast of Norway and in the southeastern Barents Sea in 1973. *Annee Biol.* **30**: 55–56.
Digby, P. S. B. 1954. The biology of the marine planktonic copepods of Scoresby Sound, east Greenland. *J. Anim. Ecol.* **23**: 298–338.
Dugdale, R. C. & J. J. Goering. 1967. Uptake of new and regenerated forms of nitrogen in primary productivity. *Limnol. Oceanogr.* **12**: 196–206.
El-Sayed, S. Z. 1985. Plankton of the Antarctic seas. *In* "Antarctica (W. N. Bonner & D. W. H. Walton, eds.), pp. 135–153. Pergamon, Oxford.
El-Sayed, S. Z. & S. Taguchi. 1981. Primary production and standing crop of phytoplankton along the ice edge in the Weddell Sea. *Deep-Sea Res.* **28**: 1017–1032.
English, T. S. & R. Horner. 1977. Beaufort Sea plankton studies. *Environ. Assess. Alaskan Cont. Shelf, Annu. Rep.* **9**: 275–627.
Ettershank, G. 1984. A new approach to the assessment of longevity in the Antarctic krill *Euphausia superba*. *J. Crustacean Biol.* **4**: 295–305.
———. 1985. Population age structure in males and juveniles of the Antarctic krill, *Euphausia superba* Dana. *Polar Biol.* **4**: 199–201.
Evans, M. S. & E. H. Grainger. 1980. Zooplankton in a Canadian Arctic estuary. *In* "Estuarine Comparisons" (V. Kennedy, ed.), pp. 199–210. Academic Press, New York.
Everson, I. 1977. "The Living Resources of the Southern Ocean." FAO, Rome.

———. 1982. Diurnal variations in mean volume backscattering strength of an Antarctic krill (*Euphausia superba*) patch. *J. Plankton Res.* **4:** 155–162.

———. 1983. Variations in vertical distribution and density of krill swarms in the vicinity of South Georgia. *Mem. Natl. Inst. Polar Res.* **27:** 84–92.

———. 1984. Zooplankton. *In* "Antarctic Ecology" (R. M. Laws, ed.), Vol. 2, pp. 463–490. Academic Press, London.

———. 1987. Some aspects of the small scale distribution of *Euphausia crystallorophias. Polar Biol.* **8:** 9–15.

Everson, I. & E. Murphy. 1987. Mesoscale variability in the distribution of krill *Euphausia superba. Mar. Ecol.: Prog. Ser.* **40:** 53–60.

Everson, I. & P. Ward. 1980. Aspects of Scotia Sea zooplankton. *Biol. J. Linn. Soc.* **14:** 93–101.

Falk-Petersen, S., J. R. Sargent, C. C. E. Hopkins & B. Vaja. 1982. Ecological investigations on the zooplankton community of Balsfjorden, northern Norway: Lipids in the euphausiids *Thysanoessa raschi* and *T. inermis* during spring. *Mar. Biol. (Berlin)* **68:** 97–102.

Falk-Petersen, S., J. R. Sargent & K. S. Tande. 1987. Lipid composition of zooplankton in relation to the subarctic food web. *Polar Biol.* **8:** 115–120.

Ferrari, F. & M. Dojiri. 1987. The calanoid copepod *Euchaeta antarctica* from Southern Ocean Atlantic sector midwater trawls, with observations on spermatophore dimorphism. *J. Crustacean Biol.* **7:** 458–480.

Fevolden, S. E. 1980. Krill off Bouvetoya and in the southern Weddell Sea with a description of larval stages of *Euphausia crystallorophias. Sarsia* **65:** 149–162.

Fevolden, S. E. & F. J. Ayala. 1981. Enzyme polymorphism in Antarctic krill (Euphausiacea); genetic variation between populations and species. *Sarsia* **66:** 167–181.

Fleminger, A. & K. Hulsemann. 1977. Geographical range and taxonomic divergence in North Atlantic *Calanus* (*C. helgolandicus, C. finmarchicus,* and *C. glacialis*). *Mar. Biol. (Berlin)* **40:** 233–248.

Foster, B. A. 1987. Composition and abundance of zooplankton under the spring sea-ice of McMurdo Sound, Antarctica. *Polar Biol.* **8:** 41–48.

Foxton, P. 1956. The distribution of the standing crop of zooplankton in the Southern Ocean. *Discovery Rep.* **28:** 191–236.

———. 1964. Seasonal variations in the plankton of Antarctic waters. *In* "Biologie Antarctique" (R. Carrick, M. W. Holgate & J. Prevost, eds.), pp. 311–318. Hermann, Paris.

———. 1966. The distribution and life history of *Salpa thompsoni*–Foxton with observations on a related species *Salpa gerlachei*–Foxton. *Discovery Rep.* **34:** 1–116.

Fraser, F. C. 1936. On the development and distribution of the young stages of krill. *Discovery Rep.* **14:** 1–192.

Frost, B. W. 1971. Taxonomic status of *Calanus finmarchicus* and *C. glacialis* (Copepoda), with special reference to adult males. *J. Fish. Res. Board Can.* **28:** 23–30.

———. 1987. Grazing control of phytoplankton stock in the open subarctic Pacific Ocean: A model assessing the role of mesozooplankton, particularly the large calanoid copepods *Neocalanus* spp. *Mar. Ecol.: Prog. Ser.* **39:** 49–68.

———. 1989. A taxonomy of the marine calanoid copepod genus *Pseudocalanus. Can. J. Zool.* **67:** 525–551.

Frost, B. W., M. R. Landry & R. P. Hassett. 1983. Feeding behavior of large calanoid copepods *Neocalanus cristatus* and *N. plumchrus* from the subarctic Pacific Ocean. *Deep-Sea Res.* **30:** 1–13.

Fukuchi, M. 1977. Regional distribution of Amphipoda and Euphausiacea in the northern North Pacific and Bering Sea in summer of 1969. *Res. Inst. North Pac. Fish., Hokkaido Univ. Spec. Vol.:* 439–458.

Fukuchi, M. & H. Sasaki. 1981. Phytoplankton and zooplankton standing stocks and down-

ward flux of particulate material around fast ice edge of Lutzow–Holm Bay, Antarctica. *Mem. Natl. Inst. Polar Res.* **34**: 13–36.
Fukuchi, M., A. Tanimura & H. Ohtsuka. 1985. Zooplankton community conditions under sea ice near Syowa Station, Antarctica. *Bull. Mar. Sci.* **37**: 518–528.
Grainger, E. H. 1961. The copepods *Calanus glacialis* Jaschnov and *Calanus finmarchicus* (Gunnerus) in Canadian Arctic–subarctic waters. *J. Fish. Res. Board Can.* **18**: 663–678.
———. 1963. Copepods of the genus *Calanus* as indicators of eastern Canadian waters. *R. Soc. Can., Spec. Publ.* **5**: 68–94.
———. 1965. Zooplankton from the Arctic Ocean and adjacent Canadian waters. *J. Fish. Res. Board Can.* **22**: 543–564.
Greene, C. H. & M. R. Landry. 1988. Carnivorous suspension feeding by the subarctic calanoid *Neocalanus cristatus. Can. J. Fish. Aquat. Sci.* **45**: 1069–1074.
Grice, G. D. 1962. Copepods collected by the nuclear submarine *Seadragon* on a cruise to and from the North Pole, with remarks on their geographic distribution. *J. Mar. Res.* **20**: 97–109.
Grigg, H. & S. J. Bardwell. 1982. Seasonal observations on moulting and maturation in stage V copepodites of *Calanus finmarchicus* from the Firth of Clyde. *J. Mar. Biol. Assoc. U. K.* **62**: 315–327.
Grønvik, S. & C. C. E. Hopkins. 1984. Ecological investigations of the zooplankton community of Balsfjorden, northern Norway: Generation cycle, seasonal vertical distribution, and seasonal variations in body weight and carbon and nitrogen content of the copepod *Metridia longa* (Lubbock). *J. Exp. Mar. Biol. Ecol.* **80**: 93–107.
Hagen, W. 1985. On distribution and population structure of Antarctic Chaetognatha. *Meeresforschung* **30**: 280–281.
———. 1988. Zur Bedeutung der Lipide im antarktischen Zooplankton. *Ber. Polarforsch.* **49**: 1–129.
Hallberg, E. & H.-J. Hirche. 1980. Differentiation of mid-gut in adults and overwintering copepodids of *Calanus finmarchicus* (Gunnerus) and *C. helgolandicus* Claus. *J. Exp. Mar. Biol. Ecol.* **48**: 283–295.
Hamner, W. M. 1984. Aspects of schooling in *Euphausia superba. J. Crustacean Biol.* **4**: 67–74.
———. 1988. Biomechanics of filter feeding in the Antarctic krill *Euphausia superba:* Review of past work and new observations. *J. Crustacean Biol.* **8**: 149–163.
Hamner, W. M., P. P. Hamner, S. W. Strand & R. W. Gilmer. 1983. Behavior of Antarctic krill, *Euphausia superba:* Chemoreception, feeding, schooling, and molting. *Science* **220**: 433–435.
Haq, S. M. 1967. Nutritional physiology of *Metridia lucens* and *M. longa* from the Gulf of Maine. *Limnol. Oceanogr.* **12**: 40–51.
Hardy, A. C. 1936. General account. The Arctic plankton collected by the *Nautilus* expedition. 1931. *J. Linn. Soc. London Zool.* **39**: 391–404.
Hardy, A. C. & E. R. Gunther. 1935. The plankton of the South Georgia whaling grounds and adjacent waters, 1926–1927. *Discovery Rep.* **11**: 1–456.
Head, E. J. H. 1986. Estimation of Arctic copepod grazing rates *in vivo* and comparison with *in-vitro* methods. *Mar. Biol. (Berlin)* **92**: 371–379.
Head, E. J. H. & L. R. Harris. 1985. Physiological and biochemical changes in *Calanus hyperboreus* from Jones Sound NWT during the transition from summer feeding to overwintering conditions. *Polar Biol.* **4**: 99–106.
Head, E. J. H., L. R. Harris & C. Abou Debs. 1985. Effect of day length and food concentration on *in situ* diurnal feeding rhythms in Arctic copepods. *Mar. Ecol.* **24**: 281–288.
Heinrich, A. K. 1962. The life history of plankton animals and seasonal cycles of plankton communities in the oceans. *J. Cons., Cons. Int. Explor. Mer* **27**: 15–24.

_____. 1968. Seasonal phenomena in the plankton of the northeast Pacific Ocean. *Oceanology (Engl. Transl.)* **8**: 231–239.

Hempel, G. 1985. On the biology of polar seas, particularly the Southern Ocean. *In* "Marine Biology of Polar Regions and Effects of Stress on Marine Organisms" (J. S. Gray & M. E. Christiansen, eds.), pp. 3–33. Wiley, Chichester.

Hempel, I. & G. Hempel. 1977/1978. Larval krill (*Euphausia superba*) in the plankton and neuston samples of the German Antarctic Expedition 1975/76. *Meeresforschung* **26**: 206–216.

Hempel, I. & G. Hempel. 1982. Distribution of euphausiid larvae in the southern Weddell Sea. *Meeresforschung* **29**: 253–266.

_____. 1985. Field observations on the developmental ascent of larval *Euphausia superba* (Crustacea). *Polar Biol.* **6**: 121–126.

Hempel, I. & E. Marschoff. 1980. Euphausiid larvae in the Atlantic sector of the Southern Ocean. *Meeresforschung* **28**: 32–47.

Hempel, I., G. Hempel & A. de C. Baker. 1979. Early life stages of krill (*Euphausia superba*) in Bransfield Strait and Weddell Sea. *Meeresforschung* **27**: 267–281.

Herman, A. 1983. Vertical distribution patterns of copepods, chlorophyll and production in northeastern Baffin Bay. *Limnol. Oceanogr.* **28**: 709–719.

Herman, Y. 1974. Topography of the Arctic Ocean. *In* "Marine Geology and Oceanography of the Arctic Seas" (Y. Herman, ed.), pp. 73–81. Springer-Verlag, Berlin.

Hewes, C. D., O. Holm-Hansen & E. Sakshaug. 1985. Alternate carbon pathways at lower trophic levels in the Antarctic food web. *In* "Antarctic Nutrient Cycles and Food Webs" (W. R. Siegfried, P. R. Condy & R. M. Laws, eds.), pp. 277–283. Springer-Verlag, Berlin.

Heywood, R. B., I. Everson & J. Priddle. 1985. The absence of krill from the South Georgia zone, winter 1983. *Deep-Sea Res.* **32**: 369–378.

Hicks, G. R. F. 1974. Variation in zooplankton biomass with hydrological regime beneath the seasonal ice, McMurdo Sound, Antarctica. *N. Z. J. Mar. Freshwater Res.* **8**: 67–77.

Hirche, H.-J. 1983. Overwintering of *Calanus finmarchicus* and *Calanus helgolandicus*. *Mar. Ecol.: Prog. Ser.* **11**: 281–290.

_____. 1987. Temperature and plankton. II. Effect on respiration and swimming activity in copepods from the Greenland Sea. *Mar. Biol. (Berlin)* **94**: 347–356.

_____. 1989. Spatial distribution of digestive enzyme activities of *Calanus finmarchicus* and *C. hyperboreus* in Fram Strait/Greenland Sea. *J. Plankton Res.* **11**: 431–443.

Hirche, H.-J. & R. N. Bohrer. 1987. Reproduction of the Arctic copepod *Calanus glacialis* in Fram Strait. *Mar. Biol. (Berlin)* **94**: 11–17.

Holm-Hansen, O. & M. Huntley. 1984. Feeding requirements of krill in relation to food sources. *J. Crustacean Biol.* **4**: 156–173.

Hopkins, C. C. E., K. S. Tande & S. Grønvik. 1984. Ecological investigations of the zooplankton community of Balsfjorden, Norway: An analysis of growth and overwintering tactics in relation to niche and environment in *Metridia longa* (Lubbock), *Calanus finmarchicus* (Gunnerus), *Thysanoessa inermis* (Krøyer) and *T. raschi (M. Sars)*. *J. Exp. Mar. Biol. Ecol.* **82**: 77–99.

Hopkins, C. C. E., K. S. Tande, S. Grønvik, J. R. Sargent & T. Schweder. 1985. Ecological investigations of Balsfjorden, northern Norway: Growth and quantification of condition in relation to overwintering and food supply in *Metridia longa, Calanus finmarchicus, Thysanoessa inermis,* and *Thysanoessa raschi. In* "Marine Biology of Polar Regions and Effects of Stress on Marine Organisms" (J. S. Gray & M. E. Christiansen, eds.), pp. 83–100. Wiley, Chichester.

Hopkins, T. L. 1969. Zooplankton standing crop in the Arctic basin. *Limnol. Oceanogr.* **14**: 80–95.

———. 1971. Zooplankton standing crop in the Pacific sector of the Antarctic. *Antarct. Res. Ser.* **17**: 347–362.

———. 1985a. Food web of an Antarctic midwater ecosystem. *Mar. Biol. (Berlin)* **89**: 197–212.

———. 1985b. The zooplankton community of Croker Passage, Antarctic Peninsula. *Polar Biol.* **4**: 161–170.

———. 1987. Midwater food web in McMurdo Sound, Ross Sea, Antarctica. *Mar. Biol. (Berlin)* **96**: 93–106.

Horner, R. & D. Murphy. 1985. Species composition and abundance of zooplankton in the nearshore Beaufort Sea in winter–spring. *Arctic* **38**: 201–209.

Hosie, G. W., T. Ikeda & M. Stolp. 1988. Distribution, abundance and population structure of the Antarctic krill (*Euphausia superba* Dana) in the Prydz Bay region, Antarctica. *Polar Biol.* **8**: 213–224.

Hubold, G. & I. Hempel. 1987. Seasonal variability of zooplankton in the southern Weddell Sea. *Meeresforschung* **31**: 185–192.

Hubold, G., I. Hempel & M. Meyer. 1988. Zooplankton communities in the southern Weddell Sea (Antarctica). *Polar Biol.* **8**: 225–233.

Huntley, M. 1981. Nonselective, nonsaturated feeding by three calanoid copepod species in the Labrador Sea. *Limnol. Oceanogr.* **26**: 831–842.

Huntley, M., K. W. Strong & A. T. Dengler. 1983. Dynamics and community structure of zooplankton in the Davis Strait and northern Labrador Sea. *Arctic* **36**: 143–161.

Huntley, M., K. Tande & H. C. Eilertsen. 1987. On the trophic fate of *Phaeocystis pouchetii* (Hariot). II. Grazing rates of *Calanus hyperboreus* (Kroyer) on diatoms and different size categories of *Phaeocystis pouchetii*. *J. Exp. Mar. Biol. Ecol.* **110**: 197–212.

Ikeda, T. 1984. Sequences in metabolic rates and elemental composition (C, N, P) during the development of *Euphausia superba* Dana and estimated food requirements during its life span. *J. Crustacean Biol.* **4**: 273–284.

Ikeda, T. & P. Dixon. 1982. Body shrinkage as a possible overwintering mechanism of the Antarctic krill, *Euphausia superba* Dana. *J. Exp. Mar. Biol. Ecol.* **62**: 143–151.

———. 1984. The influence of feeding on the metabolic activity of Antarctic krill (*Euphausia superba* Dana). *Polar Biol.* **3**: 1–9.

Ikeda, T. & P. G. Thomas. 1987a. Longevity of the Antarctic krill (*Euphausia superba*) based on a laboratory experiment. *Proc. NIPR Symp. Polar Biol.* **1**: 56–62.

———. 1987b. Moulting interval and growth of juvenile Antarctic krill *Euphausia superba* fed different concentrations of the diatom *Phaeodactylum tricornutum* in the laboratory. *Polar Biol.* **7**: 339–343.

Ishii, H. 1986. Feeding behavior of the Antarctic krill, *Euphausia superba* Dana. II. Effects of food condition on particle selectivity. *Mem. Natl. Inst. Polar Res.* **44**: 96–106.

Ishii, H., M. Omori & M. McMurano. 1985. Feeding behavior of the Antarctic krill, *Euphausia superba* Dana. I. Reaction to size and concentration of food particles. *Trans. Tokyo Univ. Fish.* **6**: 117–124.

Jackowska, H. 1980. Krill monitoring in Admiralty Bay (King George Island, South Shetland Islands) in summer 1979/1980. *Pol. Polar Res.* **1**: 117–125.

Jaschnov, W. A. 1946. The distribution of the autochthon pelagic fauna of the Arctic region. *Biul. Mosk. Obshchestva Ispytatele Prir. Otd. Biol.* (Engl. Summary) **51**: 40–50.

———. 1955. Morphology, distribution and systematics of *Calanus finmarchicus* s.1. *Zool. Zh.* **34**: 1210–1223.

———. 1970. Distribution of *Calanus* species in the seas of the northern hemisphere. *Int. Rev. Gesamten Hydrobiol.* **55**: 197–212.

———. 1972. On the systematic status of *Calanus glacialis, Calanus finmarchicus* and *Calanus helgolandicus*. *Crustaceana* **22**: 279–284.

Jażdżewski, K., W. Kittel & K. Lotocki. 1982. Zooplankton studies in the southern Drake Passage and in the Bransfield Strait during austral summer (BIOMASS–FIBEX, February–March 1981). *Pol. Polar Res.* **3**: 203–242.

John, D. D. 1936. The southern species of the genus *Euphausia*. *Discovery Rep.* **14**: 195–325.

Johnson, M. W. 1958. Observations on inshore plankton collected during summer 1957 at Point Barrow, Alaska. *J. Mar. Res.* **17**: 272–281.

———. 1963. Zooplankton collections from the high polar basin with special reference to the Copepoda. *Limnol. Oceanogr.* **8**: 89–102.

Kaczmaruk, B. Z. 1983. Occurrence and distribution of the Antarctic copepods along the ice shelves in the Weddell Sea in summer 1979/80. *Meeresforschung* **30**: 25–41.

Kalinowski, J. & Z. Witek. 1980. Diurnal vertical distribution of krill aggregations in the western Antarctic. *Pol. Polar Res.* **1**: 127–146.

Kalland, G. A. 1972. Oxygen consumption rates of four species of adult female Arctic copepods. *Sci. Alaska* **23**: 32–33.

Kanda, K., K. Takagi & Y. Seki. 1982. Movement of the larger swarms of Antarctic krill *Euphausia superba* population off Enderby Land during 1976–1977 season. *J. Tokyo Univ. Fish.* **68**: 25–42.

Kane, J. E. 1966. The distribution of *Parathemisto gaudichaudii* (Guer.), with observations on its life history in the 0° to 20°E sector of the Southern Ocean. *Discovery Rep.* **34**: 163–198.

Kato, M., S. Segawa, E. Tanoue & M. Murano. 1982. Filtering and ingestion rates of the Antarctic krill, *Euphausia superba* Dana. *Trans. Tokyo Univ. Fish.* **5**: 167–175.

Kattner, G. & M. Krause. 1987. Changes in lipids during the development of *Calanus finmarchicus* sensu lato from copepodid I to adult. *Mar. Biol. (Berlin)* **96**: 511–518.

Kawaguchi, K., S. Ishikawa & O. Matsuda. 1986. The overwintering strategy of Antarctic krill (*Euphausia superba* Dana) under the coastal fast ice off the Ongul Islands in Lutzow–Holm Bay, Antarctica. *Mem. Natl. Inst. Polar Res.* **44**: 67–85.

Kellermann, A. 1987. Food and feeding ecology of postlarval and juvenile *Pleuragramma antarcticum* (Pisces; Notothenioidei) in the seasonal pack ice zone off the Antarctic Peninsula. *Polar Biol.* **7**: 307–315.

Kils, U. 1981a. Swimming behavior, swimming performance and energy balance of Antarctic krill, *Euphausia superba*. *Biomass Sci. Ser.* **3**: 1–233.

———. 1981b. Size dissociation in krill swarms. *Kiel. Meeresforsch., Sonderh.* **5**: 262–263.

———. 1983. Swimming and feeding of Antarctic krill, *Euphausia superba*—some outstanding energetics and dynamics, some unique morphological details. *Ber. Polarforsch., Sonderh.* **4**: 130–155.

Kinder, T. H. & L. K. Coachman. 1978. The front overlaying the continental slope in the eastern Bering Sea. *J. Geophys. Res.* **83**: 4551–4559.

Kittel, W. & K. Jażdżewski. 1982. Studies on the larval stages of *Euphausia superba* Dana (Crustacea, Euphausiacea) in the southern Drake Passage and in the Bransfield Strait in February and March 1981 during the BIOMASS—FIBEX expedition. *Pol. Polar Res.* **3**: 273–280.

Kittel, W. & R. Stepnik, 1983. Distribution of *Euphausia crystallorophias, E. frigida, E. triacantha* and *Thysanoessa macrura* (Crustacea, Euphausiacea) in the southern Drake Passage and Bransfield Strait in February and March 1981. *Pol. Polar Res.* **4**: 7–19.

Kittel, W., Z. Witek & H. Czykieta. 1985. Distribution of *Euphausia frigida, Euphausia crystallorophias, Euphausia triacantha* and *Thysanoessa macrura* in the southern part of Drake Passage and in the Bransfield Strait during the 1983–1984 austral summer (BIOMASS–SIBEX). *Pol. Polar Res.* **6**: 133–149.

Klindt, H. & F. Zwack. 1984. A method for acoustic estimation of krill *Euphausia superba* Dana) abundance applied to FIBEX data. *Arch. Fischereiwiss.* **34**: 121–144.

Køgeler, J. W., S. Falk-Petersen, A. Kristensen, F. Petterson & J. Dalen. 1987. Density and sound speed contrasts in sub-Arctic zooplankton. *Polar Biol.* **1**: 231–235.

Kolosova, E. G. 1975. Temperature effects upon the distribution of the mass zooplankton groups in the White Sea. *Okeanologiya (Moscow)* **15**: 129–133.

———. 1978. Some regularities in the vertical distribution of the White Sea zooplankton established with the aid of the correlation pleiads method. *Okeanologiya (Moscow)* **18**: 320–326.

Kosobokova, K. N. 1978. Diurnal vertical distribution of *Calanus hyperboreus* Krøyer and *Calanus glacialis* Jaschnov in the central polar basin. *Oceanology (Engl. Transl.)* **18**: 476–480.

———. 1982. Composition and distribution of the biomass of zooplankton in the central Arctic basin. *Oceanology (Engl. Transl.)* **22**: 744–750.

Lee, R. F. 1974. Lipid composition of the copepod *Calanus hyperboreus* from the Arctic Ocean. Changes with depth and season. *Mar. Biol. (Berlin)* **26**: 313–318.

———. 1975. Lipids of Arctic zooplankton. *Comp. Biochem. Physiol.* **51**: 263–266.

Lee, R. F., J. Hirota & A. M. Barnett. 1971. Distribution and importance of wax esters in marine copepods and other zooplankton. *Deep-Sea Res.* **18**: 1147–1165.

Lee, R. F., J. C. Nevenzel & G. A. Paffenhofer. 1972. The presence of wax esters in marine planktonic copepods. *Naturwissenschaften* **59**: 406–411.

Lie, U. 1965. Quantities of zooplankton and propagation of *Calanus finmarchicus* at permanent stations on the Norwegian coast and at Spitsbergen, 1959–1962. *Fiskeri dir. Skr., Ser. Havunders.* **13**: 5–19.

Littlepage, J. L. 1964. Seasonal variation in lipid content of two Antarctic marine crustacea. In Biologie Antarctique (R. Carrick, M. Holgate & J. Prevost, eds.), pp. 463–470. Hermann, Paris.

Longhurst, A., D. Sameoto & A. Herman. 1984. Vertical distribution of Arctic zooplankton in summer: Eastern Canadian Archipelago. *J. Plankton Res.* **6**: 137–168.

Macaulay, M. C., T. S. English & O. A. Mathisen. 1984. Acoustic characterization of swarms of Antarctic krill (*Euphausia superba*) from Elephant Island and Bransfield Strait. *J. Crustacean Biol.* **4**: 16–44.

MacDonald, C. M., R. Williams & M. Adams. 1986. Genetic variation and population structure of krill (*Euphausia superba* Dana) from the Prydz Bay region of Antarctic waters. *Polar Biol.* **6**: 233–236.

Mackintosh, N. A. 1934. Distribution of macrozooplankton in the Atlantic sector of the Antarctic. *Discovery Rep.* **9**: 65–160.

———. 1937. The seasonal circulation of the Antarctic macrozooplankton. *Discovery Rep.* **16**: 365–412.

———. 1972. Life cycle of Antarctic krill in relation to ice and water conditions. *Discovery Rep.* **36**: 1–94.

———. 1973. Distribution of post-larval krill in the Antarctic. *Discovery Rep.* **36**: 95–156.

Maclellan, D. C. 1967. The annual cycle of certain calanoid species in west Greenland. *Can. J. Zool.* **45**: 101–115.

Makarov, R. R. 1977. Distribution of the larvae and some questions of the ecology of reproduction of the euphausiid *Euphausia frigida* Hansen, 1911 (Crustacea: Euphausiacea) in the southern part of the Scotia Sea. *Oceanology (Engl. Transl.)* **17**: 208–213.

———. 1978. Vertical distribution of euphausiid eggs and larvae off the northeastern coast of South Georgia Island. *Oceanology (Engl. Transl.)* **15**: 708–711.

———. 1979. Larval distribution and reproductive ecology of *Thysanoessa macrura* (Crustacea: Euphausiacea) in the Scotia Sea. *Mar. Biol. (Berlin)* **52**: 377–386.

Marchant, H. J. 1985. Choanoflagellates in the Antarctic marine food chain. *In* "Antarctic Nutrient Cycles and Food Webs" (W. R. Siegfried, P. R. Condy & R. M. Laws, eds.), pp. 271–276. Springer-Verlag, Berlin.
Marin, V. 1987. The oceanographic structure of the eastern Scotia Sea. IV. Distribution of copepod species in relation to hydrography in 1981. *Deep-Sea Res.* **34:** 105–121.
Markhaseva, E. L. 1984. Aetideidae copepods (Copepoda: Calanoida) of the eastern sector of the central Arctic basin. *Oceanology (Engl. Transl.)* **24:** 391–393.
Marr, J. W. S. 1962. The natural history and geography of the Antarctic krill (*Euphausia superba* Dana). *Discovery Rep.* **32:** 33–464.
Marshall, S. M. & A. P. Orr. 1972. "The Biology of a Marine Copepod." Springer-Verlag, Berlin.
Maruyama, T., H. Toyoda & S. Suzuki. 1982. Preliminary report on the biomass of macroplankton and micronekton collected with a bongo net during the Umitaka Maru FIBEX cruise. *Trans. Tokyo Univ. Fish.* **5:** 145–153.
Matthews, J. B. L. 1967. *Calanus finmarchicus* s.l. in the North Atlantic. The relationships between *Calanus finmarchicus* s.str., *C. glacialis*, and *C. helgolandicus*. *Bull. Mar. Ecol.* **6:** 159–179.
Matthews, J. B. L., L. Hestad & J. L. W. Bakke. 1978. Ecological studies in Korsfjorden, western Norway. The generations and stocks of *Calanus hyperboreus* and *C. finmarchicus* in 1971–1974. *Oceanol. Acta* **1:** 277–284.
Mauchline, J. 1980. Studies on patches of krill, *Euphausia superba*. *Biomass Handb.* **6:** 1–36.
Mauchline, J. & L. R. Fisher. 1969. The biology of euphausiids. *Adv. Mar. Biol.* **7:** 1–454.
Mayzaud, P., A. von Wormhoudt & O. Roche-Mayzaud. 1987. Spatial changes in the concentrations and activities of amylase and trypsin in *Euphausia superba*. A comparison between activity measurement and immunoquantitation. *Polar Biol.* **8:** 73–80.
McClatchie, S. 1988. Food-limited growth of *Euphausia superba* in Admiralty Bay, South Shetland Islands, Antarctica. *Cont. Shelf Res.* **8:** 329–345.
McClatchie, S. & C. M. Boyd. 1983. Morphological study of sieve efficiencies and mandibular surfaces in the Antarctic krill, *Euphausia superba*. *Can. J. Fish. Aquat. Sci.* **40:** 955–967.
McLaren, I. A., J.-M. Sevingy & C. J. Corkett. 1988. Body sizes, development rates and genome sizes among *Calanus* species. *Hydrobiologia* **167:** 275–284.
Miller, C. B. 1988. *Neocalanus flemingeri*, a new species of Calanidae (Copepoda: Calanoida) from the subarctic Pacific Ocean, with a comparative redescription of *Neocalanus plumchrus* (Marukawa) 1921. *Prog. Oceanog.* **20:** 223–273.
Miller, C. B. & M. J. Clemons. 1988. Revised life history analysis for large grazing copepods in the subarctic Pacific Ocean. *Prog. Oceanogr.* **20:** 293–313.
Miller, C. B., B. W. Frost, H. P. Batchelder, M. J. Clemons & R. E. Conway. 1984. Life histories of large, grazing copepods in a subarctic ocean gyre: *Neocalanus plumchrus*, *Neocalanus cristatus*, and *Eucalanus bungii* in the northeast Pacific. *Prog. Oceanogr.* **13:** 201–243.
Miller, D. G. M. 1986. Results from biological investigations of krill (*Euphausia superba*) in the southern Indian Ocean during SIBEX I. *Mem. Natl. Inst. Polar Res.* **40:** 117–139.
Miller, D. G. M., I. Hampton, J. Henry, R. W. Abrams & J. Cooper. 1985. The relationship between krill food requirements and phytoplankton production in a sector of the southern Indian Ocean. *In* "Antarctic Nutrient Cycles and Food Webs" (W. R. Siegfried, P. R. Condy & R. M. Laws, eds.), pp. 362–371. Springer-Verlag, Berlin.
Minoda, T. 1967. Seasonal distribution of Copepoda in the Arctic Ocean from June to December, 1964. *Rec. Oceanogr. Works Jpn.* **9:** 161–168.
———. 1971. Pelagic Copepoda in the Bering Sea and the northwestern North Pacific with special reference to their vertical distribution. *Mem. Fac. Fish., Hokkaido Univ.* **18:** 1–74.

Minoda, T. & R. Marumo. 1974. Regional characteristics of distribution of phyto- and zooplankton in the eastern Bering Sea and Chukchi Sea in June-July, 1972. *In* "Bering Sea Oceanography: An Update" (Y. Takenouti & D. W. Hood, eds.), pp. 83-95. Univ. of Alaska Press, Fairbanks.

Montù, M. & I. R. Oliveira. 1986. Zooplanktonic associations, trophic relations and standing stock of krill and other groups of the community near Elephant Island (February-March 84/85). *Nerita* **1**: 111-129.

Morris, D. J. & A. Keck. 1984. The time course of the moult cycle and growth of *Euphausia superba* in the laboratory. *Meeresforschung* **30**: 94-100.

Morris, D. J. & J. Priddle. 1984. Observations on the feeding and moulting of the Antarctic krill, *Euphausia superba* Dana, in winter. *Br. Antarct. Surv. Bull.* **65**: 57-63.

Morris, D. J. & C. Ricketts. 1984. Feeding of krill around South Georgia. I. A model of feeding activity in relation to depth and time of day. *Mar. Ecol.: Prog. Ser.* **16**: 1-7.

Morris, D. J., P. Ward & A. Clarke. 1983. Some aspects of feeding in the Antarctic krill, *Euphausia superba*. *Polar Biol.* **2**: 21-26.

Motoda, S. & T. Minoda. 1974. Plankton of the Bering Sea. *In* "Oceanology of the Bering Sea with Emphasis on Renewable Resources" (D. W. Hood & E. J. Kelly, eds.), pp. 207-241. Univ. of Alaska Press, Fairbanks.

Mujica, A. R. & V. V. Asencio, 1985. Fish Larvae, euphausiids and community structure of zooplankton in the Bransfield Strait (SIBEX-Phase I). 1984. *Ser. Cient. INACH* **33**: 131-154.

Mujica, A. R. & A. G. Torres, 1982. Qualitative and quantitative analysis of the Antarctic zooplankton. *Ser. Cient. INACH* **28**: 165-174.

Mullin, M. M. 1969. Production of zooplankton in the ocean: The present status and problems. *Oceanogr. Mar. Biol.* **7**: 293-314.

Nansen, F. 1902. The oceanography of the North Polar Basin. Norwegian North Polar Expedition, 1893-1896. *Scientific Results* **9**: 1-427.

Nast, F. 1978/79. The vertical distribution of larval and adult krill (*Euphausia superba* Dana) on a time station south of Elephant Island, South Shetlands. *Meeresforschung* **27**, 103-118.

Nast, F. 1982. The assessment of krill (*Euphausia superba* Dana) biomass from a net sampling programme. *Meeresforschung* **29**: 154-165.

―――. 1986. Changes in krill abundance and in other zooplankton relative to the Weddell-Scotia Confluence around Elephant Island in November 1983, November 1984 and March 1985. *Arch. Fischereiwiss.* **37**: 73-94.

Nast, F. & W. Gieskes. 1986. Phytoplankton observations relative to krill abundance around Elephant Island in November 1983. *Arch. Fischereiwiss.* **37**: 95-106.

Nast, F., K. H. Kock, D. Sahrhage, M. Stein & J. E. Tiedtke. 1988. Hydrography, krill and fish and their possible relationships around Elephant Island. *In* "Antarctic Ocean and Resources Variability" (D. Sahrhage, ed.), pp. 183-198. Springer-Verlag, Berlin.

Nemoto, T. 1967. Feeding pattern of euphausiids and differentiations in their body characters. *Inf. Bull. Planktol. Jpn:* 157-171.

Nesmelova, V. A. 1968. Quantitative dynamics of zooplankton in the Dal Nezelenetskaya region of the Barents Sea in 1964. *Tr. Murm. Morsk. Biol. Inst.* **17**: 22-29.

Niebauer, H. J. 1983. Multiyear sea ice variability in the eastern Bering Sea: An update. *J. Geophys. Res.* **88**: 2733-2742.

O'Brien, D. P. 1988. Direct observations of the behavior of *Euphausia superba* and *Euphausia crystallorophias* (Crustacea; Euphausiacea) under pack ice during the Antarctic spring of 1985. *J. Crustacean Biol.* **7**: 437-448.

Ommaney, F. D. 1936. *Rhincalanus gigas* (Brady), a copepod of the southern macroplankton. *Discovery Rep.* **13**, 277-384.

Ottestad, P. 1932. On the biology of some southern copepods. *Hvalradets Skr.* **5**: 1–61.
Parsons, T. R. & C. M. Lalli. 1988. Comparative oceanic ecology of the plankton communities of the subarctic Atlantic and Pacific oceans. *Oceanogr. Mar. Biol.* **26**: 317–359.
Pavshtiks, E. A. 1969. The effect of currents on seasonal variations in the zooplankton of Davis Strait. *Gidrobiol. Zh.* **5**: 58–63.
Pavshtiks, E. A. 1971. Seasonal variations in the number of zooplankton in the region of the North Pole. *Doklady Biol. Sci.* **96**: 59–62.
Pertsova, N. M. 1981. Number of generations and their span in *Pseudocalanus elongatus* (Copepoda, Calanoida) in the White Sea. *Zool. Zh.* **60**: 673–684.
Perueva, E. G. 1976. On feeding of IV copepodid state of *Calanus glacialis* Jaschnov from the White Sea on colonial alga *Chaetoceros crinitus* Schutt. *Okeanologiya (Moscow)* **16**: 1087–1091.
———. 1977a. Some experimental information on the feeding of IV copepodid stage of *Calanus glacialis* Jaschnov; qualitative composition of food. *Okeanologiya (Moscow)* **17**: 890–895.
———. 1977b. The diurnal feeding rhythmn of IV copepodite stage of *Calanus glacialis* Jaschnov. *Oceanology (Engl. Transl.)* **17**: 1085–1089.
———. 1983. Daily feeding rhythmn of *Metridia longa* (Copepoda: Crustacea) in the White Sea. *Oceanology (Engl. Transl.)* **23**: 100–103.
———. 1984. A comparison of the feeding of two abundant copepods of the White Sea. *Oceanology (Engl. Transl.)* **24**: 613–617.
Piatkowski, U. 1985. Distribution, abundance and diurnal migration of macrozooplankton in Antarctic surface waters. *Meeresforschung* **30**: 264–279.
———. 1987. Zoogeographische Untersuchungen und Gemeinschaftsanalysen an antarktischen Makroplankton. *Ber. Polarforsch.* **34**: 1–150.
Pires, A. M. S. 1986. Vertical distribution of euphausiid larvae (crustacea) in the Bransfield Strait during the First Brazilian Antarctic Expedition (Summer 1982/1983). *An. Acad. Brasil. Cienc.* **58**: 44–51.
Pirozhnikov, P. L. 1985. Brackish water calanoids of Kara and Laptev seas and pecularities of their distribution areas. *Biol. Morya* **5**: 64–66.
Poleck, T. & C. F. Denys. 1982. Effects of temperature on the molting, growth and maturation of the Antarctic krill *Euphausia superba* (Crustacea: Euphausiacea) under laboratory conditions. *Mar. Biol. (Berlin)* **70**: 255–265.
Price, H. J., K. R. Boyd. & C. M. Boyd 1988. Omnivorous feeding behavior of the Antarctic krill *Euphausia superba*. *Mar. Biol. (Berlin)* **97**: 67–77.
Priddle, J., J. P. Croxall, I. Everson, R. B. Heywood, E. J. Murphy, P. A. Prince & C. B. Sear. 1988. Large-scale fluctuations in distribution and abundance of krill—a discussion of possible causes. *In* "Antarctic Ocean and Resources Variability" (D. Sahrhage, ed.), pp. 169–182. Springer-Verlag, Berlin.
Prygunkova, R. V. 1985. Some causes of the interannual changes in the distribution of zooplankton in Kandalaksha Bay, White Sea. *Biol. Morya* **4**: 179–185.
Quetin, L. B. & R. M. Ross. 1984. School composition of the Antarctic krill *Euphausia superba* in the waters west of Antarctic Peninsula in the austral summer of 1982. *J. Crustacean Biol.* **4**: 96–106.
———. 1985. Feeding by Antarctic krill, *Euphausia superba:* Does size matter? *In* "Antarctic Nutrient Cycles and Food Webs" (W. R. Siegfried, P. R. Condy & R. M. Laws, eds.), pp. 372–377. Springer-Verlag, Berlin.
Rakusa-Suszczewski, S. 1984. Krill larvae in the Atlantic sector of the Southern Ocean during FIBEX 1981. *Polar Biol.* **3**: 141–147.
Reinhardt, S. B. & E. S. Van Vleet. 1986. Lipid composition of twenty-two species of Antarctic midwater zooplankton and fish. *Mar. Biol. (Berlin)* **91**: 149–159.

Rosenberg, A. A., J. R. Beddington & M. Basson. 1986. Growth and longevity of krill during the first decade of pelagic whaling. *Nature (London)* **324**: 152-154.
Runge, J. A. 1985. Egg production rates of *Calanus finmarchicus* in the sea off Nova Scotia. *Arch. Hydrobiol.* **21**: 33-40.
Runge, J. A. & R. G. Ingram. 1988. Under-ice grazing by planktonic, calanoid copepods in relation to a bloom of ice microalgae in southeastern Hudson Bay. *Limnol. Oceanogr.* **33**: 280-286.
Runge, J. A., I. A. McLaren, C. J. Corkett, R. N. Bohrer & J. A. Koslow. 1985. Molting rates and cohort development of *Calanus finmarchicus* and *C. glacialis* in the sea off southwest Nova Scotia. *Mar. Biol. (Berlin)* **86**: 241-246.
Sambrotto, R. N., J. J. Goering & C. P. McRoy. 1984. Large yearly production of phytoplankton in western Bering Strait. *Science* **225**: 1147-1150.
Sameoto, D. 1987. Vertical distribution and ecological significance of chaetognaths in the Arctic environment of Baffin Bay. *Polar Biol.* **7**: 317-328.
Sargent, J. R. & K. J. Whittle. 1981. Lipids and hydrocarbons in the marine food web. *In* "Analysis of Marine Ecosystems" (A. R. Longhurst, ed.), pp. 491-533. Academic Press, London.
Sargent, J. R., R. R. Gatten & R. McIntosh. 1977. Wax esters in the marine environment — their occurrence, formation, transformation and ultimate fates. *Mar. Chem.* **5**: 573-584.
Sargent, J. R., R. R. Gatten & R. J. Henderson. 1981. Lipid biochemistry of zooplankton from high latitudes. *Oceanis* **7**: 623-632.
Sars, G. O. 1900. Crustacea. Norwegian North Pole Expedition, 1893-1896. *Sci. Results* **5**: 1-141.
Schnack, S. B. 1985. Feeding by *Euphausia superba* and copepod species in response to varying concentrations of phytoplankton. *In* "Antarctic Nutrient Cycles and Food Webs" (W. R. Siegfried, P. R. Condy & R. M. Laws, eds.), pp. 311-323. Springer-Verlag, Berlin.
Schnack, S. B., S. Marschall & E. Mizdalski. 1985a. On the distribution of copepods and larvae of *Euphausia superba* in Antarctic waters during February 1982. *Meeresforschung* **30**: 251-263.
Schnack, S. B., V. Smetacek, B. von Bodungen & P. Stegmann. 1985b. Utilization of phytoplankton by copepods in Antarctic waters during spring. *In* "Marine Biology of Polar Regions and Effects of Stress on Marine Organisms" (J. S. Gray & M. E. Christiansen, eds.), pp. 65-81. J Wiley, Chichester.
Schneppenheim, R. & C. M. MacDonald. 1984. Genetic variation and population structure of krill (*Euphausia superba*) in the Atlantic sector of Antarctic waters and off the Antarctic Peninsula. *Polar Biol.* **3**: 19-28.
Seno, J., Y. Komaki & A. Takeda. 1963. Reports on the biology of the "Umitaka-Maru" expedition. Plankton collected by the "Umitaka-Maru" in the Antarctic and adjacent waters, with special references to copepoda. *J. Tokyo Univ. Fish.* **49**: 53-62.
Siegel, V. 1982. Investigations on krill (*Euphausia superba*) in the southern Weddell Sea. *Meeresforschung* **29**: 244-252.
———. 1986. Untersuchungen zur Biologie des antarktischen Krill, *Euphausia superba*, im Bereich der Bransfield Strasse und angrenzender Gebiete. *Mitt. Inst. Seefisch., Hamburg* **38**: 1-244.
———. 1987. Age and growth of Antarctic Euphausiacea (Crustacea) under natural conditions. *Mar. Biol. (Berlin)* **96**: 483-495.
———. 1988. A concept of seasonal variation of krill (*Euphausia superba*) distribution and abundance west of the Antarctic Peninsula. *In* "Antarctic Ocean and Resources Variability" (D. Sahrhage, ed.), pp. 219-230. Springer-Verlag, Berlin.

Simard, Y., G. Lacroix & L. Legendre. 1985. *In situ* twilight grazing rhythm during diel vertical migrations of a scattering layer of *Calanus finmarchicus*. *Limnol. Oceanogr.* **30:** 598–606.
Smith, S. L. 1988. Copepods in Fram Strait in summer: Distribution, feeding, and metabolism. *J. Mar. Res.* **46:** 145–181.
———. 1990. Egg production and feeding by copepods prior to the spring bloom of phytoplankton in the Fram Strait area of the Greenland Sea. *Mar. Biol. (Berlin)* **104** (in press).
Smith, S. L., W. O. Smith, Jr., L. A. Codispoti & D. L. Wilson. 1985. Biological observations in the marginal ice zone of the East Greenland Sea. *J. Mar. Res.* **43:** 693–717.
Smith, S. L. & J. Vidal. 1984. Spatial and temporal effects of salinity, temperature and chlorophyll on the communities of zooplankton in the southeastern Bering Sea. *J. Mar. Res.* **42:** 221–257.
———. 1986. Variations in the distribution, abundance, and development of copepods in the southeastern Bering Sea in 1980 and 1981. *Cont. Shelf Res.* **5:** 215–239.
Smith, S. L., J. Vidal & P. V. Z. Lane. 1983. "PROBES 1981 Zooplankton Data Report," Rep. PDR82-017. Univ. of Alaska, Fairbanks.
Smith, S. L., P. V. Z. Lane & E. M. Schwarting. 1986. "Zooplankton Data Report: The Marginal Ice Zone Experiment MIZEX, 1984." Tech. Rep. BNL 37972. Brookhaven Natl. Lab., Upton, New York.
Smith, S. L., P. V. Z. Lane, E. M. Schwarting & B. Beck. 1988. "Zooplankton Data Report: Winter MIZEX, 1987," Tech. Rep. BNL 42225. Brookhaven Natl. Lab., Upton, New York.
Soboleva, M. S. 1972. Distribution and abundance of euphausiids in the autumn winter period of 1971–1972 in the southern Barents Sea. *Annee Biol.* **29:** 40–41.
———. 1975. The distribution and abundance of euphausiids in the Barents Sea in autumn and winter 1972–1973. *Annee Biol.* **30:** 58–60.
Sømme, J. D. 1934. Animal plankton of the Norwegian coast waters and the open sea. *Fiskeridir. Skr., Ser. Havunders.* **4:** 1–163.
Springer, A. M. 1988. The paradox of pelagic food webs on the Bering–Chukchi continental shelf. Ph.D. Dissertation, Univ. of Alaska, Fairbanks.
Springer, A. M. & D. G. Roseneau. 1985. Copepod-based food webs: Auklets and oceanography in the Bering Sea. *Mar. Ecol.: Prog. Ser.* **21:** 229–237.
Springer, A. M., E. C. Murphy, D. G. Roseneau, C. P. McRoy & B. A. Cooper. 1987. The paradox of pelagic food webs in the northern Bering Sea. I. Seabird food habits. *Cont. Shelf Res.* **7:** 895–911.
Stein, M. & S. Rakusa-Suszczewski. 1984. Meso-scale structure of water masses and bottom topography as the basis for krill distribution in the SE Bransfield Strait February–March 1981. *Meeresforschung* **30:** 73–81.
Stepnik, R. 1982. All-year populational studies of Euphausiacea (Crustacea) in the Admiralty Bay (King George Island, South Shetland Islands Antarctic). *Pol. Polar Res.* **3:** 49–68.
Stretch, J. J., P. P. Hamner, W. M. Hamner, W. C. Michel, J. Cook & C. W. Sullivan. 1988. Foraging behavior of Antarctic krill *Euphausia superba* on the sea ice microalgae. *Mar. Ecol.: Prog. Ser.* **44:** 131–139.
Tande, K. S. 1982. Ecological investigations on the zooplankton community of Balsfjorden, northern Norway: Generation cycles, and variations in body weight and body content of carbon and nitrogen related to overwintering and reproduction in the copepod *Calanus finmarchicus* (Gunnerus). *J. Exp. Mar. Biol. Ecol.* **62:** 129–142.
Tande, K. S. & U. Båmstedt. 1985. Grazing rates of the copepods (*Calanus glacialis* and *C. finmarchicus* in Arctic waters of the Barents Sea. *Mar. Biol. (Berlin)* **87:** 251–258.

Tande, K. S. & S. Grønvik. 1983. Ecological investigations of the zooplankton community of Balsfjorden, northern Norway: Sex ratio and gonad maturation cycle in the copepod *Metridia longa* (Lubbock). *J. Exp. Mar. Biol. Ecol.* **71**: 43–54.

Tande, K. S. & C. C. E. Hopkins. 1981. Ecological investigations of the zooplankton community of Balsfjorden, northern Norway: The genital system in *Calanus finmarchicus* and the role of gonad development in overwintering strategy. *Mar. Biol. (Berlin)* **63**: 159–164.

Tande, K. S. & D. Slagstad. 1982. Ecological investigations on the zooplankton community of Balsfjorden, northern Norway. Seasonal and short-time variations in enzyme activity in copepodite stage V and VI males and females of *Calanus finmarchicus* (Gunnerus). *Sarsia* **67**: 63–68.

Tande, K. S., A. Hassel & D. Slagstad. 1985. Gonad and maturation and possible life cycle strategies in *Calanus finmarchicus* and *Calanus glacialis* in the northwestern part of the Barents Sea. *In* "Marine Biology of Polar Regions and Effects of Stress on Marine Organisms" (J. S. Gray & M. E. Christiansen, eds.), pp. 141–155. Wiley, Chichester.

Tanoue, E. 1985a. Organic chemical composition of fecal pellet of the krill *Euphausia superba* Dana. I. Lipid composition. *Trans. Tokyo Univ. Fish.* **6**: 125–134.

———. 1985b. Organic chemical composition of fecal pellet of the krill *Euphausia superba* Dana. II. Amino acid composition. *Trans. Tokyo Univ. Fish.* **6**: 135–138.

Thomas, P. G. & K. Green. 1988. Distribution of *Euphausia crystallorophias* within Prydz Bay and its importance to the inshore marine ecosystem. *Polar Biol.* **8**: 327–331.

Timofeev, S. F. 1987. Habitats of eggs and early larvae of Barents Sea euphausiids. *Okeanologiya (Moscow)* **27**: 237–238.

Totton, A. K. & H. E. Bargmann. 1965. "A Synopsis of the Siphonophora." Br. Mus. Nat. Hist., London.

Tranter, D. J. 1982. Interlinking of physical and biological processes in the Antarctic ocean. *Oceanogr. Mar. Biol.* **20**: 11–35.

Uribe, E. 1982. Influence of the phytoplankton and primary production of the Antarctic waters in relationship with the distribution and behavior of krill. *Ser. Cient. INACH* **28**: 147–163.

Ussing, H. H. 1938. The biology of some important plankton animals in the fjords of east Greenland. *Medd. Groenl.* **100**: 1–108.

Vervoort, W. 1965. Note on the biogeography and ecology of freeliving marine copepods. *In* "Biogeography and Ecology in Antarctica" (J. von Mieghem, P. von Oye & J. Schell, eds.), pp. 381–400. Junk Publ., The Hague.

Vidal, J. 1971. Taxonomy and distribution of the Arctic species of *Lucicutia* (Copepoda: Calanoida). *Bull. S. Calif. Acad. Sci.* **70**: 23–30.

Vidal, J. & S. L. Smith. 1986. Biomass, growth, and development of populations of herbivorous zooplankton in the southeastern Bering Sea during spring. *Deep-Sea Res.* **33**: 523–556.

Vidal, J. & T. E. Whitledge. 1982. Rates of metabolism of planktonic crustaceans as related to body weight and temperature of habitat. *J. Plankton Res.* **1**: 77–83.

Vinogradov, M. E. 1970. "Vertical Distribution of the Oceanic Zooplankton," Pp. 93–114 and 260–267. Israel Program Sci. Transl., Jerusalem.

Vinogradov, M. E. & A. G. Naumov. 1964. Quantitative distribution of plankton in Antarctic waters of the Indian and Pacific oceans. *Inf. Bull. Sov. Antarct. Exped.* **1**: 110–112.

Vladimirskaya, Y. V. 1975. Distribution of zooplankton in Scotia Sea in fall and winter. *Oceanology (Engl. Transl.)* **15**: 359–368.

———. 1978. Age composition of winter population of abundant copepod species in the southern part of the Scotia Sea. *Oceanology (Engl. Transl.)* **18**: 202–204.

von Bodungen, B. 1986. Phytoplankton growth and krill grazing during spring in the Bransfield Strait, Antarctica – Implications from sediment trap collections. *Polar Biol.* **6**: 153–160.

von Bodungen, B., E. M. Nöthig & Q. Sui. 1988. New production of phytoplankton and sedimentation during summer 1985 in the southeastern Weddell Sea. *Comp. Biochem. Physiol.* **90**: 475–487.
von Brökel, K. 1981. The importance of nanoplankton within the pelagic Antarctic ecosystem. *Kiel. Meeresforsch., Sonderh.* **5**: 61–67.
Voronina, N. M. 1968. The distribution of zooplankton in the Southern Ocean and its dependence on the circulation of water. *Sarsia* **34**: 277–284.
———. 1971. The annual cycle of plankton in the Antarctic. *In* "Fundamentals of the Biological Productivity of the Ocean and Its Exploration" (K. V. Beklemisev, ed.), pp. 64–71. Nauka, Moscow.
———. 1972a. The spatial structure of interzonal copepod populations in the Southern Ocean. *Mar. Biol. (Berlin)* **15**: 336–343.
———. 1972b. Vertical structure of a pelagic community in the Antarctic. *Oceanology (Engl. Transl.)* **12**: 415–420.
Ward, P. & D. B. Robins. 1987. The reproductive biology of *Euchaeta antarctica* Giesbrecht (Copepoda: Calanoida) at South Georgia. *J. Exp. Mar. Biol. Ecol.* **108**: 127–145.
Watkins, J. L. 1986. Variations in the size of Antarctic krill *Euphausia superba* Dana, in small swarms. *Mar. Ecol.: Prog. Ser.* **31**: 67–73.
Watkins, J. L., D. J. Morris, C. Ricketts & J. Priddle. 1986. Differences between swarms of Antarctic krill and some implications for sampling krill populations. *Mar. Biol. (Berlin)* **93**: 137–146.
Weber, L. H. & S. Z. El-Sayed. 1987. Contributions of the net, nano- and picoplankton to the phytoplankton standing crop and primary productivity in the Sourthern Ocean. *J. Plankton Res.* **9**: 973–994.
Weigmann-Haass, R. & G. Haass. 1980. Geographische Verbreitung und vertikale Verteilung der Euphausiacea (Crustacea) während der Antarktis-Expedition 1975/76. *Meeresforschung* **28**: 19–31.
Wiborg, K. F. 1954. Investigations on zooplankton in coastal and offshore waters of western and northwestern Norway. *Fiskeridir. Skr., Ser. Havunders.* **11**: 1–212.
———. 1976. Fishery and commercial exploitation of *Calanus finmarchicus* in Norway. *J. Cons., Cons. Int. Explor. Mer* **36**: 251–258.
Witek, Z. & W. Kittel. 1985. Larvae of the species of the genus *Euphausia* (Euphausiacea, Crustacea) in the southern part of Drake Passage and the Bransfield Strait during the BIOMASS–SIBEX (December 1983–January 1984). *Pol. Polar Res.* **6**: 117–132.
Witek, Z., A. Koronkiewicz & G. J. Soszka. 1980. Certain aspects of the early life history of krill *Euphausia superba* Dana (Crustacea). *Pol. Polar Res.* **1**: 97–115.
Witek, Z., J. Kalinowski, A. Grelowski & N. Wolnomiejski. 1981. Studies of aggregations of krill (*Euphausia superba*). *Meeresforschung* **28**: 228–243.
Witek, Z., M. Pastuszak & A. Grelowski. 1982. Net-phytoplankton abundance in western Antarctic and its relation to environmental conditions. *Meeresforschung* **29**: 166–180.
Witek, Z., W. Kittel, H. Czykieta, M. I. Zmijewska & E. Presler. 1985. Macrozooplankton in the southern Drake Passage and in the Bransfield Strait during BIOMASS–SIBEX (December 1983–January 1984). *Pol. Polar Res.* **6**: 95–225.
Yamada, S. & A. Kawamura. 1986. Some characteristics of the zooplankton distribution in the Prydz Bay region of the Indian sector of the Antarctic ocean in the summer of 1983/84. *Mem. Natl. Inst. Polar Res.* **44**: 86–95.
Zelikman, E. A., I. Lukashevich & S. Drobysheva. 1978. Aggregative distribution of *Thysanoessa inermis* (Krøyer) and *Thysanoessa raschi* (M. Sars) (Euphausiacea) in the Barents Sea. *Okeanologiya (Moscow)* **18**: 709–713.

———. 1979. Year-round vertical migrations of the euphausiids *Thysanoessa inermis* and *Thysanoessa raschi* in the Barents Sea. *Okeanologiya (Moscow)* **19**: 132–138.

Zelikman, E. A., I. Lukashevich, S. Drobysheva & A. Degtereva. 1980. Fluctuations of the quantities of eggs and larvae in the Barents Sea of the euphausiids *Thysanoessa inermis* and *Thysanoessa raschi*. *Okeanologiya (Moscow)* **20**: 1090–1097.

Zmijewska, M. I. 1985. Copepoda in the southern part of Drake Passage and in Bransfield Strait during early summer 1983–1984 (BIOMASS-SIBEX, December–January). *Pol. Polar Res.* **6**: 79–93.

Zwally, H. J., J. C. Comiso, C. L. Parkinson, W. J. Campbell, F. D. Carsey & P. Gloersen. 1983. "Antarctic Sea Ice, 1973–1973: Satellite Passive-Microwave Observations," NASA, Spec. Publ. 459. Natl. Aeron. Space Admin., Washington, D.C.

11 The Upper Trophic Levels in Polar Marine Ecosystems

David G. Ainley
Pt. Reyes Bird Observatory
Stinson Beach, California

Douglas P. DeMaster
National Marine Fisheries Service
La Jolla, California

I. Introduction 599
II. The Arctic and Antarctic Environments 600
III. Polar Communities and Food Webs 602
   A. Northern Hemisphere 603
   B. Southern Hemisphere 606
IV. Factors That Concentrate Upper-Level Trophic Interactions in Polar Waters 609
   A. Marginal Ice Zone 610
   B. Shelf-Related Fronts 613
   C. Insular Fronts 615
   D. Oceanic Fronts 616
   E. Freshwater Plumes 618
V. Competitive Interactions 620
VI. Summary and Future Directions 623
   References 625

I. Introduction

Marine food webs in polar regions are short, with often only two to three carbon transfers between diatoms and apex predators (Dunbar, 1981; Sanger, 1987). In the present contribution we will show how the mega- and mesoscale patterns of occurrence of the top trophic level consumers in polar regions reflect the physical environment as well as the biological characteristics of food webs. The apex predators with which we are concerned include baleen whales (Mysticetes), toothed whales (Odontocetes), seals (Phocidae), fur seals (Otariidae), walruses (Odobenidae), bears (Ursidae), and seabirds of several families, the important ones being penguins (Spheniscidae), petrels

(Procellariidae), skuas, gulls, and terns (Laridae), auks (Alcidae), and oldsquaw and eider ducks (Anatidae).

Recently, much effort has been directed toward relating occurrence patterns of top level consumers to the physical and biological characteristics of marine environments. Most of the work has been accomplished in polar, and especially Arctic, regions. The motivation behind this activity has been to learn about and perhaps predict the effects of mineral development on the biota of the wide continental shelves of the Arctic. Much less work has been completed in the Antarctic, where our understanding of oceanographic and trophic interactions is rudimentary at best. A large portion of this work has been on seabirds and has been reviewed from several perspectives (e.g., Nettleship *et al.,* 1984; Nettleship and Birkhead, 1986; Croxall, 1987). In contrast, analogous information on marine mammals is relatively scarce, although much work on mammals has been completed (Laws, 1984). This difference in the extent of our knowledge about the ecology of apex predators in marine systems is certainly a function of visibility and population size: seabirds offer a much better chance to quantify interactions with marine processes.

Our task here is to synthesize the information mentioned above from an oceanographic perspective rather than one of wildlife ecology and to add details from the most recent work.

II. The Arctic and Antarctic Environments

In polar seas a unique habitat feature, at least in relation to apex predators, is ice (Divoky, 1979, 1984). Otherwise, the oceanographic processes and habitat features that affect the trophic pathways to top predators are similar in polar, temperate, and tropical regions (Brown, 1980a). Sea ice, however, is not the same in the north and south. In the Arctic it is greatly consolidated except at its periphery, and in much of the Arctic basin it never melts. In the Antarctic it is very divergent with many internal leads, and during summer most of it melts except for several "refugia" near the Antarctic continent (Gilbert and Erickson, 1977; Zwally *et al.,* 1983). Important effects of ice in the north, thus, are negative ones: preventing access of predators to the water column and inhibiting light penetration and, therefore, the photosynthetic processes necessary for pelagic food webs (Divoky, 1979, 1984; Dunbar, 1981). The same would certainly be true for the Antarctic, though only during winter and early spring or for limited areas of permanent and persistent fast ice. A direct consequence of the more divergent Antarctic pack ice may be the observed richer biota that resides year round in the Antarctic compared to the Arctic. Another consequence of the difference between

Arctic and Antarctic ice conditions is the central importance of polynyas to pelagic marine organisms in the Arctic and their seeming lesser importance in the Antarctic (polynyas are areas within the ice pack that are almost always clear of ice; see below). In both polar regions positive effects of ice include the plankton blooms associated with pack ice margins (see Chapter 9), the provision of substrate from which or in which both micro- and macronekton seek refuge against predators, and a rich epontic community that provides important inputs to pelagic food webs (Bradstreet, 1988).

Another habitat feature that strongly affects the upper trophic level biota of polar seas is the physiographic setting, and in this respect the Arctic and Antarctic are completely opposite. In the north, with broad, relatively shallow continental shelves and the majority of the surface area ice covered year-round, the Arctic Ocean is relatively unproductive (Chapter 2). The Bering Sea provides a narrow connection to the Pacific basin, and the East Greenland and Labrador seas connect the Arctic with the Atlantic Basin. These connections provide important inputs of productive waters to small portions of the Arctic Ocean (see below). They also exert a major influence on the regional zoogeography of the Arctic marine biota; i.e., there is marked longitudinal as well as latitudinal change in faunal composition (King, 1964; Hunt and Nettleship, 1988).

In contrast, the Antarctic or Southern Ocean is deep and surrounds a land mass. Connection to the Pacific, Atlantic, and Indian basins is virtually continuous, and the fauna changes mainly only in the latitudinal dimension (King, 1964; Hunt and Nettleship, 1988; see below). Very importantly, the Southern Ocean is much more productive than the Arctic Ocean. It has several "embayments" or seas, the most oceanographically distinctive of which are the Ross and Weddell seas, but even these are broadly connected with adjacent waters.

We consider "polar seas" to be those influenced by sea ice. This leads to another distinction between the northern and southern polar oceans that is strongly reflected in the upper trophic level biota. It is related to the "horizontal," oceanographic complexity of surface waters and is partly a function of the much higher latitude of marine waters in the Arctic compared to the Antarctic, as well as factors such as the age and consequent low salinity of Arctic ice, the narrow connections to the major ocean basins, and, probably, the intense cold that overlies the region during winter as a result of land masses. In the north, waters influenced by sea ice include Polar Water as well as North Atlantic and Bering Sea water. Following the scheme of Nettleship and Evans (1986), Polar Water corresponds to the high Arctic zone, where water is unmixed and in August (the warmest month) has a surface temperature of $0°-3°C$ and a salinity $<31‰$. The low salinity is likely to be a function of the relative freshness of Arctic ice as well as the input of fresh water from very large river systems in several areas. North Atlantic Water that is influenced by ice is designated by Nettleship and Evans (1986) as low Arctic. In August, surface temperatures are 4° – 10°C and salinity ranges from 31 to 34‰. Ice even occurs, during winter, in the very northern edge of "boreal waters," i.e., waters that in August are 10° – 19°C and have salinities > 34‰. The situation is not as complex in the Bering Sea region. Besides the Polar Water in the Arctic basin, by and large sea ice influences only Bering Sea Shelf Water of the northern and eastern shelf; during August these waters reach 6°C with a salinity < 34‰. Thus, as is by now apparent, winter brings ice to surface waters in the northern hemisphere that during summer warm up appreciably and rapidly (see Preface). As a result, seasonally there are dramatic large-scale east – west and north – south migrations of many species (see below).

The large-scale oceanographic climate of "polar seas" in the southern hemisphere is far less complex (Foster, 1984). Sea ice covers only Antarctic Surface Water, i.e., water south of the Antarctic Convergence. These waters during February have temperatures ≤ 3°C and salinities of 33 – 35‰. Except in the Drake Passage area, only the northernmost extent of Antarctic Surface Water is not covered by sea ice during winter. The waters that are overlain by ice can be identified even during summer by their lower temperature and salinity and are known as "winter water." Surface waters in the shallow Ross and Weddell seas are very similar to those over greater depths, except that in the latter salinities never exceed 34.51‰, but in the shallow seas they can reach 35‰. Thus, compared to the Arctic, Antarctic seas are relatively homogeneous seasonally as well as spatially. Compared to the complex situation in the Arctic, there is little east – west, seasonal movement in the Antarctic biota, which itself is spatially and seasonally homogeneous. Any invasions of species may well have much more to do with the advance and retreat of ice rather than changed water-column characteristics (Ainley *et al.*, 1984; see below).

III. Polar Communities and Food Webs

In temperate or subpolar waters many of the fish and invertebrates that are the important links between primary consumers and higher levels of the food web are also species that are of great commercial importance, e.g., sardines (*Sardinops* spp.), anchovies (*Engraulis* spp.), and squid (*Loligo* spp.) of the eastern boundary currents; capelin (*Mallotus villosus*), herring (*Clupea harengus*), sandlance (*Ammodytes hexapterus*), and squid (*Illex* spp.) in the North Atlantic low Arctic zone; and walleye pollock (*Theragra chalcogramma*) and herring in the Bering Sea low Arctic (Belopol'skii, 1961;

Brown and Nettleship, 1984; Hunt et al., 1981; Lowry, 1984). Thus, a great deal is known about the life cycles, abundance, and distribution of these important species. In high Arctic and Antarctic waters, however, the important secondary consumers are relatively less well known, and studies of the top trophic level predators, used in a sense as sampling devices, have supplied much information about the natural history of their prey.

Most of what is known about Arctic cod (*Boreogadus saida*), a key species in Arctic food webs, has come through studies of predators, especially ringed seals (*Phoca hispida*) and thick-billed murres (*Uria lomvia*) (Bradstreet et al., 1986). In the Antarctic, studies of predators have provided important information about the natural history of the Antarctic krill (*Euphausia superba*) (Marr, 1962; Croxall et al., 1985; Ainley et al., 1986, 1988; Croxall and Prince, 1987; Daly and Macaulay, 1988), also a key species in the food web. Directed fisheries research has more recently yielded a great deal of information on krill biology (e.g., Everson, 1984a; George, 1984). Most of what is known about cephalopods in the Antarctic is the result of predator studies (e.g., Clarke, 1980). As for fish, little is known about the natural history of species that are important in pelagic and midwater neritic food webs of the Antarctic, but much is known about insular, demersal fish of lower latitudes, except for their role in food webs (e.g., Everson, 1984b).

A. Northern Hemisphere

In the Arctic Ocean two trophic webs have been identified, one associated with the shallow nearshore and the other with the pelagic offshore habitats (Fig. 11.1). Arctic cod, the pagophilic amphipod (*Apherusa glacialis*), euphausiids (*Thysanoessa* spp.), and copepods (*Calanus* spp.) are central to the pelagic web and are important in that of the nearshore, but in the latter, organisms of the epibenthos contribute substantially to energy flow. Arctic cod in their first year feed principally on copepods, but as juveniles and adults they switch more to amphipods such as *Parathemisto* sp. (Bradstreet et al., 1986). Although a number of mammals are components of the Arctic marine food webs year-round, especially polar bear (*Ursus maritimus*), ringed seal, and beared seal (*Erignathus barbatus*), only two avian predators are: black guillemot (*Cepphus grylle*) and ivory gull (*Pagophila eburnea*) (Stirling et al., 1981; Divoky, 1979; Brown and Nettleship, 1981). Polynyas are particularly important to development of these Arctic Ocean food webs (Stirling and Cleator, 1981; see below).

Several apex predators winter in large numbers in the marginal annual ice of the Bering, Labrador, East Greenland, and Barents seas and in Davis Strait. Where large polynyas occur, such as the one at St. Lawrence Island in the Bering Sea, these predators can occur within the pack ice as well. In-

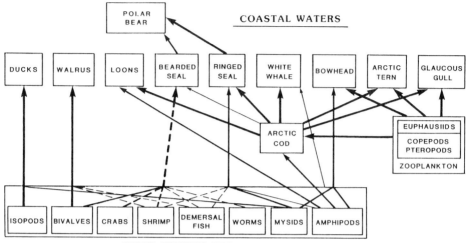
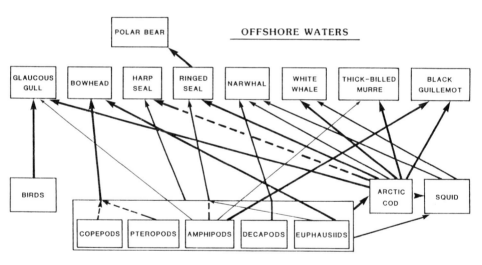

Figure 11.1 Summary of major energy flows in the upper trophic levels of coastal and offshore waters in the high Arctic. Data from Bradstreet and Cross (1982), Bradstreet et al. (1986), Divoky (1984), Frost and Lowry (1984.)

cluded among the mammals are the bowhead (*Balaena mysticetus*), narwhal (*Monodon monoceros*), white whale (*Delphinapterus leucas*), ribbon seal (*Histriophoca fasciata*), and walrus (*Odobenus rosmarus*) (King, 1964; Stirling et al., 1981). Major components of the avian fauna are the ivory gull, as well as glaucous, Iceland, slaty-backed, and Ross's gulls (*Larus glaucescens*,

L. leucopterus, L. schistisagus, and *Rhodostethia rosea*); common and especially thick-billed murres (*Uria aalge* and *U. lomvia*); and the dovekie (*Plautus alle*) (Divoky, 1979, also personal communication; Brown and Nettleship, 1981). At the Bering Sea ice edge walleye pollock, capelin, euphausiids, and a pelagic amphipod (*Parathemisto libellula*) are the mainstays of the midwater food web; walrus feed on mollusks of the benthos, and ribbon seals feed on demersal fishes (Divoky, 1979; Frost and Lowry, 1981; Lowry and Frost, 1981). Copepods, again, are important primary consumers. At high Arctic ice edges, Arctic cod, copepods (*Calanus* spp.), *Parathemisto,* and pagophilic amphipods are important in the intermediate trophic levels (Bradstreet and Cross, 1982). Farther south, e.g., in the Davis Strait and off Newfoundland, Arctic cod, capelin, and *P. libellula* are the major prey of upper-level predators (Bradstreet and Brown, 1986).

During summer a number of migrants increase the numbers of predators, but most do so only in ice-free waters (Fig. 11.2). Harp seals (*Pagophilus groenlandicus*) move north to feed in the epibenthos on the shelves of the eastern Canadian Arctic, as do gray whales (*Esrichtius robustus*) in the Bering region; fin whales (*Balaenoptera physalus*) and other baleen whales, as well as fur seals (*Callorhinus ursinus*) in the Bering region, frequent oceanic waters along continental slopes (King, 1964; Frost and Lowry, 1981; Harry and Hartley, 1981). Also moving into polar waters are many species of seabirds, the most abundant of which are oldsquaw (*Clangula hyemalis*) and eider ducks (*Somateria* spp.) in the nearshore (where they feed in the epibenthos), and shearwaters (*Puffinus* spp.; Bering region), northern fulmars (*Fulmarus glacialis*), kittiwakes (*Rissa tridactyla* and *R. brevirostris*), murres, auklets (*Aethia* spp., *Cychlorhynchus psittaculus*), and dovekies.

Arctic cod are central to the food web of high Arctic areas, but key prey vary regionally in low Arctic areas. For example, in the eastern Barents Sea during summer, Arctic cod is the main prey of seabirds, but in the western Barents Sea, which is influenced strongly by the warm North Atlantic Current at that time, sandlance, capelin, and herring are important food items to predators (Belopol'skii, 1961). In the Chukchi Sea, Bering Sea Shelf Water advected through the Bering Strait imports sandlance and capelin; in years when advection is weak, these important prey species do not appear (Springer *et al.,* 1984, 1987). The occurrence of certain predators in the Chukchi during summer is also a function of the strength of the advected warmer waters (Divoky, 1984). In the Labrador Sea and on the Grand Banks capelin are exceedingly important as food to the upper trophic level predators during summer (Brown and Nettleship, 1984). On the northwest Bering Sea shelf, because of strong advection of central Bering Sea water toward the Bering Strait, the food web is oceanic with euphausiids and large copepods predominating (Springer and Roseneau, 1985). On the southeastern Bering

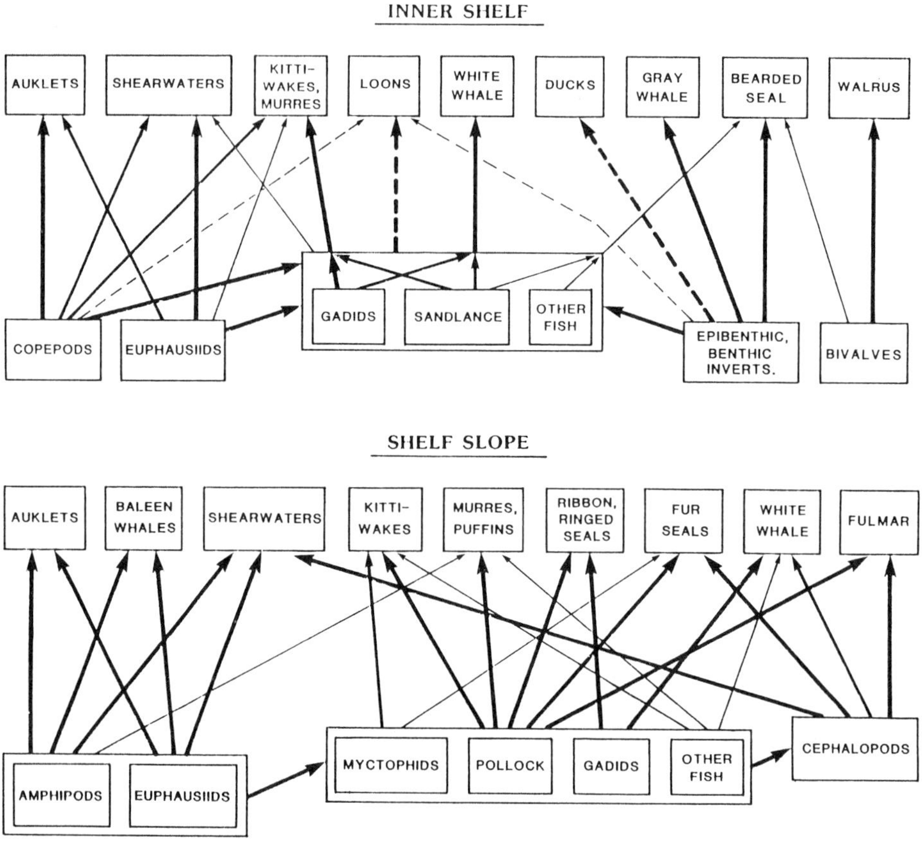

Figure 11.2 Summary of major energy flows in the upper trophic levels of ice-influenced, inner shelf and slope waters of the Bering Sea. Data from Divoky (1984), Hunt et al. (1981), Frost and Lowry (1981), Lowry and Frost (1981).

Sea shelf pollock are important to most pelagic predators; myctophids and euphausiids are important to some (Hunt et al., 1981).

B. Southern Hemisphere

In the Antarctic, continental shelves are unusually deep and, thus, there is no nearshore food web analogous to that of the Arctic. Nevertheless, epipelagic food webs in the "shallow" Ross and Weddell seas differ from those of deeper, oceanic water.

Compared to the Arctic, a much larger, more diverse biota overwinters in the Antarctic due to the relatively less severe weather over the ocean and the more divergent pack ice. The Antarctic organisms that do overwinter are truly pagophilic; most retreat in summer to remain with the pack ice that resides in several refugia, the most important of which are those of the Ross and Weddell seas. Key species of the intermediate levels of the food web over the shelves include various crustacea and other small organisms (Hopkins, 1987), and above them in the food web are a euphausiid (*E. crystallorophias*), the silverfish (*Pleuragramma antarcticum*), some pagotheniid fishes, and a squid (*Psychroteuthis glacialis*) (Ainley *et al.*, 1984; Offredo and Ridoux, 1986; Plötz, 1986; Thomas and Green, 1988; Ridoux and Offredo, 1989).

Permanently resident predators of the highest trophic level in the pack ice over the shelves include the minke whale (*Balaenoptera acutorostrata*), four seals (crabeater, *Lobodon carcinophagus;* Ross, *Ommatophoca rossi;* Weddell, *Leptonychotes weddelli;* and leopard, *Hydrurga leptonyx*), and five birds (emperor and Adélie penguin, *Aptenodytes forsteri* and *Pygoscelis adelia;* snow and Antarctic petrel, *Pagodroma nivea* and *Thalassoica antarctica;* and south polar skua, *Catharacta maccormicki;* Fig. 11.3). During winter at the pack ice edge, which occurs over deep, oceanic waters hundreds of kilometers north of the continent, the predators characteristic of the shelves are present and so are beaked whales (Ziphiidae), Antarctic fur seal (*Arctocephalus gazella*) and elephant seal (*Mirounga leonian*), chinstrap penguin (*Pygoscelis antarctica*), and several species of petrels (Procellariidae). In these waters, *E. superba,* myctophid fishes, pasiphaeid shrimp, and two squid species (*Gonatus antarcticus* and *Galiteuthis glacialis*) are important at intermediate levels in the food web (Ainley *et al.*, 1984, 1986, 1988; also D. G. Ainley and W. R. Fraser unpublished; Laws, 1984; Siniff and Stone, 1985). Within the micronekton and macrozooplankton, the ecological role of mesopelagic fish (many of which rise to the surface) has until recently been underestimated. Lancraft *et al.* (1989) estimate that mesopelagic fish are perhaps the most important consumers of krill in oceanic waters.

During summer, as the pack ice retreats, Antarctic surface waters are invaded by a host of predators that feed on the abundant micronekton and macrozooplankton. These predators, including many species that resided at the ice edge during winter, move as far south as the pack ice will allow, sometimes right to the Antarctic continent. Included among the ecologically most important predators are several large baleen whales (Balaenopteridae and Balaenidae); male sperm whales (*Pyseter macrocephalus*), which feed in the mesopelagic zone; killer whales (*Orcinus orca*); and about 10 species of seabird, almost all of the order Procellariiformes (Laws, 1977; Ainley *et al.*, 1984). The same prey listed above for oceanic, epipelagic waters remain numerically important during the summer. Unlike most other predators, which feed on larger micronekton, right whales (*Eubalaena australis*),

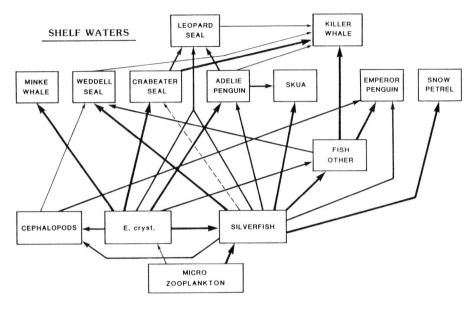

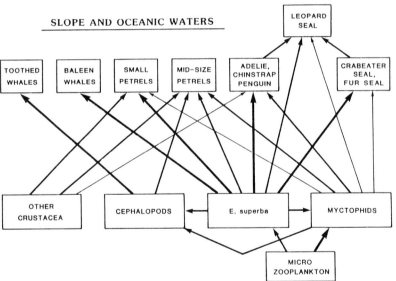

Figure 11.3 Summary of major energy flows in the upper trophic levels of ice-influenced, shelf and slope/oceanic waters in the Antarctic. Data from Ainley et al. (1984), D. G. Ainley and W. R. Fraser (unpublished), DeWitt and Hopkins (1977), Hopkins (1987), Nemoto et al. (1985), Offredo and Ridoux (1986), Siniff and Stone (1985), Williams (1985.)

prions (*Pachyptila* spp.), and diving petrels (*Pelecanoides* spp.) feed heavily on small copepods as well (Ainley et al., 1984; Croxall, 1984; S. G. Brown and Lockyer, 1984).

In oceanic waters during summer, myctophids, *E. superba,* and the squid species mentioned above are important prey of top trophic level predators (Ainley et al., 1984; Siniff and Stone, 1985; Ridoux and Offredo, 1989; D. G. Ainley, W. R. Fraser, and C. A. Ribic, unpublished). At shelf breaks the prevalence of *E. superba* increases in diets, and over the continental shelves the important intermediate links in the food web are the species mentioned above for the winter. For most food webs of the Antarctic pelagic zone, copepods are important primary consumers, but euphausiids act as primary consumers as well, especially during summer (Hopkins and Torres, 1989; Fig. 11.3).

IV. Factors That Concentrate Upper-Level Trophic Interactions in Polar Waters

Brown (1980b, 1986) has pointed out that upper trophic level predators need to forage where prey can be captured efficiently. Such a requirement is particularly true during the parts of the year when energy demands on predators are high, e.g., when feeding young or before or during migration. Observations on predators foraging in and out of areas where prey are dense are all too few. Brown has repeatedly cited two examples. First, Brodie *et al.* (1978) demonstrated that in order to forage efficiently, fin whales off Nova Scotia sought areas where the density of euphausiids was 175 times greater than average densities in surrounding waters. Second, Zelikman and Golovkin (1972) discovered that 76% of the dovekies in the vicinity of Novaya Zemlya foraged where copepod concentrations were 10 times the average in those waters. Brown (1986) proposed that the fall, winter, and early spring movements of dovekies, as well as thick-billed murres, can be viewed as a series of "jumps" from one area to another where copepods and capelin, respectively, become seasonally abundant. Thus, highly visible predators such as seabirds and large marine mammals can serve to demonstrate areas in the ocean that not only are highly productive but also have concentrations of organisms of the intermediate trophic levels. The movements of these predators can perhaps provide a better view of the temporal and spatial changes in secondary and tertiary production than is possible through expensive fisheries research, particularly for large marine systems. The following is a review of the kinds of mesoscale marine processes and phenomena where trophic interactions and energy/carbon flux involving the upper levels of the pelagic food web are maximal.

A. Marginal Ice Zone

1. Large-scale ice edges

For a number of years, ice edges have been known as productive regions but the responsible mechanisms have been determined only recently (Dunbar, 1981; Niebauer and Alexander, 1985; Smith and Nelson, 1985; Wilson et al., 1986). Smith et al. (1988) have shown that productivity at the ice edge over large distances contributes significantly to the annual production budget of Antarctic Surface Water. Quetin and Ross (1984) have hypothesized that enhanced productivity at such localities as ice edges and shelf-break fronts is necessary for successful reproduction by krill. Additional work has identified the Antarctic marginal ice zone as a nursery area for krill and other nekton (Ainley et al., 1988; Daly and Macaulay, 1988; L. Quetin and R. Ross, personal communication).

Marr (1962), Laws (1977), and most recently Mizroch et al. (1986) noted that some species of baleen whales concentrate at the Antarctic marginal ice zone and follow its retreat southward. These whales migrate in order to maintain themselves in areas where their crustacean prey are concentrated. Interestingly, the right whale, which feeds more on tiny copepods and may not require the dense concentrations of prey needed by balaenopterid whales for efficient foraging, does not move south with the ice but remains well to the north (Nemoto, 1970; S. G. Brown and Lockyer, 1984). Pinnipeds are also attracted to the pack ice edge (Ribic et al., 1990), as are seabirds (Ainley and Jacobs, 1981; Ainley et al., 1984; Fraser and Ainley, 1986; Fig. 11.4). In Figure 11.4 a series of parameter peaks in the vicinity of the ice edge correspond to bands of ice separated by bands of open water. Erickson et al. (1971) described the affinity of different seal species for particular ice floe characteristics (percent ice cover, floe size, floe age) and then estimated seal abundances by the distribution of ice types. Revised estimates, which include the now-recognized importance of the marginal ice zone and of shelf-break fronts to the distribution of seals, would be of interest. In the Arctic, seals seem to segregate by only one characteristic of sea ice (whether it is fast or pack ice), and this pattern may be largely influenced by predation by polar bears and the absence of predator–prey interactions among the seal species themselves. In the Antarctic there are no polar bears (nor any ecological equivalent), but leopard seals are important predators on other seals (Siniff and Bengtson, 1977).

One species of seabird, the Arctic tern (*Sterna paradisaea*), migrates annually to the edge of the Antarctic pack ice, where it finds prey to be so abundant that it becomes almost flightless during a month-long period of molting and replacement of its feathers. (Antarctic species, such as penguins, also molt at the edge of the Antarctic ice pack; Ainley et al., 1984.) Baleen

11 Upper Trophic Levels in Polar Marine Ecosystems 611

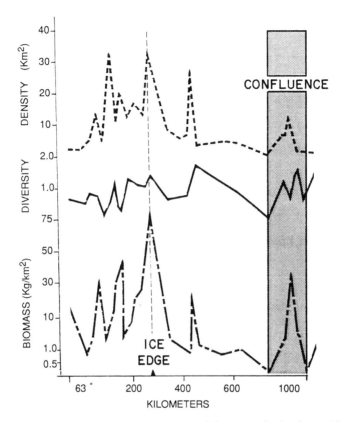

Figure 11.4 A section in the Weddell Sea sector of the Antarctic showing seabird density, species diversity, and biomass at the edge of the pack ice as well as the Scotia-Weddell Confluence. Note change in horizontal scale. From Fraser and Ainley (1986) by permission of the authors.

whales, and especially the blue whale (*Balaenoptera musculus*), also undertake long, annual migrations to reach the Antarctic ice edge, where they feed and accumulate fat reserves for sustenance the rest of the year. The smallest baleen whale, the minke (*Balaenoptera acutorostrata*), has lower fat reserves than its congeners, and it exhibits the closest association with the pack ice edge than any other whale (Laws, 1977). This perception, though, may change when more whale surveys are completed in the pack ice (Ribic et al. 1990).

The marginal ice zone in the Arctic may be relatively more important to regional production budgets than that in the Antarctic. Many species of seabirds and mammals use Arctic ice edges as migration routes during spring (Divoky, 1979; Brown, 1980a, 1986; Stirling *et al.*, 1981). Without the reliable availability of prey, these migrations would probably fail. The breeding efforts of Arctic seabirds would fail as well (Bradstreet, 1988). Divoky (1979) estimated the density of seabirds within the marginal ice zone of the Bering Sea during winter to be 500 individuals km^{-2}, compared to 0.1 birds km^{-2} in the adjacent ice to the north and 10 birds km^{-2} in the adjacent open water. Species that feed by diving for their prey, which is the most energy-demanding type of seabird feeding behavior, constituted over 90% of the marginal ice zone avifauna.

2. Polynyas

Dunbar (1981) and Bradstreet (1988) have pointed out that predators and their prey are attracted to polynyas, not just because of the physical existence of open water but also because of the enhanced productivity at their ice edges. The distribution of marine mammals in the high Arctic zone, particularly in the winter, is, in fact, determined by where recurring polynyas exist (Stirling *et al.*, 1981; Bradstreet, 1982). Polynyas and their ice edges are also critically important to seabirds. At least in the Canadian Arctic, every major seabird rookery, except one, occurs adjacent to or within bird flight range of a recurring polynya or of persistently unconsolidated pack ice that usually occurs around polynya boundaries (Brown and Nettleship, 1981). Judging from the comments of Divoky (1979), the same can be said of seabird colonies in the western North American Arctic, although a much greater limitation on the availability of nesting cliffs renders polynyas relatively less important in this regard. Whether or not the ice at the edge of Arctic polynyas is consolidated or dispersed can bear importantly on how attractive it is to predators (Bradstreet, 1982).

Polynyas are apparently of only limited importance to Antarctic seabirds, in part because many species nest far from the ocean anyway (i.e. 100 km or more; Broady *et al.* 1989). The apparently minimal role of polynyas also may be the result of limited scientific work in the Antarctic during winter and early spring. A few studies have indicated that Antarctic polynyas do affect the occurrence patterns of some birds, but no observations are available on mammals (Stonehouse, 1967; Parmelee *et al.*, 1977). Ainley *et al.* (1990) analyzed the demography of south polar skuas at Cape Crozier, Ross Island, and, after finding a low emmigration but a high immigration rate of skuas to the population, hypothesized that skuas were attracted to the site by the large polynya that forms annually adjacent to Cape Crozier. The polynya, described by Jacobs and Comiso (1989) on the basis of satellite imagery, provides open water and access to food for surface-feeding species, like the skua, weeks before waters are free of ice adjacent to other skuaries in the Ross Sea. Jacobs and Comiso (1989) also described an annually persistent loosening of ice, in a sense a polynya, along the continental slope of the Ross Sea during winter. The shelf break front is one of the biologically richest areas of the Ross Sea region (see next section, Fig. 11.5). The existence of open water there during winter would provide the habitat in which the immense penguin and seal populations of the Ross Sea could over-winter. Otherwise, the pack ice edge would be hundreds of kilometers farther north.

B. Shelf-Related Fronts

1. Shelf-break fronts

In the Antarctic the continental shelf-break front is exceedingly important in pelagic biological production and carbon flux to the upper trophic levels. Little attention, however, has been given to Antarctic shelf-break fronts (Foster, 1984). Ainley and Jacobs (1981) and Jacobs (1988) described the front physically, and Ainley (1985) and Veit and Braun (1984) have shown the marked concentration of predators along the shelf break of the Ross Sea (Fig. 11.5). In turn, the proximity of land to the front has contributed to the immense concentrations of seabird breeding populations at the eastern and western boundaries of the Ross Sea (Ainley *et al.*, 1984). Various sources of information indicate the biological importance of the shelf-break front elsewhere in the Antarctic. The concentrations of predators described by Cline *et al.* (1969) in the pack ice of the Weddell Sea coincide with the shelf break there. A similar concentration of predators coincides with the shelf break along the western boundary of the Antarctic Peninsula (D. G. Ainley and W. R. Fraser, unpublished), and crabeater seals appear to concentrate within the pack ice immediately seaward of the shelf break in the Indian Ocean (K. R. Kerry and P. Ensor, personal communication).

In the Arctic Ocean continental shelf-break fronts seem to be of less biological importance due to the shallow topography and sluggish circulation. In the low Arctic zone, however, such fronts are of measurable importance but they have not been well investigated. Kinder and Coachman (1978) described the shelf-break front in the Bering Sea, and Iverson *et al.* (1979) discussed the elevated carbon flux associated with it. The latter authors' attention with regard to highest trophic levels was directed toward seabirds, but mammals are also known to concentrate along the shelf break, presumably because of enhanced feeding opportunities (Nasu, 1974; Harry and Harltey, 1981).

Off Newfoundland and particularly in the Grand Banks area, copepod-feeding dovekies concentrate in winter along the southern edge of the bank, which rises directly across the flow of the North Atlantic Current; on the eastern, downstream boundary of the bank, few birds congregate (Brown, 1980b). Capelin is the main prey of seabirds and other predators in the

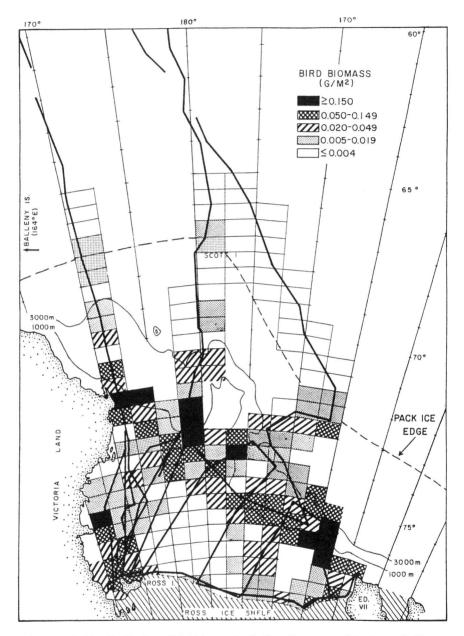

Figure 11.5 The distribution of bird biomass in the Ross Sea sector of the Antarctic. From Ainley (1985) by permission of the author.

southern Labrador Current area of eastern Canada (Brown and Nettleship, 1984). These fish are residents of the continental slope, and it is during their annual spawning migrations upslope and inshore that they become critical to the energy budgets of many predators (see also Carscadden, 1984). In eastern Canada, where the continental shelf is narrow and the rich shelf-break front is close to shore, the largest breeding concentrations of seabirds in Canada occur (Brown, 1986). The shelf break in this area is also the area where cetaceans occur in high densities (Sutcliffe and Brodie, 1977).

2. Midshelf fronts

Iverson et al. (1979) described a series of three fronts on the southeastern Bering Sea shelf. These fronts partition the shelf into inner and middle shelf domains and, except for the outermost, shelf-break front (discussed above), they tend to occur when pack ice does not overlie surface waters. Hunt et al. (1981), Schneider and Hunt (1982), and Schneider et al. (1987) noted a 50% higher carbon flux to pelagic predators over the slope and outer shelf compared to the middle shelf region. The strong midshelf front restricts oceanic herbivores to the outer shelf region and helps to account for the elevated carbon flux to higher trophic levels in the outer shelf zone (Iverson et al., 1979). In the middle and inner shelf zones, much of the phytoplankton biomass settles to the benthos, where the food web is much more productive than in surface waters. On the inner shelf occur two predators unique among related taxa because both feed in the rich substrate of the shelf benthos: a baleen whale, the gray whale, and a pinniped, the walrus.

In the northern Bering Sea a different frontal system appears to exist and is a product of the juxtaposition of water masses that flow from different origins. Shelf-break and slope waters are advected toward the Bering Strait, bringing a rich food web dominated by large, oceanic copepods across the northern Bering Sea shelf (Springer and Roseneau, 1985; Springer et al., 1987). As a result, predator species found only in the outer shelf of the southeastern Bering Sea are found over the inner shelf in the north.

In the Antarctic, there has been little if any work that has identified or investigated midshelf fronts. The much deeper continental shelves may work against the formation of such fronts.

C. Insular Fronts

Fronts associated with islands enhance production and trophic transfer rates in local areas (Wolanski and Hamner, 1988). Some of these fronts are analogous to large-scale counterparts, e.g., continental shelf- and insular shelf-break fronts.

Kinder et al. (1983) described a front coincident with the 50-m isobath that encircles each of the Pribilof Islands in the Bering Sea. The front separated vertically homogeneous and colder waters inshore from stratified waters offshore. Diving seabirds were particularly associated with the fronts, and they were thought to contribute substantially to the energy/carbon budgets of the Pribilof seabird communities. Springer and Roseneau (1985) considered these to be upwelling fronts and described similar features around other Bering Sea islands (St. Matthew and St. Lawrence) that are located on the Bering Sea continental shelf. They noted that large copepods are abundant in the insular upwelling fronts and — because fin whales and pollock, which are major copepod predators, are absent from the northern Bering Sea shelf — auklets, which also prey on copepods, are exceedingly dense around the northern Bering Sea islands. G. L. Hunt and N. M. Harrison (unpublished) have also investigated the fronts around St. Lawrence Island and have noted the role of water column stratification in concentrating prey within the frontal zones.

An insular, upwelling front likely occurs at South Georgia in the Scotia Sea. Croxall (1984) and Croxall et al. (1985) note that penguins and fur seals from Bird Island, an adjacent islet, feed principally on euphausiids off the northwest coast and shelf break of South Georgia (Fig. 11.6). This would be the downstream end of the island, where upwelling would be expected, or perhaps the enhanced feeding opportunity indicated is a function of an island "wake" phenomenon (Wolanski and Hamner, 1988). The latter phenomenon has been proposed by Abrams and Miller (1988) to explain enhanced trophic interactions near Gough Island, an oceanic island in the subantarctic Atlantic. Hunt et al. (1988) provide greater descriptive detail of the situation at South Georgia and note that in a southerly direction from Trinity Island, which is adjacent to Bird Island, predators and prey are inshore over the shelf rather than concentrating along the shelf break. Thus, in the vicinity of South Georgia it is not yet clear what the oceanographic mechanisms may be that concentrate nekton in certain areas. Increased avian biomass has been reported over the shelf break around the South Orkney Islands, also in the Scotia Sea (Hunt and Veit, 1983), as well as near King George Island in the Drake Passage (Veit and Braun, 1984).

D. Oceanic Fronts

Where two water masses or currents meet, trophic transfer rates to upper levels of the food web are often enhanced substantially by the entrapment and concentration of plankton. Nettleship and Evans (1986) note that the northern edge of the North Atlantic Current is particularly important to the natural history of alcids in the North Atlantic. These diving seabirds depend on the plentiful food found along this boundary during summer if breeding

11 Upper Trophic Levels in Polar Marine Ecosystems 617

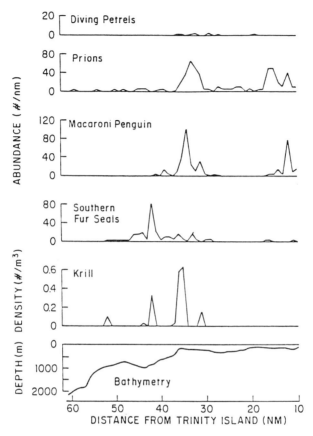

Figure 11.6 Bathymetry and the abundance of krill, fur seals, and birds along a section northward from Trinity Island, South Georgia. From Hunt et al. (1988) by permission of the authors.

sites are close by, as in Newfoundland and Iceland, and especially during winter, when birds are no longer constrained to remain near breeding sites. Brown (1980a) discussed briefly the high densities of seabirds and cetaceans that occur where a branch of the North Atlantic Current, the West Greenland Current, moves alongside the cold Labrador Current. The predators involved were attracted by the high abundance of gonatid squid and euphausiids. Convergences can also be important on a smaller scale. Brown (1980b) demonstrated that murres tend to concentrate at water-type boundaries in Hudson Strait. G. L. Hunt and N. M. Harrison (unpublished) discuss the attraction of planktivorous auklets to convergences in the northern Bering Sea. They propose that these convergences are most attractive if one water mass is highly stratified, thereby trapping prey near the surface.

In the Antarctic, convergences appear to contribute less to the enhancement of trophic transfer on the large scale. There is a difference of opinion on whether biological activity is enhanced significantly at the Antarctic Convergence, where cold and saline Antarctic Surface Water sinks beneath Subantarctic Surface Water. The perception of enhancement may be relative to the background productivity of various regions. Ainley and Boekelheide (1983) and Ainley et al. (1984) found little change in seabird biomass at the convergence in the southwestern Pacific, and there is no biomass maximum in the Drake Passage (Brown et al., 1975a; D. G. Ainley, unpublished). In contrast, Griffiths et al. (1982) detected a peak in seabird biomass at the convergence in the African sector of the Antarctic. In the areas surveyed by Ainley and co-workers and Brown et al., the oceanography may be much more complex, with a number of feeding opportunities attracting the attention of predators. In the southwest Pacific, upper trophic level predators concentrate in the upwelling zone along the edge of the Campbell Plateau (a subsea feature), which is only a few hundred kilometers north of the convergence; in the narrow Drake Passage, the South American and Antarctic slope fronts are areas where predator biomass is very high. South of Africa, such features do not occur near the convergence.

On a smaller scale, the convergence between Scotia and Weddell sea surface water is an area where a significant peak in the biomass of euphausiids and other nekton, as well as seabirds and fur seals, has been observed (Brinton, 1984; Ainley et al., 1986; Fraser and Ainley, 1986; also D. G. Ainley and W. R. Fraser unpublished). The peak in seabird biomass, however, when observed in November 1983, was much smaller than that in the marginal ice zone 200 km farther south (Fig. 11.4).

E. Freshwater Plumes

Though insignificant in the Antarctic, where relatively little seasonal melt of terrestrial ice and snow occurs, some of the largest river systems on earth drain into the Arctic Ocean and Bering Sea. During spring the relatively warm fresh water encourages early disintegration of sea ice near river mouths. This early appearance of open water alone is attractive to migrating marine mammals and birds. Additional factors make freshwater plumes attractive in other ways. During summer, when runoff is at its peak, the boundaries of freshwater plumes in the Arctic Ocean become areas where biological production is enhanced. The rivers bring nutrients and organic matter to the coast, and stability of the water column at the salinity front encourages phytoplankton growth (Parsons et al., 1988, 1989). Two food webs actually develop, one based on heterotrophic production within the plume and the other based on autotrophic, oceanic production at the plume boundary. Arctic cod are especially attracted to freshwater plumes during the summer (Bradstreet *et al.*, 1986) and thus also are various predators such as birds, seals, and cetaceans (Divoky, 1979, 1984; Nettleship and Gaston, 1978; Bradstreet *et al.*, 1987).

In the case of large freshwater features such as the Mackenzie River plume in the Beaufort Sea, the abundance of micronekton appeas to be enhanced by coastal upwelling at nearby capes and islands, but it is then the marked discontinuity between plume and Arctic marine waters which concentrates the prey sufficiently to attract bowhead whales (Borstad, 1985; Bradstreet *et al.*, 1987; Bradstreet and Fissel, 1987). Figure 11.7 shows a typical occurrence pattern of whales which are attracted to the plume boundaries and are largely absent from vast stretches of adjacent Arctic Waters. Similar large-scale plumes occur in the Asian Arctic, and the same phenomenon might well be expected to exist, although predators other than the bowhead would be involved.

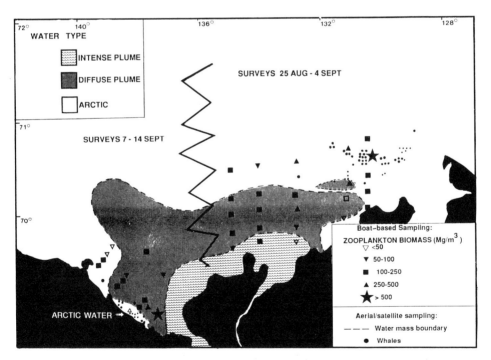

Figure 11.7 Zooplankton biomass, whale distribution, and the location of water types (intense and diffuse plume water from the Mackenzie River, and Arctic Water) in the Beaufort Sea, August–September 1986 (aerial transects for whales extend from the coast to north of 71°). From Bradstreet *et al.* (1986) by permission of the authors.

V. Competitive interactions

We mentioned above the work of Springer and Roseneau (1985), who hypothesized that the high abundance of auklets around islands on the northern Bering Sea shelf is a result of the lack of competition from other slope-water copepod predators. One predator thus far not mentioned is man, whose commercial exploitation of marine resources in polar regions has significantly affected local food webs. We are not concerned here with the usually insignificant effects of aboriginal harvests on various marine mammal populations.

Sherman and Alexander (1986) classify large marine ecosystems into two general types. The first type includes relatively closed systems, usually shallow and enclosed by land masses, in which predation is a major structuring factor. The Bering Sea, North Sea, and Grand Banks are given by Sherman and Alexander (1986) as examples; most other low and high Arctic regions we have mentioned here are likely similar. The other major type of system is structured by oceanic processes; primary production is exceedingly high and often not entirely utilized in the pelagic food web; the population dynamics of the nekton are driven by production. Eastern boundary currents are the main examples offered by Sherman and Alexander (1986). Not enough work has been done in the Antarctic to classify its pelagic ecosystems in this scheme. The high productivity and the very short trophic links (phytoplankton – krill – top predator) argue for similarity to eastern boundary currents.

In the Arctic, commercial exploitation of organisms intermediate in food webs has had major impacts on the populations of top trophic level predators both off Newfoundland and in the Bering Sea. The resultant changes in populations are indicative of the closed, predation-dominated structuring proposed by Sherman and Alexander (1986). On the northeastern Grand Banks, seabirds consume an estimated 250,000 tons of capelin annually, a figure that is of similar magnitude to the take by seals and whales but is 10% that of cod (Brown and Nettleship, 1984). The proportion of the capelin biomass taken by all natural predators is unknown, but of a total 10.3 million tons taken off eastern Newfoundland in 1976, the fishery took an estimated 3.5% and natural predators 37% (seabirds, 2.4%; Winters and Carscadden, 1978; Brown and Nettleship, 1984). The seabirds cannot dive as deeply as other predators and thus the proportion of capelin biomass available to them, and which will allow efficient foraging during the nesting season, is limited. For seabirds, the critical quantity is likely not to be the total prey biomass but, rather, the density within range of their nesting colonies (i.e., for nesting species) that allows successful reproduction. The offshore capelin fishery expanded rapidly during the 1970s and early 1980s but was terminated because of its effects on capelin recruitment. At that time, however, an inshore fishery for roe was further developed. Comparing reproductive success of seabirds in Witless Bay, Newfoundland, between the late 1960s and the early 1980s, Brown and Nettleship (1984) noted a relationship between success and the proportion of capelin in the diet, with success in the later period being very low. They expressed a need for caution and a need to monitor closely the fishery and its effects on the food web.

At Bluff in Norton Sound, Alaska, just south of the Bering Strait, the common murre population declined from the mid-1970s through the 1980s (Murphy *et al.*, 1986). The birds feed on saffron cod (*Eleginus gracilis*) and sandlance during nesting. Except for a period of cold water in the mid-1970s, which led to temporarily reduced prey biomass and reduced nesting success (Springer *et al.*, 1984, 1987), success has remained sufficiently high during the period that it cannot be the cause of the population decline. Murphy *et al.* (1986) point out, however, that the population decline in murres tracks the decline in catch per unit effort (CPUE) of the pollock fishery in the southern Bering Sea, the area where the murres likely spend their fall-to-spring nonbreeding period (Fig. 11.8). They hypothesize that the pollock fishery may be

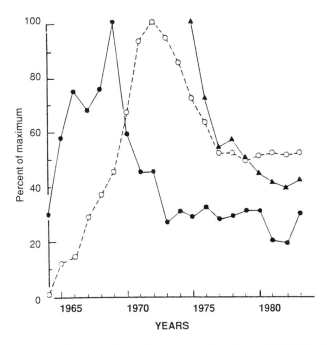

Figure 11.8 Murre numbers at Bluff, Alaska, and annual catches and CPUE of pollock in the southeastern Bering Sea, expressed as percentages of maximum values. (O) Pollock catch; (●) pollock CPUE; (▲) murre numbers. From Murphy *et al.* (1986) by permission of the authors.

affecting the survival of subadult murres. Interestingly, Wespestad and Terry (1984) have proposed that the fish stock can sustain even higher yields than current ones.

According to Springer and Roseneau (1985), the pollock fishery may have had or may be having other effects on the Bering Sea avifauna. Their hypothesis is based on the assumption that planktivorous pollock, acting as prey, can enhance the populations of piscivorous predators but, acting as predators (on copepods), can depress other planktivore populations. In the southeastern Bering Sea, where the pollock CPUE in the commercial fishery indicates a decline in fish stocks or at least a change in their availability, the nesting success of black-legged kittiwakes, a pollock predator (Hunt et al., 1981), has remained low since the late 1970s. Nesting success in other Alaska areas, however, has been much higher. At the same time, the Bering Sea populations of auklets, which feed heavily on copepods, have increased markedly since the mid-1970s.

In the Arctic many species of marine mammals were reduced to low levels by commercial exploitation in the 1800s and early 1900s, but since then several species have been increasing in response to cessation of harvesting. For example, walrus in the Bering Sea sector are thought to have increased from 70,000 animals in 1955 to over 150,000 in 1975 and are now considered to be near their carrying capacity (DeMaster, 1984). Such a large change in population size has not been without ecological repercussions. In the Bering Sea the diet of walrus overlaps considerably with that of the bearded seal (*Erignathus barbatus*), and recovery of the walrus population is thought to be related to a change in the seal's diet. In the 1950s and 1960s, the seals consumed large quantities of clams, whereas in the late 1970s clams were only a minor fraction of the seal diet (Lowry et al., 1980). Numbers of bearded seals did not declined as the walrus herd grew, probably as a result of the catholic food habits of bearded seals.

In the Antarctic no substantial fisheries on pelagic micronekton have yet developed, although extensive experimental fishing for krill has been attempted in recent years (Everson, 1984a). The overexploitation of baleen whale populations, however, is well known. As reviewed by S. G. Brown and Lockyer (1984) and Laws (1977, 1984), a marked decrease in whale numbers has led to changes in demographic parameters among their inter- and intraspecific competitors for krill. These authors note, however, that interpretation of the data is controversial. If the change in whale numbers has led to changes in the populations of their competitors, this might argue for a closed, predation-structured system in the Sherman and Alexander (1986) sense. The Antarctic fur seal and southern elephant seal were also exploited to the verge of extinction (reviewed in Laws, 1984). Their populations have since increased manyfold in response to the cessation of the commercial harvest, but it is not known if any ecological adjustments have taken place on the part of prey or competitors.

VI. Summary and Future Directions

Polar waters in the north are defined as those which are influenced by sea ice during at least part of the year; in the north, thus, polar waters include those of the high Arctic, low Arctic, and boreal zone. In the south, sea ice occurs only over a single zone, Antarctic Surface Water. Because of this and other environmental differences between the north and south polar seas, and particularly the greater severity of conditions and lower productivity in the Arctic, the composition of upper trophic level communities is much more variable regionally in the Arctic than in the Antarctic. The high Arctic community, particularly during winter, is relatively depauperate in species when compared to the high-latitude Antarctic. Furthermore, species composition in the low Arctic is dissimilar between areas adjacent to the Pacific basin and areas adjacent to the Atlantic basin, whereas in the Antarctic, community composition is mostly homogeneous. In both regions, there are differences between communities in oceanic and shelf waters, and a number of upper trophic level organisms move poleward with the retreating ice during summer.

The intermediate trophic levels of polar food webs are dominated by one or at most two species of micronekton or macrozooplankton, and the diets of upper trophic level predators overlap extensively. Arctic cod (nearshore and offshore high Arctic), capelin (Atlantic, low Arctic pelagic), sandlance (Atlantic and Bering, low Arctic shelf), amphipods (*Calanus* spp.) and copepods (*Parathemisto* sp.) (Atlantic and Bering, low Arctic pelagic), pollock and myctophids (Bering, low Arctic pelagic), euphausiids (*Thysanoessa* spp.: Bering, low Arctic pelagic; *Meganyctiphanes* sp.: Atlantic, low Arctic pelagic; *Euphausia crystallorophias:* Antarctic shelf; *E. superba:* Antarctic pelagic), squid (*Gonatus* sp., *Illex* sp. and *Psychroteuthis* sp.: Antarctic and low Arctic), silverfish (Antarctic shelf), and myctophids (Antarctic pelagic) are the key prey of most predators in respective regions. Many of these organisms are not exploited commercially, and thus little is known of their natural history. Much information has been supplied by studies of predators.

Carbon flux and energy transfer are enhanced and upper trophic level organisms are concentrated in both polar regions by shelf-break fronts, midshelf fronts, the marginal ice zone, convergences, and insular fronts. On a regional scale, shelf-break fronts appear to be relatively more important in the Antarctic than the Arctic, whereas midshelf fronts, convergences, and insular fronts are more important in the Arctic. In the polar regions of both hemispheres marginal ice zones are very important, but polynyas are of major significance perhaps only in the Arctic. In the latter region freshwater plumes are also consequential in the life history patterns of upper trophic level organisms.

Competition for prey has affected the occurrence patterns of a number of upper trophic level consumers, especially in the Arctic. Included in competitive interactions are commercial fisheries, acting as predators. The fisheries for capelin off Newfoundland, for pollock in the southeastern Bering Sea (a direct and an indirect effect), for walrus in the Arctic, and for baleen whales in the Antarctic are examples of fisheries that have caused major changes in marine communities.

Marine biological research in the Antarctic is highly fragmented and is confined in both the temporal and spatial scales. Rarely has there been a truly multidisciplinary effort, especially one seeking to learn of trophic interactions above the microzooplankton level. Our understanding of the patterns of production and trophic transfer in some Arctic marine systems is much better. A model for future work is that accomplished during the 1970s and 1980s in the eastern Bering Sea shelf (PROBES), but even that multidisciplinary work is limited in temporal and geographic scope at least from the perspective of upper trophic level predators. To use an analogy, we have snapshots when we really need cinematic coverage. As noted by Brown (1986), the lives of organisms in the upper trophic levels are played out at vast temporal and spatial scales; the migrations of these organisms tend to link areas of high biological production at the mesoscales where pelagic marine biologists are most comfortable. Marine biologists, however, have not linked mesoscale areas or temporal patterns of production as well as have upper trophic level organisms. Much more multidisciplinary effort would increase manyfold our knowledge of the important marine processes. The use of satellite imagery, with ground-truth work, will help us to develop a more integrated and synoptic view of polar marine ecosystems and to unravel interactions simultaneously at several scales and at all seasons. Such work would also help us to understand the relative seasonal importance of pack ice processes as opposed to dynamic water-column processes in affecting the abundance and spatial distribution of upper trophic level organisms.

Acknowledgments

Preparation of this chapter was possible through support from the National Science Foundation, Division of Polar Programs, grant DPP 84–19894 to D.B.A. R. Allen assisted in the preparation of the figures. This is contribution number 392 of the Point Reyes Bird Observatory.

References

Abrams, R. W. & D. G. M. Miller. 1988. The distribution of pelagic seabirds in relation to the oceanic environment of Gough Island. *S. Afr. J. Mar. Sci.* **4**: 125–137.
Ainley, D. G. 1985. The biomass of birds and mammals in the Ross Sea. *In* "Antarctic Nutrient Cycles and Food Webs" (W. R. Siegfried, P. R. Condy & R. M. Laws, eds.), pp. 498–515. Springer-Verlag, Berlin.
Ainley, D. G. & R. J. Boekelheide. 1983. An ecological comparison of oceanic seabird communities of the South Pacific Ocean. *Stud. Avian Biol.* **8**: 2–23.
Ainley, D. G. & S. S. Jacobs. 1981. Seabird affinities for ocean and ice boundaries in the Antarctic. *Deep-Sea Res.* **28**: 1173–1185.
Ainley, D. G., E. F. O'Connor & R. J. Boekelheide. 1984. The marine ecology of birds in the Ross Sea, Antarctica. *Ornithol. Monogr.* **32**. 97 pp.
Ainley, D. G., W. R. Fraser, C. W. Sullivan, J. J. Torres, T. L. Hopkins & W. O. Smith. 1986. Antarctic mesopelagic micronekton: Evidence from seabirds that pack ice affects community structure. *Science* **232**: 847–849.
Ainley, D. G., W. R. Fraser & K. L. Daly. 1988. Effects of pack ice on the composition of micronektonic communities in the Weddell Sea. *In* "Antarctic Ocean and Resources Variability" (D. Sahrhage, ed.), pp. 140–146. Springer-Verlag, Berlin.
Ainley, D. G., C. A. Ribic & R. C. Wood. 1990. A demographic study of the south polar skua *Catharacta maccormicki* at Cape Crozier. *J. Anim. Ecol.* **59**: 1–20.
Belopop'skii, L. O. 1961. "Ecology of Sea Colony Birds of the Barents Sea." Israel Program Sci. Transl., Jerusalem.
Borstad, G. A. 1985. Water colour and temperature in the southern Beaufort Sea: Remote sensing in support of ecological studies of the bowhead whale. *Can. Tech. Rep. Fish. Aquat. Sci.* **1350**: 1–68.
Bradstreet, M. S. W. 1982. Occurrence, habitat use, and behaviour of seabirds, marine mammals, and Arctic cod at the Pond Inlet ice edge. *Artic* **35**: 28–40.
―――. 1988. Importance of ice edges to high-Arctic seabirds. *Acta Congr. Int. Ornithol. 19th,* Vol. 1: 998–1000.
Bradstreet, M. S. W. & R. G. B. Brown. 1986. Feeding ecology of the Atlantic alcidae. *In* "The Atlantic Alcidae" (D. N. Nettleship & T. M. Birkhead, eds.), pp. 264–318. Academic Press, Orlando, Florida.
Bradstreet, M. S. W. & W. E. Cross. 1982. Trophic relationships at high Arctic ice edges. *Arctic* **35**: 1–12.
Bradstreet, M. S. W. & D. B. Fissel. 1987. Zooplankton of a bowhead whale feeding area off the Yukon coast, 1985. *Environ. Stud. (Can., Dept. Indian Aff. North. Dev.)* **50**: 205–360.
Bradstreet, M. S. W., K. J. Finley, A. D. Sekerak, W. B. Griffiths, C. R. Evans, M. F. Fabijan & H. E. Stallard. 1986. Aspects of the biology of Arctic cod (*Boreogadus saida*) and its importance in Arctic marine food chains. *Can. Tech. Rep. Fish. Aquat. Sci.* **1491**: 1–193.
Bradstreet, M. S. W., D. H. Thomson & D. B. Fissel. 1987. Zooplankton and bowhead whale feeding in the Canadian Beaufort Sea, 1986. *Environ. Stud. (Can., Dept. Indian Aff. North. Dev.)* **50**: 1–204.
Brinton, E. 1984. Observations of plankton organisms obtained by bongo nets during the November–December 1983 ice-edge investigations. *Antarct. J. U.S.* **19**: 113–115.
Broady, P. A., C. J. Adams, P. J. Cleary & S. D. Weaver. 1989. Ornithological observations at Edward VII Peninsula, Antarctic, in 1987–88. *Notornis* **36**: 53–61.
Brodie, P. F., D. D. Sameoto & R. W. Sheldon. 1978. Population densities of euphausiids off Nova Scotia as indicated by net samples, whale stomach contents, and sonar. *Limnol. Oceanogr.* **23**: 1264–1267.

Brown, R. G. B. 1980a. Seabirds as marine organisms. *In* "Behavior of Marine Animals" (J. Burger, B. L. Olla & H. E. Winn, eds.), Vol. 4, pp. 1–39. Plenum, New York.
———. 1980b. The pelagic ecology of seabirds. *Trans. Linn. Soc. N. Y.* **9**: 15–22.
———. 1986. The Atlantic alcidae at sea. *In* "The Atlantic Alcidae" (D. N. Nettleship & T. M. Birkhead, eds.), pp. 384–427. Academic Press, Orlando, Florida.
Brown, R. G. B. & D. N. Nettleship. 1981. The biological significance of polynyas to Arctic colonial seabirds. *In* "Polynyas in the Canadian Arctic" (I. Stirling and H. Cleator, eds.), Occas. Pap. No. 45, pp. 59–66. Can. Wildl. Serv., Ottawa.
———. 1984. Capelin and seabirds in the northwest Atlantic. *In* "Marine Birds: Their Feeding Ecology and Commercial Fisheries Relationships" (D. N. Nettleship, G. A. Sanger & P. F. Springer, eds.), pp. 184–195. Can. Wildl. Serv., Ottawa.
Brown, R. G. B., F. Cooke, P. K. Kinnear & E. L. Mills. 1975a. Summer seabird distributions in the Drake Passage, Chilean fjords and off southern South America. *Ibis* **117**: 339–356.
Brown, R. G. B., D. N. Nettleship, P. Germain, C. E. Tull & T. Davis. 1975b. "Atlas of Eastern Canadian Seabirds." Can. Wildl. Serv., Ottawa.
Brown, S. G. & C. H. Lockyer. 1984. Whales. *In* "Antarctic Ecology" (R. M. Laws, ed.), Vol. 2, pp. 717–782. Academic Press, Orlando, Florida.
Carscadden, J. E. 1984. Capelin in the northwest Atlantic. *In* "Marine Birds: Their Feeding Ecology and Commercial Fisheries Relationships" (D. N. Nettleship, G. A. Sanger & P. F. Springer, eds.), pp. 170–183. Can. Wildl. Serv., Ottawa.
Clarke, M. R. 1980. Cephalopoda in the diet of sperm whales of the southern hemisphere and their bearing on sperm whale biology. *'Discovery' Rep.* **37**: 1–324.
Cline, D. R., D. B. Siniff & A. W. Erickson. 1969. Summer birds of the pack ice in the Weddell Sea, Antarctica. *Auk* **86**: 701–716.
Croxall, J. P. 1984. Seabirds. *In* "Antarctic Ecology" (R. M. Laws, ed.), Vol. 2, pp. 533–620. Academic Press, Orlando, Florida.
———. (ed.). 1987. "Seabirds: Feeding Ecology and Role in Marine Ecosystems." Cambridge Univ. Press, Cambridge.
Croxall, J. P. & P. A. Prince. 1987. Seabirds as predators on marine resources, especially krill, at South Georgia. *In* "Seabirds: Feeding Ecology and Role in Marine Ecosystems" (J. P. Croxall, ed.), pp. 347–368. Cambridge Univ. Press, Cambridge.
Croxall, J. P., P. A. Prince & C. Ricketts. 1985. Relationship between prey life-cycles and the extent, nature and timing of seal and seabird predation in the Scotia Sea. *In* "Antarctic Nutrient Cycles and Food Webs" (W. R. Siegfried, P. R. Condy & R. M. Laws, eds.), pp. 516–533. Springer-Verlag, Berlin.
Daly, K. L. & M. C. Macaulay. 1988. Abundance and distribution of krill in the ice edge zone of the Weddell Sea, austral spring 1983. *Deep-Sea Res.* **35**: 21–41.
DeMaster, D. P. 1984. An analysis of a hypothetical population of walruses. *NOAA Tech. Rep., NMFS* **12**: 77–80.
DeWitt, H. H. & T. L. Hopkins. 1977. Aspects of the diet of the Antarctic silverfish, *Pleuragramma antarcticum. In* "Adaptations Within Antarctic Ecosystems" (G. A. Llano, ed.), pp. 557–567. Gulf Publ. Co., Houston, Texas.
Divoky, G. J. 1979. Sea ice as a factor in seabird distribution and ecology in the Beaufort, Chukchi, and Bering seas. *In* "Conservation of Marine Birds of Northern North America" (J. C. Bartonek & D. N. Nettleship, eds.), Wildl. Res. Rep. 11, pp. 9–18. U.S. Fish Wildl. Serv., Washington, D.C.
———. 1984. The pelagic and nearshore of the Alaskan Beaufort Sea: Biomass and trophics. *In* "The Alaskan Beaufort Sea: Ecosystems and Environments" (P. W. Barnes, D. M. Schell & E. Reimnitz, eds.), pp. 417–437. Academic Press, Orlando, Florida.
Dunbar, M. J. 1981. Physical causes and biological significance of polynyas and other open

water in sea ice. *In* "Polynyas in the Canadian Arctic" (I. Stirling & H. Cleator, eds.), Occas. Pap. No. 45, pp. 29–44. Can. Wildl. Serv., Ottawa.

Erickson, A. W., D. B. Siniff, D. R. Cline & R. J. Hofman. 1971. Distributional ecology of Antarctic seals. *In* "Symposium on Antarctic Ice and Water Masses" (G. Deacon, ed.), pp. 55–76. Sci. Comm. Antarct. Res., Cambridge, Massachusetts.

Everson, I. 1984a. Marine zooplankton. *In* "Antarctic Ecology" (R. M. Laws, ed.), Vol. 2, pp. 463–490. Academic Press, Orlando, Florida.

———. 1984b. Fish. *In* "Antarctic Ecology" (R. M. Laws, ed.), Vol. 2, pp. 491–532, Academic Press, Orlando, Florida.

Foster, T. D. 1984. The marine environment. *In* "Antarctic Ecology" (R. M. Laws, ed.), Vol. 2, pp. 345–372. Academic Press, Orlando, Florida.

Fraser, W. R. & D. G. Ainley. 1986. Ice edges and seabird occurrence in Antarctica. *BioScience* **36:** 258–263.

Frost, K. J. & L. F. Lowry. 1981. Foods and trophic relationships of cetaceans in the Bering Sea. *In* "The Eastern Bering Sea Shelf: Oceanography and Resources" (D. W. Hood & J. A. Calder, ed.), Vol. 2, pp. 825–836. Univ. of Washington Press, Seattle.

———. 1984. Trophic relationships of vertebrate consumers in the Alaskan Beaufort Sea. *In* "The Alaskan Beaufort Sea: Ecosystems and Environments" (P. W. Barnes, D. M. Schell & E. Reimnitz, eds.), pp. 381–402. Academic Press, Orlando, Florida.

George, R. Y. 1984. The biology of the Antarctic krill *Euphausia superba. Crustacean Biol.* **4,** 1–338.

Gilbert, J. R. & A. W. Erickson. 1977. Distribution and abundance of seals in the pack ice of the Pacific sector of the Southern Ocean. *In* "Adaptations Within Antarctic Ecosystems" (G. A. Llano, ed.), pp. 703–748. Gulf Publ. Co., Houston, Texas.

Griffiths, A. M., W. R. Siegfried & R. W. Abrams. 1982. Ecological structure of a pelagic seabird community in the Southern Ocean. *Polar Biol.* **1:** 39–46.

Harry, G. Y. & J. R. Hartley. 1981. Northern fur seals in the Bering Sea. *In* "The Eastern Bering Sea Shelf: Oceanography and Resources" (D. W. Hood & J. A. Caler, eds.), Vol. 2, pp. 847–867. Univ. of Washington Press, Seattle.

Hopkins, T. L. 1987. Midwater food web in McMurdo Sound, Ross Sea, Antarctica. *Mar. Biol. (Berlin)* **96:** 93–106.

Hopkins, T. L. & J. J. Torres. 1989. Midwater food web in the vicinity of a marginal ice zone in the western Weddell Sea. *Deep-Sea Res.* **36:** 543–560.

Hunt, G. L., Jr. & D. N. Nettleship. 1988. Seabirds of high-latitude northern and southern environments. *Acta Congr. Int. Ornithol., 19th,* Vol. 1: 1143–1155.

Hunt, G. L., Jr. & R. R. Veit. 1983. Marine bird distribution in Antarctic waters. *Antarct. J. U.S.* **28:** 167–169.

Hunt, G. L., Jr., B. Burgeson & G. A. Sanger. 1981. Feeding ecology of seabirds of the eastern Bering Sea. *In* "The Eastern Bering Sea Shelf: Oceanography and Resources" (D. W. Hood & J. A. Calder, eds.), Vol. 2, pp. 629–647. Univ. of Washington Press, Seattle.

Hunt, G. L., Jr., I. Everson, D. Heinemann & R. R. Veit. 1988. Distribution of krill, marine birds, and Antarctic fur seals in the waters off the western end of South Georgia Island. *Antarct. J. U.S.* **21:** 201–203.

Iverson, I., L. K. Coachman, R. T. Cooney, T. S. English, J. J. Goering, G. L. Hunt, Jr., M. C. Macaulay, C. P. McRoy, W. S. Reeburgh & T. E. Whitledge. 1979. Ecological significance of fronts in the southeastern Bering Sea. *In* "Ecological Processes in Coastal and Marine Systems" (R. J. Livingstone, ed.), pp. 437–466. Plenum, New York.

Jacobs, S. 1988. The Antarctic slope front. *Antarct. J. U.S.* **21:** 123–124.

Jacobs, S. S. & J. C. Comiso. 1989. Sea ice and oceanic processes on the Ross Sea continental shelf. *J. Geophys. Res.* **94(C12):** 18195–18211.

Kinder, T. H. & L. K. Coachman. 1978. The front overlying the continental slope in the eastern Bering Sea. *J. Geophys. Res.* **83:** 4551-4559.

Kinder, T. H., G. L. Hunt, Jr., D. Schneider & J. D. Schumacher. 1983. Correlations between seabirds and oceanic fronts around the Pribilof Islands, Alaska. *Estuarine, Coastal Shelf Sci.* **16:** 309-319.

King, J. E. 1964. "Seals of the World." Br. Mus. (Nat. Hist.), London.

Lancraft, T. M., J. J. Torres & T. L. Hopkins. 1989. Micronekton and macrozooplankton in the ocean waters near Antarctic ice edge zones (AMER[174]IEZ 1983 and 1986). *Polar Biol.* **9:** 225-234.

Laws, R. M. 1977. The significance of vertebrates in the Antarctic marine ecosystem. *In* "Adaptations Within Antarctic Ecosystems" (G. A. Llano, ed.), pp. 411-438. Gulf Publ. Co., Houston, Texas.

———. 1984. Seals. *In* "Antarctic Ecology" (R. M. Laws, ed.), Vol. 2, pp. 621-716. Academic Press, Orlando, Florida.

Lowry, L. F. 1984. A conceptual assessment of biological interactions among marine mammals and commercial fisheries in the Bering Sea. *In* "Proceedings of the Workshop on Biological Interactions Among Marine Mammals and Commercial Fisheries in the Southeastern Bering Sea" (B. R. Melteff & D. H. Rosenberg, eds.), Alaska Sea Grant, pp. 101-118. Univ. of Alaska, Fairbanks.

Lowry, L. F. & K. J. Frost. 1981. Feeding and trophic relationships of phocid seals and walruses in the eastern Bering Sea. *In* "The Eastern Bering Sea Shelf: Oceanography and Resources" (D. W. Hood & J. A. Calder, ed.), Vol. 2, pp. 813-824. Univ. of Washington Press, Seattle.

Lowry, L. F., K. J. Frost & J. J. Burns. 1980. Feeding of bearded seals in the Bering and Chukchi seas and trophic interaction with Pacific walruses. *Arctic* **33:** 330-342.

Marr, J. W. S. 1962. The natural history and geography of the Antarctic krill (*Euphausia superba* Dana). *'Discovery' Rep.* **32:** 33-464.

Mizroch, S. A., D. W. Rice & J. L. Bengtson. 1986. Prodromus of an Atlas of Balaenopterid Whale Distribution in the Southern Ocean Based on Pelagic Catch Data." U.S. Natl. Mar. Fish. Serv., Natl. Mar. Mammal Lab., Seattle, Washington.

Murphy, E. C., A. M. Springer & D. G. Roseneau. 1986. Population status of common guillemots *Uria aalge* at a colony in western Alaska: Results and simulations. *Ibis* **128:** 348-363.

Nasu, K. 1974. Movement of baleen whales in relation to hydrographic conditions in the northern part of the North Pacific Ocean and the Bering Sea. *In* "Oceanography of the Bering Sea with Emphasis on Renewable Resources" (D. W. Hood & E. J. Kelly, eds.), pp. 345-360. Univ. of Alaska Press, Fairbanks.

Nemoto, T. 1970. Feeding pattern of baleen whales in the ocean. *In* "Marine Food Chains" (J. H. Steele, ed.), pp. 241-252. Oliver & Boyd, Edinburg.

Nemoto, T., M. Okiyama & M. Takahashi. 1985. Aspects of the roles of squid in food chains of marine Antarctic ecosystems. *In* "Antarctic Nutrient Cycles and Food Webs" (W. R. Siegfried, P. R. Condy & R. M. Laws, eds.), pp. 415-420. Springer-Verlag, Berlin.

Nettleship, D. N. & T. M. Birkhead (eds.). 1986. "The Atlantic Alcidae." Academic Press, Orlando, Florida.

Nettleship, D. N. & P. G. H. Evans. 1986. Distribution and status of the Atlantic alcidae. *In* "The Atlantic Alcidae" (D. N. Nettleship & T. M. Birkhead, eds.), pp. 54-155. Academic Press, Orlando, Florida.

Nettleship, D. N. & A. J. Gaston. 1978. "Patterns of Pelagic Distribution of Seabirds in Western Lancaster Sound and Barrow Strait, NWT," Occas. Pap. 39. Can. Wildl. Serv., Ottawa.

Nettleship, D. N., G. A. Sanger & P. A. Springer (eds.). 1984. "Marine Birds: Their Feeding Ecology and Commercial Fisheries Relationships." Can. Wildl. Serv., Ottawa.

Niebauer, H. J. & V. Alexander. 1985. Observations and significance of ice-edge oceanographic frontal structure in the Bering Sea. *Cont. Shelf Res.* **4**: 367–388.

Offredo, C. & V. Ridoux. 1986. The diet of emperor penguins *Aptenodytes forsteri* in Adelie Land, Antarctica. *Ibis* **128**: 409–412.

Parmelee, D. F., W. R. Fraser & D. R. Neilson. 1977. Birds of the Palmer Station area. *Antarct. J. U.S.* **12**: 14–21.

Parsons, T. R., D. G. Webb, H. Dovey, R. Haigh., M. Lawrence & G. E. Hopky. 1988. Production studies in the Mackenzie River–Beaufort Sea estuary. *Polar Biol.* **8**: 35–39.

Parsons, T. R., D. G. Webb, B. E. Rokeby, M. Lawrence, G. E. Hopky & D. B. Chiperzak. 1989. Autotrophic and heterotrophic preduction in the Mackenzie River/Beaufort Sea estuary. *Polar Biol.* **9**: 261–266.

Plötz, J. 1986. Summer diet of Weddell seals (*Leptonychotes weddelli*) in the eastern and southern Weddell Sea, Antarctica. *Polar Biol.* **6**: 97–102.

Quetin, L. B. & R. M. Ross. 1984. School composition of the Antarctic krill *Euphausia superba* in the waters west of the Antarctic Peninsula in the austral summer of 1982. *Crustacean Biol.* **4**: 96–106.

Ribic, C. A., D. G. Ainley & W. R. Fraser. 1990. Marine mammal habitat associations along the pack ice edge in the Antarctic. *Antarct. Sci.* (submitted for publication).

Ridoux, V. & C. Offredo. 1989. The diets of five summer breeding seabirds in Adélie Land, Antarctica. *Polar Biol.* **9**: 137–145.

Sanger, G. A. 1987. Trophic levels and trophic relationships of seabirds in the Gulf of Alaska. *In* "Seabirds: Feeding Ecology and Role in Marine Ecosystems" (J. P. Croxall, ed.), pp. 229–258. Cambridge Univ. Press, Cambridge.

Schneider, D. & G. L. Hunt. 1982. Carbon flux to seabirds in waters with different mixing regimes in the southeastern Bering Sea. *Mar. Biol. (Berlin)* **67**: 337–344.

Schneider, D., N. M. Harrison & G. L. Hunt, Jr. 1987. Variation in the occurrence of marine birds at fronts in the Bering Sea. *Estuarine, Coastal Shelf Sci.* **25**: 135–141.

Sherman, K. & L. M. Alexander. 1986. "Variability and Management of Large Marine Ecosystems." Am. Assoc. Adv. Sci., Washington, D.C.

Siniff, D. B. & J. L. Bengtson. 1977. Observations and hypotheses concerning the interactions among crabeater seals, leopard seals, and killer whales. *J. Mammal.* **58**: 414–416.

Siniff, D. B. & S. Stone. 1985. The role of the leopard seal in the trophodynamics of the Antarctic marine ecosystem. *In* "Antarctic Nutrient Cycles and Food Webs" (W. R. Siegfried, P. R. Condy & R. M. Laws, eds.), pp. 555–560. Springer-Verlag, Berlin.

Smith, W. O., Jr. & D. M. Nelson. 1985. Phytoplankton bloom produced by a receding ice edge in the Ross Sea: Spatial coherence with the density field. *Science* **227**: 163–166.

Smith, W. O., Jr., N. K. Keene & J. C. Comiso. 1988. Interannual variability in estimated primary productivity of the Antarctic marginal ice zone. *In* "Antarctic Ocean and Resources Variability" (D. Sahrhage, ed.), pp. 131–139. Springer-Verlag, Berlin.

Springer, A. M. & D. G. Roseneau. 1985. Copepod-based food webs: Auklets and oceanography in the Bering Sea. *Mar. Ecol.: Prog. Ser.* **21**: 229–237.

Springer, A. M., D. G. Roseneau, E. C. Murphy & M. I. Springer. 1984. Environmental controls of marine food webs: Food habits of seabirds in the eastern Chukchi Sea. *Can. J. Fish. Aquat. Sci.* **41**: 1202–1215.

Springer, A. M., E. C. Murphy, D. G. Roseneau, D. P. McRoy & B. A. Cooper. 1987. The paradox of pelagic food webs in the northern Bering Sea. I. Seabird food habits. *Cont. Shelf Res.* **8**: 895–911.

Stirling, I. & H. Cleator (eds.). 1981. "Polynyas in the Canadian Arctic" Occas. Pap. No. 49. Can. Wildl. Serv., Ottawa.
Stirling, I., H. Cleator, & T. G. Smith. 1981. Marine mammals. In "Polynyas in the Canadian Arctic" (I. Stirling & H. Cleator, eds.), Occas. Pap. No. 49, pp. 45–58. Can. Widl. Serv., Ottawa.
Stonehouse, B. 1967. Occurrence and effects of open water in McMurdo Sound, Antarctica, during winter and early spring. Polar Rec. **87**: 775–778.
Sutcliffe, W. H. & P. F. Brodie. 1977. Whale distribution in Nova Scotia waters. Tech. Rep.— Fish Mar. Serv. (Can.) **722**: 1–83.
Thomas, P. B. & K. Green. 1988. Distribution of Euphausia crystallorophias within Prydz Bay and its importance to the inshore marine ecosystem. Polar Biol. **8**: 327–331.
Veit, R. R. & B. M. Braun. 1984. Hydrographic fronts and marine bird distribution in Antarctic and subantarctic waters. Antarct. J. U.S. **19**: 165–167.
Wespestad, V. G. & J. A. Terry. 1984. Biological and economic yields for eastern Bering Sea walleye pollack under different fishing regimes. North Am. J. Fish. Manag. **4**: 204–215.
Williams, R. 1985. Trophic relationships between pelagic fish and euphausiids in Antarctic waters. In "Antarctic Nutrient Cycles and Food Webs" (W. R. Siegfried, P. R. Condy & R. M. Laws, eds.), pp. 452–459. Springer-Verlag, Berlin.
Wilson, D. L., W. O. Smith & D. M. Nelson. 1986. Phytoplankton bloom dynamics of the western Ross Sea ice edge. I. Primary productivity and species-specific production. Deep-Sea Res. **33**: 1375–1387.
Winters, G. H. & J. E. Carscadden. 1978. Review of capelin ecology and estimation of surplus yield from predator dynamics. Int. Comm. NW Atl. Fish Res. Bull. **13**: 21–30.
Wolanski, E. & W. H. Hammer. 1988. Topographically controlled fronts in the ocean and their biological influence. Science **241**: 177–181.
Zelikman, E. A., & A. N. Golovkin. 1972. Composition, structure and productivity of neritic plankton communities near the bird colonies of the northern shores of Novaya Zemlya. Mar. Biol. (Berlin) **17**: 265–274.
Zwally, H. J., J. C. Comisco, C. L. Parkinson, W. J. Campbell, F. D. Carsey & P. Bloersen. 1983. "Antarctic Sea Ice, 1973–1976: Satellite Passive-Microwave Observations." Natl. Aeron. Space Admin., Washington, D.C.

12 Polar Benthos

Paul K. Dayton
*Scripps Institution of Oceanography A-001
La Jolla, California*

I. Introduction   632
II. On Early Exploration and Collections   634
III. Origin, Evolution, and Historical Background of Arctic Habitats and Species   635
  A. Greenland, Jan Mayen, and Iceland   637
  B. Svalbard and the Barents Sea   639
  C. The White Sea   639
  D. Northern Soviet Seas   640
  E. The Chukchi Sea   640
  F. The Beaufort Sea   640
  G. The Canadian Arctic   640
IV. Arctic Patterns and Processes   641
  A. Land-Fast Ice Zone   643
  B. Bering and Chukchi Seas   646
V. Arctic Macroalgae   651
VI. Origin, Evolution, and Historical Background of Antarctic Habitats and Species   653
  A. Biogeography   656
  B. Sedimentation   657
VII. Antarctic Patterns and Processes   658
  A. Zonation and Benthic Assemblages   659
  B. Soft-Bottom Habitats   660
  C. Deeper Benthic Habitats   663
  D. Productivity and Benthic Communities   665
  E. Life Below Antarctic Ice Shelves   668
VIII. Antarctic Benthic Macroalgae   669
IX. Discussion   672
   References   676

I. Introduction

Apart from their polar position, low temperatures, and the seasonal trends in irradiance and ice cover, the north and south polar benthos are strikingly different from each other. Their origins and evolution are different, as are their oceanic habitats and climates and the types of terrestrial influences. They differ physiographically and oceanographically: the Artic Ocean is an enclosed ocean surrounded by land, whereas the Southern Ocean is an oceanic ring flowing around a continent (Table 12.1, see also Fig. A, page xiii).

Most benthic communities reflect their biogeographic histories and the

Table 12.1 Oceanographic Characteristics of Antarctic and Arctic Seas[a]

Antarctic	Arctic
1. Ocean area $35-38 \times 10^6$ km^2 as a circumpolar ring between 50 and 70°S	Oceanic area 14.6×10^6 km^2 as an enclosed ocean surrounded by land between 70 and 80°N
2. Permanent ice shelf 1.5×10^6 km^2 Mean maximum pack ice 22×10^6 km^2 Mean maximum pack ice 2.6×10^6 km^2	No permanent ice shelf Mean maximum pack ice $12-36 \times 10^6$ km^2 Mean pack ice $6-7 \times 10^6$ km^2
3. Free connection with major oceans, the Atlantic, Pacific, and Indian	Very limited connection with the Pacific, much greater with the Atlantic; maximum sill depth 70 m (Pacific) and 440 m (Atlantic)
4. Circumpolar currents (West Wind Drift, East Wind Drift) and large eddies (Weddell, Ross)	Transpolar current, Beaufort Gyre

5. No pronounced stratification or vertical stability. Considerable mixing with sinking of high salinity (greater than 34.5 ‰) and low temperature (less than 0.5°C) near the continent, and upwelling of high salinity (34.7 ‰) and high temperature (1-2°C) in the region of the Antarctic Divergence	Stratification pronounced. Considerable vertical stability at all times of the year.
	 a. Surface layer (100-150 m)
	    Low salinity (30-32 ‰)
	    Low temperature (-1.5 to -1.7°C)
	 b. Second layer (100-200 m)
	    Normal salinity 34 ‰
	    Low temperature (-1.0°C)
	 c. Warm Atlantic water layer (600 m)
	    High salinity (34.7-349 ‰)
	    High temperature (0-1.8°C)
	 d. Bottom water (600-800 m)
	    High salinity (34.7-34.9 ‰)
	    Low temperature (0.5 to -1.0°C)

6. Abundant large tabular icebergs	Less abundant, small irregular icebergs mainly in Greenland Sea, some in Bering Sea; none in Arctic basin

7. Continuous high nutrient levels in the euphotic zone	Lower levels and seasonal lack of nutrients in the euphotic zone
8. Extremely high levels of primary productivity nearshore; moderate offshore	Moderate levels of primary productivity
9. Narrow shelf, 400–600 m deep; open to all oceans with large exchange in deep water	Broad shelf, > 100–500 m deep, two narrow openings at Bering and Fram straits
10. Mosaic of glacial marine sediments, including muds, fine and coarse sands, and large and small boulders. Sediments generally poorly sorted. No riverborne sediments	Large input of sediment from rivers flowing into the Arctic basin. Mud and clayey sediments dominant
11. High degree of stability of the marine climate over the last 200 years	Evidence of a considerable rise of temprature in recent decades, especially over the last 40 years

[a] After Knox and Lowry (1977) and Hempel (1985).

physical stresses delimiting general patterns of distribution and abundance. On this very general format, the populations require allochthonous organic input, although this can be supplemented by *in situ* benthic primary production in some shallow habitats. The historical and physical factors determine the species composition, and the availability of organic material and biotic interaction determines their relative mass and abundance. The supply of organic material usually is a function of water column productivity, depth, and the various transport processes. Understanding such benthic systems depends on the nature, abundance, and predictability of organic inputs; their utilization by the benthic populations; and the degree to which the structure of the community is regulated by abiotic disturbance and biotic factors influencing recruitment and age-specific survivorship and fecundity.

Ultimately, the strength of the generalities or the theories and the value of the ecosystem and evolutionary hypotheses rest on the strength of the descriptive natural history. The methods of science quickly render good hypotheses invalid, but good descriptive natural history is timeless. Polar workers are lucky to have inherited solid foundations of descriptive research. I briefly review some of the early explorers who, with the scientists of their time, are the giants on whose shoulders we stand. Then I summarize pertinent aspects of the salient geological patterns which have created the habitats we see today. Starting with summaries of biogeographical patterns, I will consider processes which appear important in the various habitats. Geographically, the Antarctic is reasonably well delimited; however, the distinction between the Arctic and subArctic is not clear, and I have avoided this issue by using "Arctic" in the generic inclusive sense.

This is a general review of polar benthic communities. For expedience and textual coherence, I discuss the Arctic and Antarctic separately. Because the literature is self-contained, it seems expedient to consider the algal patterns separately. Finally, while it may be premature, I will offer contrasts and speculations regarding possible contributions of the polar systems to the interplay between ecosystem and evolutionary ecology.

II. On Early Exploration and Collections

It seems odd that the polar regions are often considered to be unexplored and virtually unknown. Perhaps this misconception is based on the fact that much of our scientific tradition comes from temperate climates, whereas polar regions are considered too cold for appropriate study. In reality, both polar regions have long and proud histories of exploration which through the 20th century easily parallel the exploration of many temperate and tropical regions. Although the travails of Scott and Shackleton have captured the imagination of the public, other Antarctic explorers and especially Arctic explorers deserve more recognition. Much of the Arctic history of exploration, from the "Thule" of Pytheas (330–325 B.C.) to Hwui Shan (a Buddhist monk who reached Alaska in A.D. 458) to the gentle Saint Brendan (early 500s) to the Norse (Ottar to the White Sea about 880 and the active but ultimately unsuccessful colonization of Greenland and the New World ending with the Little Ice Age), is very poorly known. Some of this early history as well as the aggressive European exploration of the 15th through 19th centuries through Nansen and Stefansson in the 20th century is poetically, if succinctly, summarized by Lopez (1986). Many of us have been raised on the epic survival story of Shackleton in the *James Caird* in 1916, but equally remarkable sagas and open-boat voyages exist in the Arctic. Consider, for example, Willem Barent and his crew overwintering almost 400 years ago on the northern tip of Novaya Zemlya after their ship was wrecked. Most of the crew returned in an open boat, covering more than 1600 miles. The return of Bering's crew in 1741 is also impressive, and it is particularly important because it included Steller, who died on the return but whose notebooks form a vital part of our scientific legacy. Scientific collections in those early Arctic explorations were indeed rare, but some were very important. Perhaps the most important scientific observation was based on another remarkable open-boat survivorship story of some of George DeLong's crew and their recovery of his log after the sinking of the *Jeannette* in 1881 near 77°N and 155°E (see Ellsberg, 1938). The publication of the record of their westward drift plus the serendipitous discovery and news account of some of the

Jeannette's articles on the southwest coast of Greenland demonstrated to Nansen the Arctic current patterns and stimulated his and Sverdrup's classical oceanographic research. See Kirwan (1960) for a more complete account of polar history.

Though it occupies the public imagination, the exploration of the Antarctic lacks the long history of Arctic exploration (but consider Peter Buck's 1958 account of early Polynesian discoveries of cold white islands south of New Zealand). The first successful exploration was Cook's circumnavigation in the early 1770s. Soon thereafter sealers and eventually whalers devastated the mammals of the Southern Ocean. But at the same time explorers such as Bellingshausen, Biscoe, Wilkes, Dumont d'Urville, and Ross began the scientific and geographic exploration of the Antarctic which was continued in the 20th century by Borchgrevink, Nordenskjöld, Scott, Shackleton, Mawson, and many others. A balanced and complete summary of the Antarctic history can be found in the Readers Digest (1985) "Antarctica." Even the very early expeditions resulted in voluminous natural history publications [e.g., Gaudichaud (1826), Ehrenberg (1844), Montagne (1845, 1846), Hooker (1847), and others summarized in White (1984)]. Some of the leaders, especially Wilkes, were obsessed with the publication of the reports of their expeditions. Dell's (1972) remarkably complete compendium and White's (1984) more recent review summarize some of the accomplishments of these and of the *Challenger* (1872–1876) and *Discovery* investigations (1925–1937) through the IGY years (for example, Bullivant and Dearborn, 1967). White correctly laments the early taxonomic confusion and comments on the tremendous advantages of scuba techniques. Nevertheless, I doubt if the early collectors will ever get their just recognition. At McMurdo Station, for example, we admire the remarkable IGY era collections of Dearborn (1965) and others; however, I can but stand in awe of the prodigious accomplishments of Scott's first collector, T.V. Hodgson: even after more than two decades of scuba diving in the areas in which he worked, I have not found many large species he did not collect! To consider this and to read (Hodgson, 1907) the understated collecting techniques of this man (the oldest in the group and little appreciated in the journals of his peers) in the *Terra Nova* reports is extremely humbling.

III. Origin, Evolution, and Historical Background of Arctic Habitats and Species

The Arctic Ocean may have been in existence since Paleozoic times. The rifting that produced the Atlantic commenced in the Jurassic, at which time the Arctic was a deep-sea connection to the Pacific. The modern configuration began developing in the late Cretaceous. The derivation of much of the modern world marine fauna was influenced by the warm-water Tethys Sea; at this time [perhaps 60-70 million years before present (m.y. BP)] the Arctic Sea was nearly tropical or at least subtropical (Menzies et al., 1973; McKenna, 1980; Francis, 1988).

The Arctic Ocean was strongly affected by many Cenozoic events. The Miocene cooling resulted in a major expansion of the Antarctic ice sheet and concomitant lowering of the sea level and establishment of the Bering land bridge which isolated the Arctic and Pacific oceans. Because the ocean circulation was latitudinal, there was probably little interchange between the Arctic and Atlantic oceans; a sudden appearance of Pacific mollusks in Iceland about 3.5 m.y. ago implies that the Bering land bridge was flooded at that time (Herman and Hopkins, 1980). This and the emergence of the Isthmus of Panama resulted in a very different oceanographic regime and affected the Arctic Ocean by causing a persistent flow of low-salinity Pacific water and saline Atlantic water into the Arctic Ocean. The inception of permanent sea ice cover was only about 0.7 million years ago (Herman and Hopkins, 1980).

The modern biogeography of the Arctic Ocean is mostly a result of events occurring during the Quaternary period (Denton and Hughes, 1981). Eventually, an Arctic estuarine water mass was formed by the freshwater inputs of the Siberian rivers; this water mass developed a faunistic complex of freshwater immigrants and local euryhaline species. During the last sea level transgression, this estuarine complex distributed itself along the polar basin and became an important ecological barrier to the dispersal of stenohaline marine species (Golikov and Averincev, 1977). There are various biogeographic schemes for the Arctic region. Zenkevitch's (1963) classical work may still be the standard (Knox and Lowry, 1977; Golikov and Averincev, 1977). The autochthonous groups are from preglacial times and have been able to respond to the reduced temperatures of the ice ages. There must have been considerable Pliocene penetration into the Arctic basin, especially from the Pacific; however, most of these species were eliminated in shallow water by the sea level regressions, with the survivors successfully moving into deeper water. The Pacific is poorly represented in the present fauna. These species gave rise to the high Arctic fauna, which was subsequently restricted by the salinity barriers discussed by Golikov and Averincev (1977). These Quaternary changes resulted in the interconnected complex of estuarine seas which variously connected with both the Atlantic and the Pacific. In postglacial times there has been limited exchange with the Pacific and especially the Atlantic. Knox and Lowry (1977) suggest that the Arctic basin is presently being populated by low temperature-tolerant species from the north Atlantic. Because of the predominantly nonplanktonic dispersal modes, these invasions are limited and relatively slow.

Soviet workers recently have focused on Arctic biogeographic questions, and much of this work is synthesized in an excellent review by Nesis (1983) which discusses the distinct and limited nature of the various distributional areas. Nesis suggests that the eastern Arctic and Arctic–Pacific species are stenobathic (100–200 m), whereas the western Arctic and Arctic–Atlantic species are eurybathic. He utilizes recent geological conclusions to suggest that Quaternary glaciation was responsible for causing and maintaining these biogeographic groupings. He also argues that the western Arctic shelf was covered with ice shields with very deep outlet glaciers precluding stenobathic shallow benthic species; in contrast, the eastern Arctic had no glaciers and eurybathic benthic species were abundant.

There is a long history of benthic research in the Arctic and subarctic regions. Many nations have been involved, and different objectives have brought researchers to the north. Most of the early work was done by European and Soviet investigators, and many of the early programs involved commercial or military expectation. Although there were a few early programs from Canada and the United States, it is only recently that the petroleum interests have stimulated large Canadian and American research efforts. Finally, the types of scientific questions have shifted from pure exploration (with its descriptive approach) to much more process-oriented research. The types of scientific questions are dependent on the particular countries doing the research, and the types of data, sampling methods, and types of processes under study are extremely variable. This section offers a very superficial summary of regional studies (mostly from Curtis, 1975) designed to demonstrate the histories of the research efforts and the types of data available. Any serious effort must also include the recent Soviet references such as those summarized in Nesis (1983).

The excellent review by Curtis (1975) includes summaries of different regional faunal patterns (Figure 12.1). He concludes that the highest species richness occurs in areas of hydrographic mixing, and areas of lowest species richness are those influenced by the brackish waters and the deep sea. He also summarizes recent examples of quantitative community analysis and concludes that the data are insufficient to allow much sensitivity in such analyses. Curtis also summarizes efforts to evaluate patterns of benthic biomass as roughly paralleling the regional patterns and decreasing sharply with depth. Finally, Curtis reviews the success of the Thorson and Meleikovsky life history research and argues that their observations of nonplanktonic larval development in northern seas appears to be very general. This is especially true in the Antarctic (Dell, 1972).

A. Greenland, Jan Mayen, and Iceland

The rugged Greenland coast has received a surprising amount of excellent research. West Greenland is bathed by the northerly Irminger Current,

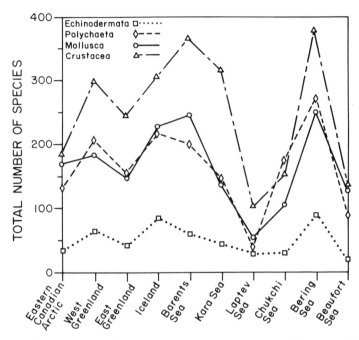

Figure 12.1 Total numbers of benthic species of crustaceans, echinoderms, mollusks, and polychaete worms known to occur in various circumpolar marine areas. After Curtis (1975).

which mixes with polar waters in the north and has long been the site of cod and shrimp fisheries. East Greenland lies in Arctic waters dominated by the cold East Greenland Current and does not have a history of commercial fisheries. The first published work was Otto Fabricus's (1780) "Fauna Groenlandica." Over a hundred years later this work was supplemented by the *Ingolf* (1895–1896) and the *Denmark* (1906–1908) expeditions, which were published in the journal *Meddelelserom Groenland* in 1898 and 1917, respectively (see Curtis, 1975). Whereas these and other early programs were mainly taxonomic, extremely thorough quantitative work was carried out in East Greenland by the "Three-Year Expedition," which was synthesized by Thorson (1936). This massive publication included an extremely thorough descriptive component and was the seminal paper establishing global marine life history studies. Continuing research (see Curtis, 1975) in Greenland has had regional focus and has obtained valuable time-series data (Petersen, 1964).

Jan Mayen is an isolated Arctic region which received early study by the Greenland expeditions as well as descriptive surveys following a 1970 volcanic eruption (Skjaeveland, 1973; Gulliksen, 1974). The initial colonization was slow, but Gulliksen *et al.* (1980) visited Jan Mayen again in 1978 and found many changes. They compared old and new grounds at several depths. Both the old and new communities between 5 and 10 m were physically stressed and populated by motile animals such as amphipods but were otherwise relatively barren. But at 20 – 30-m depths the new substrata had a much higher biomass, largely composed of the bivalve *Hiatella arctica.*The old substratum was not dominated by a single species and from the photographs appears to be an urchin – coralline "barren" *(sensu* Lawrence, 1975). Gulliksen *et al.* argue that predators such as the sea urchin *Strongylocentrotus droebachiensis* and the anemone *Tealia felina* structure the old community and prevent the monopolization of space by *Hiatella* and several species of sponges, polychaetes, etc. Though not really Arctic, Iceland also has a rich history of early 20th-century research which has continued through the recent work following volcanic eruptions (Curtis, 1975; Sigurdsson, 1972).

B. Svalbard and the Barents Sea

Commercial interests stimulated much 19th-century research by European exploratory cruises, and these were followed by extensive Soviet, German, and Norwegian research early in the 20th century. These included quantitative surveys in the Spitsbergen area (Brotzky, 1930; Idelson, 1930). Much of this early work is well summarized by Blacker (1957). The Soviets also have been very active in Novaya Zemlya and the Murmon Coast. These studies have quantified benthic biomass and found high benthic biomass in these polar front regions, where there is hydrographic mixing and open water (Zenkevitch, 1963). Shifts in long-term climate are associated with changes in the benthic community (see other reviews by Gulliksen, 1988; Nesis, 1959, 1960, 1983). Gulliksen (1979) found that the benthos of the current-swept and physically disturbed Bear Island is generally lacking many predators and between 10 and 30 m is dominated by colonial ascidians. Gulliksen and colleagues (1974, 1985) have also been active at Svalbard, where the soft-bottom habitats had rather depauperate fuanas, mostly polychaetes and bivalves. The distribution patterns seem influenced by sills and terrestrial and freshwater runoff. From the photographs the hard-bottom community varies from "urchin-barren grounds" to protected habitats with complete substrata domination by many types of filter feeders.

C. The White Sea

This small sea is isolated from the Barents Sea by a shallow sill and tends to be stratified and relatively unproductive; nevertheless, it has been extensively sampled (Zenkevitch, 1963). It is also the region where invertebrate life history patterns were studied by Mileykovsky (1970, 1971, 1972), and an active Soviet program has continued in this region (Malveyeva, 1974; Kaufman, 1974).

D. Northern Soviet Seas

This is the extensive Soviet area dominated by seasonal estuarine conditions and river runoff. Zenkevitch (1963) and Curtis (1975) summarize some of the early work from the *Fram* (1893) and *Belgica* (1907). Golikov and Averincev (1977) summarize their extensive work between Franz Josef Land and the Novosibirskye Islands which emphasizes the importance of substrata, depth, and especially salinity. They also present useful summaries of habitat types, oceanographic conditions, and biota for many areas in this region. Finally, they emphasize that the Novosibirskye Islands and the Lomonosov Ridge represent the transition area between the Atlantic and Pacific fauna.

E. The Chukchi Sea

This area was the site of relatively intensive whaling in the 19th century and has been visited by several research expeditions, including Beechey's expedition in the *Blossom* to Icy Cape in 1826. Nordenskiold's *Vega* expedition worked the west portion, and the *Yukon* and *Corwin* worked the eastern area in the late 19th century. The Chukchi Sea has received much more intensive Soviet work starting in the 1950s, with sporadic American work starting in the late 1950s. The first environmental work was conducted in the late 1950s and early 1960s (Sparks and Pereyra, 1966; Wilimovsky, 1966).

F. The Beaufort Sea

This large area received only minor attention until the U.S. Naval Arctic Research Laboratories (NARL) were established at Point Barrow after World War II. A great deal of taxonomic and ecological work was conducted in the late 1940s (MacGinitie, 1955). The Canadians were active in the eastern Beaufort Sea with the small fisheries research ship *Salvelinus* in the early years and now are sponsoring much more intensive research.

G. The Canadian Arctic

Apart from the dredging program off the *Fram* in 1898–1902, there was little benthic research until the Canadian Fisheries Research Board sponsored expeditions on the *Calanus* (1948–1961) and the *Salveninus* (1960–

1965). This and other work instigated by M. Dunbar resulted in reasonably complete surveys, especially those of Ellis (1960). Like the U.S. Beaufort Sea, northeastern Canadian seas have received a flurry of recent activity.

One of the most complete and succinct papers I have seen is Thomson's (1982) of the Baffin Bay area. A few density data of this massive survey, which also includes biomass data and a sophisticated multivariate analysis, are summarized later in Table 12.3. This area had a rather homogeneous sediment type, and the differences are probably related to availability of food. Biomass tended to be higher in this Baffin Bay region than elsewhere in the high Arctic. Thomson compared Lancaster Sound (520 g m^{-2}), northwest Baffin Bay (297 g m^{-2}), Eclipse Sound (258 g m^{-2}), and central Baffin Bay (88 g m^{-2}) with the Alaskan Beaufort Sea (41 g m^{-2}), Melville Island (94 g m^{-2}), and Brentford Bay (188 g m^{-2}). He related these differences to higher nutrient levels and primary production than are found in most high Arctic regions. Additional increases in density and biomass appear correlated with higher current regimes, where there are more large suspension feeders; areas with lower currents tended to have lower biomass and were dominated by deposit feeders. This important synthesis deserves more recognition.

IV. Arctic Patterns and Processes

The Arctic Ocean is an enclosed sea with only two connections, the Bering Strait (70 km wide by 70 m deep at the sill) and Fram Strait (400 km wide with a sill depth of 440 m). The Arctic basin is covered with a permanent but dynamic moving ice cover and is surrounded by broad shelf areas. The Arctic Ocean is strongly stratified. The large amount of fresh continental runoff and winter brine rejection (see Chapters 2,4,5) produce a cold surface layer to 200 m which overlies a layer of warmer saline Atlantic water (200–900 m), below which lies cold saline bottom water. The circulation includes two major currents: the anticyclonic Beaufort Gyre and the relatively straight current which flows from eastern Siberian waters to the Greenland Sea (Fig. 4.3), and which continues as the East Greenland Current and exits through Fram Strait. The Arctic Ocean experiences seasonal nutrient depletions in the surface waters of the continental shelf; this, in conjunction with the large proportion of permanent ice cover, results in low levels of annual primary production (see Chapter 9) and a paucity of benthic animal life relative to other broad shelf areas. Paul and Menzies (1974), working from an ice island, sampled the bottom between 1000 and 2500 m and found a low biomass of about 0.04 g m^{-2}. This is comparable to that found at about 6000 m in the oligotrophic red clay of the central Pacific and is 40 times less than that found at comparable depths off the Antarctic. The fauna was depauperate: 53% of the biomass was foraminifera, 27% bivalves, 7% sponges, and only 5% polychaetes. Considering that most oligotrophic benthic habitats are dominated by detritus feeders, it is interesting that the three most abundant species are suspension feeders.

The Beaufort Sea is one of several shelf habitats of the Arctic Ocean. In contrast to the others, it is relatively deep and saline. Although Golikov and Averincev (1977) describe many areas with different habitats marked by large numbers of filter feeders, most other shallow muddy habitats superficially, at least, seem similar to those of the Beaufort Sea. Because the Beaufort Sea is well described by the extensive survey work of the past two to three decades, I focus on this region for a general description of patterns and processes.The shelf is generally muddy with sandy areas nearshore and patches of gravel at the shelf break. Ice covers the shelf from September through June or July. A land-fast ice zone extends to depths of 15–23 m, where it impinges on the polar ice pack at the shear zone. This ice breaks up in the summer, but as it drifts and grounds, it becomes (with winter anchor ice) (Barnes *et al.,* 1982; Reimnitz *et al.,* 1986, 1987) a major predictable benthic disturbance.

A reasonably consistent picture of depth stratification or zonation has been developed in the Beaufort Sea (see Carey and Ruff, 1977; Carey *et al.,* 1984a,b; Thorsteinson, 1987). This includes littoral, nearshore (0–2 m), inshore or coastal (2–20 m), offshore (20–70 m), and slope (70–200 m) zones. These zones are strongly affected by large river systems (e.g., the Mackenzie River), which superimpose on the depth zonation estuarine, transitional, and marine associations. In addition, the countless smaller drainage systems contribute organic-rich runoff.

The heavily disturbed littoral zone is characterized by ephemeral populations of chironomid larvae and oligochaete annelids. Motile isopods *Saduria entomon)* and amphipods *(Gammarus setosus* and *Onisimus litoralis)* are the most common epifauna. The deeper inshore zone includes patches of different substrata with associated biota; here the infauna is represented by many species of polychaetes, bivalves, and isopods and the epifauna by mysids, amphipods, isopods, copepods, and euphausiids. All are patchy but often abundant. There seems to be no consistent pattern within the bivalve fauna (Carey *et al.,* 1984b). The offshore zone is characterized by polychaetes, bivalves, ophiuroids, holothurians, and many crustacea. The polychaetes constitute 32–87% of the total macrobenthos, with mollusks and arthropods contributing between 5 and 50%. Farther offshore, the species composition was roughly similar, but the biomass and density of larger macrofauna increased (Carey and Ruff, 1977).

The cross-shelf patterns have been determined with several transects across the shelf of the western Beaufort Sea (from Thorsteinson, 1987). By far the most abundant groups are polychaetes, bivalves, and gammarid amphipods, but more than 300 species of invertebrates have been collected in the Alaskan Beaufort Sea. Various hypotheses have been offered to explain the various cross-shelf patterns. The nearshore peaks may be associated with terrestrial peat from coastal and river erosion, whereas some of the offshore peaks may represent depressions or other low-velocity sites resulting in detrital accumulation.

Carey and Ruff (1977) analyzed 10 transects across the western Beaufort shelf and found several recurrent groups, which they associated with various physical characteristics. As would be expected from most benthic research, sediment or substratum type is an important variable. The disruptive influence of ice in shallow waters was another important variable influencing benthic associations. Hydrographic patterns can have a strong influence because they can be associated with large salinity and temperature variations, which can also be related to nutrient fluxes. Finally, the availability of food is always an important component of community structure. Here the primary production is associated with phytoplankton, ice algae, and patches of macroalgae; it is often strongly augmented with material from runoff.

The macroepifauna is most abundant on the outer and upper slope (Frost and Lowry, 1984; Carey et al., 1984a,b; Thoresteinson, 1987).These are most conspicuously represented by echinoderms, especially ophiuroids.Two major epifaunal communities were described qualitatively in the southwestern Beaufort Sea. West of 154° the dominant species were ophiuroids, holothurians, and a soft coral *(Gersemia* sp.), which were associated with muddy substrata. East of 154° the epifauna association was characterized by scallops, crinoids, holothurians, and echinoids, as well as ophiuroids and shrimps and a coarser substratum. Tanner crabs, very common in the Bering and Chukchi seas, were also found in parts of the Beaufort Sea but were not reproductive. This suggests that some biogeographic patterns represent regional environmental differences rather than small-scale dispersal or habitat restrictions.

A. Land-Fast Ice Zone

The zone beneath the land-fast ice includes the inshore zones and is a conspicuous general component of the Arctic Ocean (Newbury, 1983). This shallow zone is ice covered about 75% of the year along most Arctic coastlines. It is the zone inshore of the area where pieces of ice become grounded in shallow water, usually at depths of 15–23 m. The width of the land-fast ice zone averages between 15 and 60 km in the Alaskan Beaufort Sea, with similar widths being observed in the eastern Beaufort Sea and parts of the

Chukchi Sea. During the summer the area is exposed to strong wind-driven currents and freshwater flushing from the many adjacent river systems. Winter ice formation eliminates these wind-driven currents and eventually results in different hydrographic regimes in the inner and outer areas. Under the inner land-fast ice cover, current velocities up to 25 cm s^{-1} can be created by tidal pumping; these currents can cause significant erosion. The outer land-fast ice zone, however, often has no measurable currents for months.

The benthos is subjected to several types of disturbance. It is exposed to tremendous salinity extremes during the winter as the brine from the freezing surface layer sinks to the bottom. Salinity values of 80–100 per mil, and in one case to 182.8 per mil, have been measured. These unusually high values are in restricted areas; nevertheless, the brine on the bottom does raise the salinity. As the brine moves seaward it drives a modest thermohaline circulation. During the summer the high-salinity situation reverses and the shallow benthos is exposed to fresh water from the many river systems (Newbury, 1983). Another important disturbance for at least epifaunal organisms is the ubiquitous formation and uplift of anchor ice, as documented by Reimnitz and co-workers (1986). The outer zone is much less disturbed by this phenomenon. Still another related type of disturbance which is more prevalent in the outer zone is gouging by ice keels. Transects have shown more than 100 gouges km^{-1} with an average gouge width of 7.5 m; almost 75% of the transects in that zone intercept recent gauges. Finally, the shallow zone is disturbed by ice gouging during breakup and the resuspension during fall storms, which can completely overturn the muddy bottom to a depth of 30 cm. Certainly these represent major disturbances by any standard.

The marine primary production of the Barents and Beaufort seas is small, about 10–25 g C m^{-2} yr^{-1} (Horner, 1984; Walsh *et al.*, 1989). As would be expected, primary production essentially ceases during the fall and winter, when the much reduced ambient irradiance is essentially eliminated by snow, windblown debris from the shore, and especially sediments uplifted by anchor ice. Interestingly, this is the period during which the kelps elongate, utilizing stored carbon (Dunton *et al.*, 1982). In the absence of photosynthesis, the benthic primary consumers depend on stored reserves and detritus.

The carbon sources into the Beaufort Sea are very interesting (Schell, 1983). Over half ($\approx$ 30 g m^{-2} yr^{-1} peat versus $\approx$ 20 g m^{-2} yr^{-1} phytoplankton) of the carbon input is from terrestrial sources, and almost all of this is very old (several thousand years) peat being eroded directly into the sea or flushed by the rivers. But with the exception of one species of amphipod, very little of this carbon input from peat is utilized by marine macrofauna. This "fossil fuel subsidy" is very important to freshwater and terrestrial food webs but not to marine systems, despite the fact that more peat-derived carbon than phytoplankton-derived carbon is available to the shallow-water species (Schell, 1983). The fact that this carbon does not accumulate emphasizes the importance of nearshore bacterial decomposition (Griffiths and Morita, 1981). The meiofaunal component of this issue certainly needs study. Despite the very general importance of fluvial input of peat and other terrestrial materials, kelp carbon is also recycled in regions where kelp grow. Dunton and Schell (1987) have analyzed the stable carbon isotope ratios, which show that a remarkable amount of Arctic kelp productivity is incorporated into a broad array of consumers. Those with large fractions of kelp carbon include not only herbivores but also filter feeders (such as ascidians) and carnivores (such as *Polinices* and nemerteans). In addition, the mysid *Mysis littoralis* incorporates a surprising amount of kelp carbon, which mediates the transfer of macroalgal biomass to higher trophic levels because *Mysis* is a major prey item of many vertebrates, including marine mammals. It is interesting to note that 7% of the carbon in the soft coral *Gersemia rubiformis*, a carnivorous filter feeder, comes from kelps, suggesting that the kelp carbon is recycled back into the zooplankton.

The most common benthic invertebrates are gammarid amphipods and mysids, and the isopod *Saduria entomon* is an important predator. These are motile and efficient scavengers as well as deposit feeders. Epibenthic organisms are exposed to predation by the isopod and two abundant species of fish, the Arctic cod, *Boreogadus saida,* and the sculpin, *Myoxocephalus quadricornis.* These fish, along with invertebrates, are important prey for the ringed seal, *Phoca hispida,* in the land-fast ice areas.

Carey (1987) offers a good summary of sedimentation measurements in shallow waters under sea ice. The sedimentation represents carbon production from both ice algae and the water column, but almost all comes from the ice algae. The sedimentation early in the year is rather uniform and low, representing about 1% of the ice algal production. In this area the annual production is low and does not have a strong spring phytoplankton bloom. The spring snow cover is scant or melts and the ice algae tend to intercept the light and may delay or reduce the phytoplankton bloom. The ice algae also may offer an early source of nutrition to the motile benthic grazers, especially the lysianassid amphipod *Pseudolibrotus litoralis.* Large numbers of them graze the ice algae but eventually become pure carnivores and prey on the other grazers. Eventually 1–10% of the ice algal production (<0.2– 2.6 mg C m^{-2} h^{-1}) must reach the benthos. This coupling is weak but still important; the low rate of production presumably explains the low benthic standing stocks observed by Carey and Ruff (1977).

After breakup and during periods of clear ice, there is a large burst of photosynthesis by epontic algae (Horner and Schrader, 1982). This forms a brief but probably important source of production for epibenthic grazers as well as pelagic species. Eventually the ice disintegrates and the benthos is disturbed by drifting ice and resuspension from wind-driven storms and flooding rivers.

B. Bering and Chukchi Seas

With a continental shelf of more than 1.5×10^6 km^2, the combined Bering and Chukchi seas represent one of the largest and most productive shelf habitats in the world. Because of the extensive research programs related to petroleum exploitation and the important commercial fisheries, the marine ecology of the Bering Sea may be one of the best-studied marine habitats anywhere (reviewed by Hood and Calder, 1981; Lewbel, 1983). Most of the sampling has been done in the eastern and northern Bering Sea, although the western Bering Sea has a remarkably high level of primary production (Sambrotto *et al.*, 1984). Much of this production is advected northward into the Bering Strait and southern Chukchi Sea. Historically, the Chukchi Sea has had much less research attention than the Bering Sea, but this situation is changing (see Walsh *et al.*, 1989; Grebmeier, 1987; Grebmeier *et al.*, 1989; Grebmeier and McRoy, 1989). J. M. Grebmeier reports (personal communication) that the western Chukchi Sea has a very high biomass.

Stoker (1981) reviewed the benthic communities in the eastern Bering and Chukchi seas, identifying eight major assemblages and two north-to-south gradients. He evaluated several possible factors which may contribute to these patterns: patterns of primary productivity, current structure and velocity, organic flux from rivers, predator pressure, sediment type, and latitude. Recent studies by Grebmeier (Grebmeier *et al.*, 1988, 1989) indicate a strong coupling between water mass characteristics, food supply, and benthic processes in the region. Benthic community structure was primarily correlated to sediment heterogeneity and type in both studies (Stoker, 1978, 1981; Grebmeier, 1987; Grebmeier *et al.*, 1989), although food supply also had a variable influence on community composition (Grebmeier *et al.*, 1989). Stoker (1981) described a north-to-south trend in benthic biomass in which the northern Bering Sea and southern Chukchi Sea have the highest infaunal biomass, a conclusion supported by Grebmeier *et al.* (1988). In the most recent study, Grebmeier *et al.* (1988) concluded that the quality and quantity of organic carbon deposited to the benthos directly influenced benthic biomass. High benthic abundance and biomass correlated with high primary production in the overlying Bering Shelf–Anadyr water in the west, with low primary production in Alaska coastal water to the east determined to limit benthic biomass (Grebmeier *et al.*, 1988). This east–west variance in carbon flux to the benthos was supported by experiments on total sediment respiration (Grebmeier, 1987; Grebmeier and McRoy, 1989).

Bivalve mollusks are the most important infaunal species in terms of biomass in most areas of the Bering and Chukchi seas, and they are the most important food of the Pacific walrus. In addition, large areas of the shelf are dominated by ampeliscid amphipods, especially *Ampelisca* and *Byblis* species. These amphipods have their highest abundance and biomass north of St. Lawrence Island to the Bering Strait, where they furnish the major food resource to the California gray whales (Johnson and Nelson, 1984; Oliver and Slattery, 1985a; Nelson and Johnson, 1987).

Perhaps the most striking component of the Bering–Chukchi seas is the rich assemblage of large epibenthic species. These patterns have been summarized by Jewett and Feder (1981). There are at least 211 species of epifaunal invertebrates, the mollusks, arthropods, and echinoderms having the most species (76, 52, and 28, respectively). In the southeastern Bering Sea, four commercially important crabs (the king crabs *Paralithodes camtschatica* and *P. platypus* and the tanner or snow crabs *Chionoecetes opilio* and *C. bairdi*) and four species of asteroids *(Asterias amurensis, Evasterias echinosoma, Leptasterias polaris acervata,* and *Lethasterias nanimensis)* constituted some 70% of the epifaunal biomass of the entire eastern shelf. The large crabs of the eastern Bering Sea have been fished since 1930. From the early 1950s the fishing became increasingly intense through 1974, when the fishery doubled, taking almost 27,000 tons; thereafter it grew phenomenally through 1980, when almost 65,000 tons were taken. At that point the fishery began a collapse until 1983, when it was closed (Otto, 1986). Otto summarized extensive fishery literature and concluded that the exploitation was inadequate to explain the collapse, which he attributed to extremely variable year-class strength and high natural mortality of young crabs rather than inadequate management.

Feder and Jewett (1981) offer an excellent synthetic review of the feeding relationship of the dominant predators of benthic populations. Most bottom predators feed on the upper continental shelf during the winter, moving to shallower and warmer areas in the spring and summer. Organic carbon from the plankton and from rivers periodically enriches the benthos and may coincide with recruitment pulses of infauna, especially bivalves, one of the most important groups of prey for many benthic consumers. Feder and Jewett summarize the sources of organic carbon and the means by which it is distributed. The sources include regionally variable primary production, much of which is inefficiently coupled to the zooplankton and reaches the bottom in a patchy manner. This is also true of the large pulsed contribution from the many river systems and seagrass beds. The patterns of sedimentation, dispersion, and various types of water column–benthic coupling are discussed by Feder and Jewett (1981), Grebmeier *et al.* (1988, 1989).

The diets of the most important predators on benthic organisms are reviewed by Feder and Jewett (1981). Pink shrimp *(Pandalus borealis)* are carnivores on small epifauna and deposit feeders; they are also a major prey item for larger carnivores. Crabs potentially are very important predators; some, such as the king crabs *(Paralithodes* spp.) and tanner crabs *(Chionoe-*

cetes spp.) can be as large as dogs and are extremely mobile searchers which consume all other epifaunal species, especially mollusks, crustacea, and echinoderms. The crabs are eaten in turn by marine mammals, a few fishes, and other bigger crabs. Major fish predators on benthic animals include flatfishes, cods, sculpins, blennies, eelpouts, and snail fishes. Most are opportunistic predators which consume small invertebrates and fishes. Considering their high densities, effective searching abilities, and metabolic rates higher than those of invertebrates, the fin fishes certainly are extremely important predators, most of which have a major impact on the benthic community. Extremely low temperatures in the northern Bering and Chukchi seas are suggested to reduce fish predation in the region (Neiman, 1963; Jewett and Feder, 1980), which may reduce the influence of fish predation on the benthos (Jewett and Feder, 1981). In warm years flatfishes move north, where they feed on bivalves.

Asteroids are another important component of the benthic community, representing 32% of the total shelf epifaunal biomass and 68 and 45% in the northeastern Bering Sea and southeastern Chukchi Sea, respectively (Jewett and Feder, 1981). Most of these asteroids are generalized predators capable of heavily affecting benthic prey populations, especially bivalves. Feder and Jewett (1981) point out that mature asteroids channel a considerable amount of potential energy back into the system via their gametes, which are heavily preyed upon, and they speculate that, in a general sense, this can be an important part of the energy budget in the Bering Sea. Finally, MacIntosh and Somerton (1981) emphasize the ubiquity of large gastropods in the eastern Bering Sea. Indeed, these snails (mostly species of *Neptunea*) were the object of a brief fishery which collected over 3,500 metric tons in one season. Their diets are not well known, but they are probably also generalists, consuming shellfish and other invertebrates.

There are at least 25 species of marine mammals in the Bering Sea, and these have been estimated to consume $9-10 \times 10^6$ metric tons of nektonic and benthic species. This is the equivalent of about four times the commercial fisheries (Feder and Jewett, 1981). An excellent review of the general natural history of important Alaskan marine mammals has been edited by Lentfer (1988). So far as the benthos is concerned, most seals *(Phoca* spp., *Eumetopias,* and *Callorhinus)* feed predominantly on nekton, but many, especially the *Phoca* species, will consume benthic epifauna (see papers reviewed by Feder and Jewett, 1981, and in Lentfer,1988). Bearded seals, *Erignathus barbatus,* differ from other phocids by eating benthic invertebrates rather than fin fishes (see Kelly, 1988, for a recent review). *Erignathus* is ubiquitous throughout the Arctic, where it consumes essentially any epifaunal species as well as infaunal clams. The Pacific walrus *(Odobenus rosmarus divergens)* was classically considered to be a bottom feeder specializing on clams under normal conditions (Fay, 1981; Sease and Chapman, 1988). However, the population has increased considerably and the walrus have depleted their bivalve resource and have begun to eat seals (Lowry and Fay, 1984). In addition, they have acquired heavy parasite loads, suggesting a switch to fishes as well (L. Lowry, personal communication). Oliver *et al.* (1985) detail the benthic effects of walrus foraging. The walrus sometimes sees conspicuous siphons such as *Mya truncata* which it excavates (by either sucking or hydraulic jetting), leaving a deep (to 30 cm) pit. When feasible, they visually search for siphons, but when visibility is poor or when the walrus forage for bivalves with inconspicuous siphons (e.g., *Macoma* spp.), they furrow the upper few centimeters of the sediment and find the clams by touch. One such furrow was over 60 m long and resulted in the consumption of 34 clams, 19 of which were *Mya* living over 30 cm deep. Oliver *et al.* (1985) discuss the many indirect effects the walrus rooting has on the benthic community, as the pits trap floc and, with the discarded clam shells, offer habitats to many other benthic species. It is interesting to note that the walrus pits were not taken over by peracarid crustaceans, as were the much larger gray whale excavations (Oliver and Slattery, 1985a) or the ray pits of southern California (Van Blaricom, 1982).

Finally, Stoker defines a pronounced northward gradient in benthic biomass and diversity, especially of epibenthic forms. There are correlations with sediment, near-surface primary production and advected fluvial material, sediment type, and water mass, but he argues that the primary mechanism maintaining this trend is that the southern areas have very strong predation pressures, especially from fishes, crabs, and marine mammals. In each case the northern predation intensity is restricted by ice and/or cold-water temperatures. Jewett and Feder (1981) offer a great deal of supporting evidence for this hypothesis that predation limits the southern extension of the rich Chukchi benthic fauna. More recently, however, evidence has accumulated which shows a very clear relationship between primary production and benthic community composition and biomass (Grebmeier *et al.,* 1988, 1989). For example, the Bering Shelf–Anadyr water has an annual production of at least 285 g C m^{-2}, which is twice that of the southeastern Bering Sea and more than 10 times that of either the Beaufort or Barents Sea (Walsh *et al.,* 1989). These authors also point out the importance of the advection of this massive amount of primary production into the Arctic Ocean. Grebmeier *et al.* (1988, 1989) points out that the benthic biomass is an almost perfect reflection of the primary production in the overlying water mass. It would be interesting to evaluate the relationship between the benthic communities, the water-column productivity, and the roles and limitations of the large mobile predators.

Many of the same patterns have been recorded from the Soviet Arctic

Table 12.2 List of Biological and Physical Disturbance Agents in Arctic Habitats[a]

Surface disturbers

Brachyuran crabs (e.g. *Chionoecetes, Hyas*): many species, very abundant and ubiquitous. Dig in sediment; shallow surface disturbance.

Anomuran crabs (e.g., *Paralithodes, Pagurus*): many species, abundant and ubiquitous. Shallow sediment disturbance.

Crangonid shrimps (e.g., *Crangon*): common and ubiquitous; moderate disturbance of sediment surface.

Ophiuroid echinoderms (e.g., *Ophiura, Stegophiura*): very abundant and ubiquitous. Some dig after prey, leaving depressions; others skim epifauna or filter-feed.

Echinoid echinoderms (e.g., *Strongylocentrotus*): patchy on gravel and pebble bottom; can create shallow depressions.

Anchor ice: ubiquitous in shallow water; uplifts and transports surface sediment (see Reimnitz *et al.*, 1986).

Shallow bioturbators

Gastropod mollusks (e.g., *Tachyrhynchus, Natica, Neptunea*): abundant and ubiquitous. Various disturbance from surface trails to burrows.

Bivalve mollusks (e.g., *Yoldia, Nucula*): very abundant and ubiquitous. Can make various surface mounds and fecal pellets.

Amphipods (e.g., *Protomedia, Hippomedon*): abundant and ubiquitous, particularly in mud and fine sand; some make burrows.

Polychaetes (e.g., *Nepthys, Haploscoloplos, Sternaspis*): very abundant and ubiquitous. Many burrow and/or deposit feed; often make extensive tube structures or mounds and fecal pellets.

Echinoid echinoderms (*Lovenia, Echinarachinius*): heart urchins—rare in selected areas; make depressions and mounds.

Intermediate bioturbators

Bivalve mollusks (e.g., *Serripes, Macoma*): very abundant; ubiquitous in soft mud and sand sediment.

Echiuroid (*Echiurus*): common and ubiquitous in fine to coarse sand; makes mounds and fecal pellets.

Amphipods (e.g., *Ampelisca, Byblis*): can be extremely abundant in certain areas.

Polychaetes (e.g., *Myriochele, Onuphis*): extremely abundant and ubiquitous in fine sand.

Bearded seal (*Erignathus*): common, feeds on epibenthos and shallow infauna.

Deep disturbance

Holothurian (*Cucumaria*): common and ubiquitous; can make large mounds with fecal pellets.

Ascideans (*Pelonaia*): common in sand and gravel habitats.

Bivalve mollusks (e.g., *Mya, Spisula*): very common and deep burrowers; ubiquitous in hard sand and mud.

Sipunchulia (*Golfingia*): common in sand.

Polychaeta (*Lumbrineris, Ampharete, Maldane*): common; ubiquitous—some large mounds.

Walrus (*Odobenus*): abundant and ubiquitous; sucks clams from bottom, leaving pits to 0.5 m deep (see Oliver *et al.*, 1985).

Gray whales (*Eschrichtius*): seasonally common in Bering and Chukchi seas; feed on benthic invertebrates, primarily on amphipods (e.g., *Ampelisca, Pontoporeia*). Excavate pits 1 × 2 m and 0.1–0.5 m deep (see Oliver and Slattery, 1985a).

Ice disturbance: gouging and scouring very common to water depths of at least 50 m. Depth of the gouge can be over 7 m (see Barnes *et al.*, 1984).

[a] Modified from Nelson *et al.* (1981).

(Zenkevitch, 1963; Golikov and Averincev, 1977). In the appropriate habitats these reports summarize fluvial input and especially ice disturbance, with furrows and grooves common to at least 30 m. As is true in the Beaufort, the seasonal salinity fluxes can be very large near rivers, which also carry a large amount of sediment. Similar patterns of vertical zonation were also observed in many areas. However, the Beaufort Sea differs in many ways from other satellite seas of the Arctic Ocean. Perhaps most notable is that other seas seem to have much more emphasis on filter feeding, and they especially have a preponderance of bivalves, which can constitute 60–80% of the total fauna (Thorson, 1957).

Stoker (1981) summarizes benthic standing-stock estimates from East Greenland (Thorson, 1934) as 20–400 g m^{-2}, northwest Greenland (Vibe, 1939) as 160–387 g m^{-2}, Baffin Island (Ellis, 1960) as 200–300 g m^{-2}, and Sea of Okhotsk (Zenkevitch, 1963) as 200 g m^{-2}, all of which compare with his rough mean of 300 g m^{-2} for the Bering and Chukchi seas. These contrast with the impoverished White and Baltic seas, which Zenkevitch (1963) reports as averaging 20 and 33 g m^{-2} respectively. Stoker's high-biomass *Serripes groenlandicus* community means of 1–2000 g m^{-2} also compare with Vibe's (1939) estimates of 1481–3500 g m^{-2} for Greenland. These similarities suggest that many of the Bering and Chukchi patterns are generally true in similar productive shelf habitats. Certainly the most notable generalization of the Arctic benthos is that it is characterized by a remarkable diversity of disturbances (Table 12.2).

V. Arctic Macroalgae

Since the early 1800s expeditions to the Arctic have collected macroalgae, and there is a history of almost 150 years of excellent publications describing these collections. All of the giants in phycology enthusiastically published voluminous reports beginning in the 1840s; dozens of papers, all before the turn of the century, were published by Agardh, Boldt, Borge, Børgesen, Dickie, Farlow, Foslie, Harvey, Hooker, Kjellman, Lagerheim, Rosenvinge, and Setchell. This massive literature has been succinctly summarized by Taylor (1954), who somehow manages to impart an ecological and biogeographical overview. Of the ecological factors influencing the distribution and abundance of macroalgae, Taylor discounts temperature and emphasizes substrata, light, and salinity. Distinct biogeographic patterns are often related to these variables.

As with the fauna, the northern Soviet coast is relatively unproductive. In addition to the lack of an appropriate substratum for macroalgae, the region is seasonally inundated with nearly fresh water. The Murman Sea represents a transition from the relatively rich Norwegian Polar Sea to the sparse flora of the Siberian coast. With the exception of *Fucus distichus* subsp. *evenescens*, the Atlantic fucoids disappear and the more truly Arctic additions to the flora such as *Haplospora globusu, Saccorhiza dermatodea*, and *Laminaria* spp. appear in the White Sea and become more apparent to the east. The algal flora of Novaya Zemlya is also poorly developed; however, there is a richer flora to the northeast in the Matochkin Sea area, which includes polar kelps represented by *Chorda tomentosa, C. filium, Laminaria saccharina, L. solidungula, Agarum cribrosium,* and *Saccorhiza dermatodea* (luxuriant to 77°N). The Kara Sea is relatively poor but is marked by *Alaria esculenta*. The eastern Soviet area is represented by a poorly developed flora, but this may in part represent relatively low collecting effort. It is interesting to note that there is very little penetration of the rich association of Pacific species into the Arctic Ocean. More detailed distributional and biomass data can be found in Golikov and Averincev (1977).

At the time of Taylor's review the north American region was poorly described, although it was known (Dall, 1875) that the Bering Sea and perhaps parts of the Chukchi Sea had rich algal communities. I have observed richly developed kelp communities at King and Diomede islands in the Bering Strait; however, Mohr *et al.* (1957) summarize collections of algae between Points Lay and Barrow in the Barrow area and in scattered localities to longitude 144° which they characterize as being floristically very poor. These may be the extent of kelp communities in the Chukchi Sea, as K. Dunton (personal communication) reports that extensive searching has failed to find much more kelp in the Chukchi Sea and R. Wilce (personal communication) refers to it as a phycological desert. Farther north, Dunton *et al.* (1982) have described high Arctic kelp beds at the "Boulder Patch" in the western Beaufort Sea and Chapman and Lindley (1980) and Dunton and Schell (1986) have described the productivity of *Laminaria solidungula* in the high Arctic.

Taylor's summary of the macroalgal flora of the Canadian Arctic has been updated by Lee (1973, 1980), who was able to build on the work of Wilce (1959) and Ellis and Wilce (1961). Lee noted that of the 175 known species from the Canadian Arctic, 160 had Atlantic affinities; of the remaining 15 species, 12 are endemic to the Arctic and only 3 have Pacific affinities. Like Taylor, he attributes the depauperate flora to physical stress and inappropriate soft substrata. It is interesting that juxtaposed to the poor Canadian flora, Baffin Bay and west Greenland have the richest macroalgal floras in the Arctic, according to Taylor. R. T. Wilce (personal communication) characterizes all of west Greenland (north Disko Island) as subarctic and Baffin Island north to Pangnirtung as having subarctic features. Jan Mayen Island has a remarkably rich macroalgal flora, with *Phycodrys* and *Pantoneura* growing at 110–118 m depth (Taylor, 1954).

12 Polar Benthos   653

Because it reaches to such a high latitude yet has a rich flora, Spitsbergen or Svalbard is noteworthy. Taylor (1954) reviews the extensive work of Kjellman, who overwintered on the south island (R. T. Wilce, personal communication). As expected from ice-scoured shores, there are few plants in the littoral zone, although at least 14 species were reported and many characteristically littoral or shallow species were collected in deeper water. Another association which includes *Phyllophora* and *Laminaria solidungula* grows to a depth of at least 27 m. Taylor also mentions luxuriant growths of *L. solidungula* and mats of detached kelps which are continuing to grow. The Spitsbergen flora is associated with that of west Greenland and Arctic America rather than Siberia. Finally, because it is bathed by the warm current from the North Atlantic, the flora of the northern coast of Norway has a distinct North Atlantic affinity.

VI. Origin, Evolution, and Historical Background of Antarctic Habitats and Species

Unlike the Arctic coast, the Antarctic continent of today is a bleak frozen landscape with little terrestrial life. Some regions in Antarctic seas, however, may represent some of the most productive habitats in the world (but see Chapters 9 and 10). They are remarkable in their high degree of endemism. However, the continent was not always barren and frozen. In the Paleozoic and Mesozoic eras, Antarctica was part of the mild subtropical Gondwanaland. The breakup of Gondwanaland and the poleward migration of Antarctica began about 60 m.y. BP (Kennett, 1978). Sometime thereafter, the circum-Antarctic current developed and effectively isolated the continent. Recent cores from McMurdo Sound give evidence of temperate climates persisting until perhaps 22 m.y. ago (A. Pyne, personal communication). This indicates that the intense cooling came well after the isolation associated with the poleward migration and by the developing convergence, divergence, and upwelling systems. The present pattern probably became well established by 20 m.y. BP (Kennett, 1978; A. Pyne, personal communication). During the last 30 m.y. the high latitude and strong consistent divergence and convergence zones has resulted in the Southern Ocean developing into one of the most discrete marine ecosystems in the world. Although it is cold, the physical parameters are more stable and uniform than those of any other continental system; and though there is a strong seasonal pulse, the biological productivity is high and usually extremely predictable.

The origin of the Antarctic fauna and flora is still one of the most fascinating mysteries of the continent (see Lipps and Hickman, 1982). The early geological history presents a very different fauna from that of the present because the Cretaceous marine fossil fauna indicates a temperate climate

(Dell, 1972). Dell also points out that the Cretaceous mollusks and echinoderms have essentially no affinities with their modern counterparts. The lower Miocene beds in West Antarctica contain fossil penguins and a "cool" molluscan fauna distinct from the present forms. There are, however, 30-million-year-old fossils of the Notothenioid fishes that are similar to the present fish fauna. There are Eocene fossils of an echinoid, *Schizaster,* which is closely related to the living genera *Abatus, Amphipneustes,* and *Tripylus* (Dell, 1972). Newman and Ross (1971) observed that of extensive early Cirripedia fossils, only the Pleistocene *Bathylasma corolliforme* survives in deep water.

Dell (1972) elaborates on the theme of very faint biogeographic relationships with southern South America. This is especially true for the Nototheniiformes fishes that demonstrate an interesting adaptive radiation, which includes whitebloodedness and antifreeze. Andriashev (1965) argues that the Nototheniiformes and the Muraenolepidae are old endemic elements coexisting with other more recent invasions. Dell cites Andriashev's conclusion that the Antarctic fish have ancient roots with occasional links to South America and New Zealand and concludes that this pattern of an ancient element, sometimes with extensive radiation and rare recent invasions, is the general source of Antarctic species.

There now seems to be a developing consensus of biogeographers that most of the Antarctic benthic fauna is very old and representative of the original propagules which moved poleward with the continent (Picken, 1985). Although there may be a certain amount of faunal exchange from the outside, especially via the deep sea and the Scotia Shelf, this overview is supported with an extraordinarily high degree of endemism. The fishes, for example, have approximately 120 species, 100 of which are endemic with 75 in a single suborder, the Notothenioidei (DeWitt, 1971). The high degree of invertebrate endemism is reviewed by Dell (1972), Knox and Lowry (1977), and White (1984) and is summarized in Table 12.3. White (1984) and Picken (1980) have developed the theme that high levels of endemism result from isolation, which in the Antarctic is enhanced by the hydrological barriers and the fact that many of the dominant benthic invertebrate groups lack pelagic larval stages. White (1984) develops the argument that the Antarctic endemism is remarkably inconsistent within phyla and that most of the endemism is found in relatively few taxa. For example, assuming endemism is a function of time and isolation, recent colonizers should have less speciation than more ancient autochthonous groups. However, in most phyla autochthonous groups exhibit both evolutionary conservatism and extensive speciation. White offers the isopods as an excellent example: they are reasonably well known and the more ancient families are represented by very few conservative genera which are mostly monospecific, whereas only three

Table 12.3 Some Examples of Antarctic Endemism[a]

Group	Genera (%)	Species (%)	Reference
Fish	70	95	Andriashev (1965)
Isopoda and Tanaidacea	10	66	Kussakin (1967)
Pycnogonid	14	> 90	Fry (1964)
Echinodermata	27	73	Ekman (1953)
Echinoidea	25	77	Pawson (1969b)
Holothuroidea	5	58	Pawson (1969a)
Bryozoa	—	58	Bullivant (1969)
Polychaete	5	57	Knox and Lowry (1977)
Amphipoda	39	90	Knox and Lowry (1977)
Chlorophyta			
Subantarctic	—	16	Heywood and Whitaker (1984)
Low Antarctic	—	33	Heywood and Whitaker (1984)
High Antarctic	—	67	Heywood and Whitaker (1984)
Phaeophyta			
Subantarctic	—	41	Heywood and Whitaker (1984)
Low Antarctic	—	73	Heywood and Whitaker (1984)
High Antarctic	—	83	Heywood and Whitaker (1984)
Rhodophyta			
Subantarctic	—	70	Heywood and Whitaker (1984)
Low Antarctic	—	92	Heywood and Whitaker (1984)
High Antarctic	—	100	Heywood and Whitaker (1984)

[a] After Knox and Lowry (1977) and Heywood and Whitaker (1984).

families (Antarcturidae, Serolidae, and Munnidae) account for over 50% of the Antarctic species. He points out that some of these species may have Gondwanian affiliations. White also reviews Kussakin's (1967) clever use of Preston's coefficient of faunal diversity to evaluate the biogeographic affinities of the large, relatively well-known isopods and applies the coefficient to his own extensive isopod data to demonstrate the robustness of Hedgpeth's (1971) biogeographic groupings.

Picken (1985) summarizes current taxonomic and biogeographical work on prosobranch gastropods, perhaps the best-known benthic group in the Antarctic, which demonstrates a much higher level of speciation in the Scotia Arc than was previously realized. He points out that this suggests a reassessment of the hypothesis that the Scotia Arc is a pathway for gene flow between subantarctic and Antarctic regions. This hypothesis has yet to be tested. The isothermal structure of the water column allows deep-water specimens to be dredged and transferred to other areas, perhaps providing a way to test this or other biogeographic hypotheses (cf. Dayton *et al.*, 1982).

Summarizing, the Antarctic fauna appears to be derived from three sources: (1) a relic autochthonous fauna, (2) eurybathic species which have emerged from deep water into shallower coastal waters, and (3) cool-temperate species, mostly from South America, which have invaded via the Scotia Arc.

A. Biogeography

It is difficult to develop an Antarctic parallel to the regional differences between the satellite seas of the Arctic Ocean. Whereas the Arctic seas usually have distinctive physical and biological characteristics (e.g., salinity, sediments, available primary production) which can delimit the biological components, all the Antarctic benthic habitats share relatively constant physical parameters such as temperature, salinity, and substrata, with few if any important barriers. Not surprisingly, they also share many of the same species, albeit in different proportions. Most biogeographers, especially Dell (1972), have recognized the problems associated with species from the Antarctic Peninsula and the Scotia Arc partially penetrating the Antarctic polar front. White (1984) offers a clear review and supports Hedgpeth's (1969, 1971) scheme, which recognizes the large number of circumpolar species plus smaller provinces or regions. Hedgpeth's groups are (1) circumcontinental, (2) species restricted to the Bellingshausen and Weddell seas and the Antarctic Peninsula, (3) species restricted to the Scotia Arc, and (4) species restricted to the subantarctic islands. To a certain extent, the fact that they are so robust is simply a reflection that almost all the species fit into the circumcontinental group.

But such general groupings cannot deal with marked local differences. For example, in McMurdo Sound an east–west difference of less than 50 km exhibits a profound oligotrophic-to-eutrophic shift. In addition, a well-developed sponge community essentially disappears to the north, to be replaced by a starfish *(Odontaster validus)*–dominated assemblage with scattered coelenterates in less than 10 km along Hut Point Peninsula on the East Sound while the West Sound has a profound soft-bottom infaunal density gradient from high in the north to low in the southwest (Dayton and Oliver, 1977). Most coastal regions exhibit profound local shifts in community composition, usually in response to differences in wave exposure or substrata, but at McMurdo Sound it is probably related to differences in currents and ice and snow effects on light transmission, all of which have ramifications for the productivity (Dayton et al., 1986; Barry and Dayton, 1988; Barry, 1988).

A noteworthy observation about the Antarctic benthos is that photographs of sponge communities from McMurdo Station and from Mirny on the other side of the continent, as well as photographs of deep-water sponge associations from the Antarctic Peninsula, all look identical. Similarly, the picture of the *Admussium colbecki* association in Nakajima et al. (1982) from Syowa Station looks exactly like one from New Harbor, McMurdo Sound. Therefore, although major community shifts do occur over small spatial scales, given the right conditions identical species assemblages seem to occur around the entire continent. This phenomenon is unique in marine biogeography and amply justifies the very inclusive "circumcontinental" group defined by Hedgpeth.

B. Sedimentation

Because of the great depth of the continental shelf (average ca. 500 m), the absence of appreciable runoff, and ice-covered shorelines restricting wave influence, patterns of sedimentation are driven by biogenic processes and oceanographic currents and mass flow. Dunbar et al. (1985) summarize the surface sediment data from the Ross Sea west to the George V coast. The surface sediments are similar and have distributions which indicate that similar processes are active in each area, and sedimentation on the shelf is dominated by production and settlement of biogenic material and resuspension by bottom currents. The biogenic components seem especially enriched in areas characterized by open water in the summer months. The currents rework the sediments on bottoms with no obstructions to depths of about 500 m.

Leventer and Dunbar (1987) summarize their diatom flux data in McMurdo Sound. In areas below more permanent sea ice, a substantial portion of the diatom species in sediment traps were representative of ice algal species. There also were stations which received substantial amounts of pelagic species. Increased near-bottom flux at all sites indicates bottom resuspension and/or lateral advection at depth. Lateral advection is more important in some sites than in others, which are strongly influenced by the near-bottom nepheloid layer. Because of their dissolution, relatively small amounts of sea ice diatoms show up in the sedimentary record despite the high sea ice productivity. In addition, the diatom fluxes in the McMurdo Sound region vary over three orders of magnitude, with the highest fluxes at Granite Harbor and Hut Point Peninsula, areas influenced by advection from nearby open water. They also noted that fecal pellet fluxes are extremely low in McMurdo Sound compared to other areas in the Southern Ocean (e.g., von Bodungen et al., 1986).

The carbon flux at McMurdo Sound mirrors the benthic distributional pattern; Dunbar et al. (1989) found organic carbon fluxes from sediment traps to be one to two orders of magnitude higher in the eastern side of McMurdo Sound. They relate these differences to the fact that the currents bathing the southwest sound come from beneath the Ross Ice Shelf with little allochthonous carbon; in contrast, the currents along the east side tend to flow southward from a much more productive area (Barry, 1988; Barry and Dayton, 1988). This difference is exacerbated by the ice conditions (more multiyear ice and windblown debris), which probably reduce *in situ* production in the southwestern area. In addition, the absence of summer basal melting and ice breakout reduces the flux of ice algae to the bottom. Finally, Dunbar et al. (1989) make the interesting observation that resuspension events transport biogenic debris from shallow (<400 m) to shelf basins. They calculate that the organic flux to the deeper basins averages 45 mg C m^{-2} day^{-1}. This is more than an order of magnitude higher than the world average to continental margins. Because there is no evidence of anaerobic sediments, this suggests that these Antarctic deep basins may be very interesting habitats characterized by substantial benthic activity.

VII. Antarctic Patterns and Processes

The Antarctic benthic community differs from that in the Arctic in almost every regard. Rather than being an extensive shallow continental shelf, the Antarctic shelf is unique in its great depth, averaging about 500 m with troughs to over 1000 m. The Antarctic continental slope descends to some 3000 m, with the abyssal plain occurring at depths of 3700–5000 m (Knox and Lowry, 1977). Unlike the Arctic, with its massive contributions from fresh water, the Antarctic has no rivers and relatively little meltwater runoff. Whereas the Arctic shelf experiences wide fluctuations in temperature, salinity, nutrients, and dissolved gases, the Antarctic waters remain relatively constant in all of these parameters. Finally, the beaches and shallow-water areas are much less extensive and are often permanently covered with ice; there appears to be much less disturbance and transport of sediment by ice. Summarizing, the Antarctic benthic habitat is much more stable physically and has almost no terrestrial input with the exception of the runoff from penguin and seal colonies, which could be locally very important.

To understand the processes driving the benthic systems, it is important to describe the associations or communities, evaluate the nature and abundance of organic input and its utilization, and describe other forcing or regulatory functions determining the structure of the community. Despite considerable research, these characteristics of the Antarctic benthic environment are known only in the few localities which have had benthic research programs. Antarctic benthic research has dealt primarily with aspects of systematics, biogeography, reproduction, and foraging biology. Much of this research focused on the definition and causes of various zones and the definition of species assemblages and their relative densities and biomass.

A. Zonation and Benthic Assemblages

Some form of biological distribution pattern probably characterizes all vertical gradients. This is a response to the physical factors which have been well studied in marine and terrestrial habitats, where it is often found that various biological interactions modify and/or exacerbate the physical effects. The Antarctic benthos is no exception and many workers have reported zonal patterns, especially on hard and cobble substrata.

In lower latitudes the most pronounced zonal patterns are defined by the distribution of algae (Neushul, 1965; Hedgpeth, 1971; DeLaca and Lipps, 1976; Lamb and Zimmerman, 1977; Richardson, 1979; see Picken, 1985). The intertidal zone is heavily disturbed and is characterized by crustose red and brown algae and the limpet *Nacella concinna* (Hedgpeth, 1971). The shallow sublittoral region exhibits an algal zonation in which *Desmarestia menziesii* occurs above *Himantothallus* (= *Phyllogigas*) *grandifolius,* but the interface between the two zones varies from 5 to 25 m (see Picken, 1985). As is characteristic of the genus, *D. menziesii* is probably indicative of disturbed conditions. Thus *Himantothallus* is likely to be excluded from the upper zone by ice disturbance where *Desmarestia* occurs opportunistically (DeLaca and Lipps, 1976). Presumably *Himantothallus* is competitively dominant in relatively undisturbed conditions. DeLaca and Lipps offer qualitative details of the species zonation (both algal and animal) near Palmer Station, but it is unclear whether the vertical distributions relate to disturbance, the dominance relationships among the algae, or some other environmental process. Dell (1972) and Heywood and Whitaker (1984) give details of algal distribution patterns.

In higher latitudes the hard-bottom habitats also exhibit a very general zonation pattern (reviewed by Hedgpeth, 1971) which reflects a gradient of ice disturbance (see Gruzov, 1977). The shallow zone (5–15 m) is covered by ice most of the year and is relatively bare of organisms, but during the ice-free summer months there is a heavy benthic diatom growth, which attracts large populations of the asteroid *Odontaster validus* and the echinoid *Sterechinus neumayeri.* In most areas there is a second zone (15–30 m) which is periodically exposed to disturbance from grounded ice and especially the formation and uplift of anchor ice (Dayton *et al.,* 1969; Gruzov, 1977). This zone is inhibited by coelenterates such as the stoloniferan *Clavularia;* and alcyonarian *Alcyonium;* several actinarians including *Isotealia, Artemidactis,* and *Urticinopsis;* and several species of hydroids such as *Lam-*

pra, Tubularia, and *Halecium.* Several ascidians, especially *Cnemidocarpa,* are conspicuous, as are a few sponges, most notable of which is *Homaxinella balfourensis,* which has massive interdecadal population fluctuations in response to anchor ice intensity (Dayton, 1979, 1989). Many of these sedentary species have predators which are also found in this zone. Dayton *et al.* (1970) outline some of the trophic relationships. The similarities between McMurdo Sound, the Haswell Islands (Gruzov, 1977), and the Vestfold Hills (Tucker and Burton, 1987) are striking (see reviews in Hedgpeth, 1971; Gruzov, 1977; White, 1984).

The substratum below 33 m is rarely disturbed by anchor ice and at McMurdo Station is dominated by many species of long-lived sponges living on a mat of sponge spicules which may be over 1 m thick (Dayton *et al.,* 1974). However, anchor ice occurs much deeper in areas exposed to deep cold water moving up on the shelf, such as White Island or the Daily Islands, McMurdo Sound (Barry and Dayton, 1988). The distribution and abundances of most of the sponges reflect relatively intense predation by asteroids and, at least in some cases, the occasional predation by the predominantly deposit-feeding *Odontaster validus* on *Acodontaster conspicuus,* a large asteroid capable of eating the largest sponges (see Dayton *et al.,* 1974; Dayton, 1979). At McMurdo Station we have observed that the sponge zone below about 50 m is replaced by bryozoans and sabellarid polychaetes with many fewer sponges; certainly most of the sponges occur in much deeper water, but unpublished photographic transects show a marked decline in sponge abundance deeper than 45–50 m. We hypothesize that this reflects increased *Acodontaster* predation in the reduced densities of *O. validus,* which are a function of decreased primary production. I emphasize that this pattern may be idiosyncratic to the southern end of the Hut Point Peninsula. A few kilometers to the north the substratum changes to gravel or mud and the sponge zone drops below diving depths, although in one area the substratum is adequate for sponges but is covered instead by coralline algae and gorgonians. Nevertheless, as mentioned earlier, Hedgpeth's (1971) review of circum-Antarctic research emphasizes that the same species tend to occupy roughly the same zones, suggesting that this pattern is general in the appropriate habitats.

B. Soft-Bottom Habitats

Soft-bottom habitats do not reflect these generalizations because they tend much more closely to reflect local sediment type and productivity and history of disturbance (both biological and physical); zonal patterns, if they exist, are much less important than local species assemblages and the methods by which they are maintained (see Gallardo, 1987). A few, hopefully representative patterns of abundance are presented in Table 12.4. Such

Table 12.4 Comparison of Benthic Faunal Abundance at Antarctic and Arctic Locations[a]

Location	Depth (m)	Screen aperture	Abundance (individuals m^{-2})		Source
			Average	Range	
Antarctica					
McMurdo Sound, East	20	0.5	118,712		Dayton and Oliver (1977)
	20	0.5	155,573		
	30	0.5	145,781		
McMurdo Sound, West	30	0.5	2,184		
	30	0.5	45,294		
	40	0.5	10,036		
Ross Sea	500	0.5	1,960		
Arthur Harbor, Anvers Island, Bismarck Strait	5–75	1.0	18,412	3,954–34,251	Richardson and Hedgpeth (1977)
	5–75	0.5		9,440–86,514	
	300–700	1.0	1,530		
	300–700	1.0	2,891		
	300–700	0.5	1,876		
	300–700	0.5	4,362		
Arthur Harbour, Anvers Island	30	1.0	7,629	3,264–14,756	Lowry (1975)
		1.0	6,285	2,244–11,747	
King Edward Cove, South Georgia	5	1.0	31,150		Platt (1980)
	6	1.0	2,490		
	11	1.0	1,618	390–3,260	
South Shetland Islands	46–115		11,177	6,720–17,960	Mills (1975)

(*continues*)

Table 12.4 (*Continued*)

Location	Depth (m)	Screen aperture	Abundance (individuals m^{-2})		Source
			Average	Range	
Arctic					
Frustration Bay	5	2.0	243		Ellis (1960)
Northwest Baffin Bay	26–52	1.0	1,133	1,190	Thomson (1982)
	106–250	1.0	5,502	4,006	
	251–500	1.0	1,983	1,141	
		1.0	988	509	
Central Baffin Bay	5		6,193	344	
	26–52		1,730	1,092	
	106–250		867	572	
	751–1,100		231	42	
Eastern Bering Sea	20–103			330–4,414	Stoker (1973)
Southwestern Bering Sea	23–30	1.0	1,350	±195	Carey et al. (1974)
	46–52	1.0	2,120	±165	
	100–140	1.0	2,410	±593	
	360	1.0	2,380	±424	
	700	1.0	4,330	±570	
	1,700	1.0	1,730	±189	
	260	1.0	270	±54	
Nain Bay, Labrador	30–80		11,239	7,050–17,198	Mills (1975)
Northern Bering Sea					
Alaska coastal water (east)	19–42	1.0	140–7,770	2,413	Grebmeier et al. (1988)
Bering Shelf–Anadyr water (west)	20–51	1.0	188–14,365	4,718	
Chukchi Sea					
Alaska coastal water (east)	22–51	1.0	193–2,765	1,268	
Bering Shelf–Anadyr water (west)	46–54	1.0	1,080–12,115	4,461	

[a] After White (1984) and Carey and Ruff (1977).

relative density and natural history data offer more biological insight than some summary statistics such as diversity indices or cluster diagrams.

Gallardo (1987) summarizes his data (Gallardo *et al.,* 1977) from Port Foster (Deception Island, Antarctic Peninsula), which described the effects of a volcanic eruption, and from an undisturbed site at Chile Bay (Greenwich Island, Antarctic Peninsula). The Port Foster benthos appeared continually disturbed by the redistribution of the ash. There was an interesting population explosion of the echiurid *Echiurus antarcticus* which lasted 1 year. The Chile Bay samples demonstrated two very distinct groups: a deeper (>100 m) assemblage which was dominated by the polychaete *Maldane sarsi antarctic,* other polychaetes, and bivalves and a shallower assemblage in which *Maldane* was conspicuously absent and crustacea were the most abundant. Arthur Harbor, Anvers Island, Antarctic Peninsula, was studied by Lowry (1975) and Richardson and Hedgpeth (1977), reported high densities and diversity (Table 12.4) which they related to high productivity and habitat stability. We found vast differences in abundance of infauna between relative eutrophic and oligotrophic sides of McMurdo Sound (Table 12.4), which we relate to differences in benthic primary productivity and carbon supply (Dayton and Oliver, 1977; Dayton *et al.,* 1986; Barry and Dayton, 1988). The dense assemblage at McMurdo Sound is notable for both its extraordinary densities (some samples may be over 150,000 individuals m^{-2}) and its remarkable equitability (11 species maintain populations over 2000 individuals m^{-2}), which differs markedly from other, usually monospecific associations in other dense assemblages (Dayton and Oliver, 1977). Oliver and Slattery (1985b) describe some of the mechanisms by which the dense assemblage is maintained. The most important is predation by the tanaid *Nototanais dimorphus* and the phoxocephalid *Heterophoxus videns* on larvae, juveniles, and small polychaetes. The other component of the dense assemblage at McMurdo Station is a well-developed vertical structure with at least some canopy species capable of consuming larvae (Figure 12.2) such as the actinian *Edswardsi meridionalis.* Some of these canopy species sustain a modest predation from fishes. But Oliver and Slattery emphasize that the canopy structure is largely undisturbed in the dense assemblage and that this is very different from the situation in other shelf benthic habitats, which are exposed to a high intensity of disturbance from many flatfish, skates, rays, and other demersal fishes as well as marine mammals such as walrus and gray whales (see Table 12.2).

C. Deeper Benthic Habitats

The Antarctic shelf has been actively studied by many diving programs since the late 1950s and early 1960s (Bunt, 1963; Peckham, 1964; Neushul, 1965),

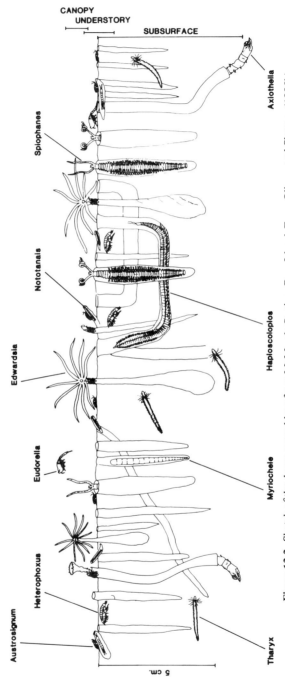

Figure 12.2 Sketch of the dense assemblage from McMurdo Station, Ross Island. From Oliver and Slattery (1985b).

but the bathyl habitats below diving depths have received relatively little attention since the pioneering work of Bullivant and Dearborn (1967). The hard substrata tend to be dominated by sponges, bryozoans, hydroids, alcyonarians, polychaetes, and bivalves. Bullivant and Dearborn (1967) review several assemblages in the Ross Sea which are associated with specific substrata. The most common is the Deep Shelf Mixed Assemblage, which occurs to 523 m on fine sediments with erratic boulders. This assemblage includes tubiculous polychaetes, bryozoans, gorgonians, ophiuroids, and crinoids. Another is the Deep Shelf Mud Bottom Assemblage, which occurs between 400 and 750 m on mud or sandy substrata; here polychaetes and ophiuroids occur with sipunculids, holothurians, asteroids, and scaphopods. Another interesting assemblage is located along the 400-m shelf break of the Pennell Bank of the Ross Sea. This is a mud–sand substratum with rocky outcrops and cliffs which support cirripeds, stylasterine corals, ascidea, bryozoans, gorgonians, ophiuroids, and pycnogonids. We tested several hypotheses regarding the distribution of the barnacle *Bathylasma corolliforme* using transplants of specimens from the Pennell Bank to McMurdo Sound. We negated hypotheses based on physiological constraints, habitat availability, dispersal, and certain types of predation which might have explained the apparent restriction of these barnacles to the deeper Pennell Bank region (Dayton *et al.*, 1982). Nevertheless, probably the most important factors influencing the various faunal patterns are substratum and sedimentation, as discussed by Dunbar *et al.* (1989).

D. Productivity and Benthic Communities

In polar marine systems the relative importance of advection and *in situ* production may be quite variable and have profound consequences for nearby shallow-water benthic communities. Polar marine systems have three *in situ* sources of primary production — phytoplankton, ice algae, and benthic microalgae — and in some Arctic systems the sea ice algal production dominates the local production (Horner and Schrader, 1982). However, in some areas advection of phytoplankton may be an equally or more important source of primary production (Thomson, 1982). This is true for some areas beneath the Ross Ice Shelf in McMurdo Sound, such as White Island (Dayton and Oliver, 1977; G. A. Knox, personal communication) or Black Island (Collen, 1979), where there are dense populations of filter feeders which live in complete darkness 30–50 km from the ice edge. In McMurdo Sound sea ice microbial communities are very productive (Palmisano and Sullivan, 1983; Kottmeier *et al.*, (1987), and detrital fallout from the sea ice may be very important to benthic communities, as it is in some Arctic habitats (Carey, 1987).

Primary production in the water column and its contribution to benthic communities potentially are very great at McMurdo Sound. Preliminary estimates of primary production from nutrient depletion calculations (spring to summer), while imprecise, indicate that productivity may have exceeded 2.0 gm C m^{-2} day^{-1} (Barry, 1988). A similar calculation from late-summer data in Littlepage (1965) corroborates this estimate, as do the early data of Bunt (1964; Bunt and Lee, 1970). That much of this material gets to the bottom in shallow waters is supported by observations of dense accumulations of algae, presumably *Phaeocystis,* on the bottom in March 1977 (J. S. Oliver, personal communication). In McMurdo Sound the benthic community assemblages have remarkably strong east–west and north–south density gradients apparently related to gradients in the availability of primary production (Dayton and Oliver, 1977; DeLaca *et al.,* 1980; Hodson *et al.,* 1981; Dayton *et al.,* 1986; Barry and Dayton, 1988; Barry, 1988).

Recent investigations of the vertical flux of lithogenic and biogenic material have both corroborated previously observed productivity patterns and revealed new insights into benthic sedimentary processes throughout McMurdo Sound (Dunbar *et al.,* 1985, 1989; Dunbar and Leventer, 1986). East–west McMurdo Sound comparisons of sedimentary processes found lower vertical fluxes for diatoms, silica, and sediments in the southwestern sound, compared to the southeastern sound. However, fluxes in the northwestern sound (Granite Harbor) were usually higher than in all other areas. Resuspension (or perhaps lateral advection) at all sites was significant, leading to a near-bottom nepheloid layer (Leventer and Dunbar, 1987). Resuspension may be caused by storm surges or short-term changes in currents, even in areas with normally sluggish flow (R. B. Dunbar, personal communication).

Summarizing, benthic communities are organized around the nature, abundance, and predictability of organic inputs, their utilization, and the degree to which the structure of the community is influenced by various pattern of disturbance. The Arctic benthos is very heavily disturbed (Table 12.4). Gallardo (1987) outlines many types of shallow-water physical disturbances, but compared to the Arctic the Antarctic benthos is relatively free of disturbance and seems to utilize the organic input very efficiently. This organic input is an important factor determining the densities and biomass of the benthic populations. Despite this importance, there seem to be few attempts to evaluate this issue. The organic carbon content of the sediments tends to be rather low (Warnke *et al.,* 1973 recorded most values as less than 1%), and although there certainly are places where pockets of organic material such as drift algae accumulate, especially along the Antarctic Peninsula reducing layers are not common. Nevertheless, Antarctic benthic communities often are characterized by high densities and biomass of epifaunal and infaunal organisms.

As Hedgpeth (1977) noted, the Antarctic seas offer a paradox: on one hand they are marked by an apparently highly productive plankton system capable of supporting great populations of whales, penguins, seals, fishes, and cephalopods, but on the other hand, as both White (1984) and Picken (1985) emphasized in their reviews, the benthic communities are characterized by low growth rates and secondary production. Most of the large epifaunal sponges along the east side of McMurdo Sound do appear to grow slowly (Dayton *et al.*, 1974; Dayton, 1979), yet there is a large amount of primary production from the ice algae, the phytoplankton, and the benthic microalgae. In other systems one would look to microbial decomposition, yet here even bacterial growth rates seem to be low (White *et al.*, 1984, 1985). Dunbar *et al.* (1989) argued that winter resuspension and apparent transport to deep basins create a tremendous organic carbon influx to the deep benthic habitat. Yet organic carbon is not accumulating, which suggests that the benthic community may be an important sink. The role of protozoans (Bowser *et al.*, 1986; DeLaca, 1986) and meiofauna needs to be quantified in this regard.

Benthic microbial and meiofaunal ecology has been neglected in Antarctic research (Picken, 1985), although there has been some benthic microbial research in the McMurdo Sound area. The microalgal and total microbial biomass is highest in the eastern sound (Hodson *et al.*, 1981; DeLaca *et al.*, 1980; Dayton *et al.*, 1986; G. A. Smith *et al.*, 1986; 1990; G. A. Smith unpublished). Bacterial biomass in both the water column and the benthos is high in some areas (Fuhrman and Azam, 1980; White *et al.*, 1984) and consists of diverse assemblages of microorganisms, but is reported to be very low in the western sound (DeLaca *et al.*, 1980; Hodson *et al.*, 1981). However, the relative importance of sedimentation versus benthic production to community energy flow through benthic macrofauna is still unknown.

Finally, the relationship between microfauna, infauna, and epifauna in temperate soft bottoms has been studied for years (Rhoads, 1974; see Gray, 1981, for review). Small-scale disturbance may or may not affect microbial populations (Alongi, 1985), but in the tropics gross primary production can be closely coupled with microbial production (Hansen *et al.*, 1987). Certainly, epibenthic species can have an important impact on the microbial community (Federle *et al.*, 1986). There may be Antarctic parallels to some of these studies. Ophiuroids, for example, are important predators on zooplankton and small infauna and epifauna in the Antarctic (Dearborn, 1977; Fratt and Dearborn, 1984; Dearborn *et al.*, 1986) and the common *Ophionotus victoriae* can ingest benthic microflora (Kellogg *et al.*, 1983). The protozoans and meiofauna are the benthic groups most sensitive to the different types of microbial production, yet except for foraminifera and some nematodes (Platt, 1979) they are virtually unstudied in Antarctic benthos.

Foraminifera are abundant components of all benthic habitats, and their study is important to an understanding of biological and geological oceanography. Carter et al. (1981), Ward et al. (1987), and Bernhard (1987) have described foraminiferal distribution in McMurdo Sound. Most deep-sea and polar research has been restricted to taxonomic and distributional studies, and little is known of the actual biology of foraminifera, much less their general importance to the benthic community. Recent results in McMurdo Sound demonstrate that these animals can be important to other benthic species. For example, the epizoic *Cibicides refulgens* settles on the shell and eventually parasitizes its scallop *(Adamussium colbecki)* host, yet with its pseudopodial net it is also able to graze algae and bacteria and suspension feed through the pseudopodial net (Alexander and DeLaca, 1987). Most important is the growing realization that foraminifera such as *Astrammina rara* are important predators of settling larvae as well as microbes and other protozoans (DeLaca, 1986). T. E. DeLaca, W. L. Stockton, S. S. Bowser, S. P. Alexander and co-workers (unpublished) have shown that in some areas such as Explorers Cove (New Harbor, McMurdo Sound) the carnivorous foraminifera occur in densities sufficient to cover most of the substratum, and they are capable of consuming almost all the benthic invertebrate larvae tested. All ecologists since Thorson (1957, 1966) have recognized the theoretical importance of predation on settling or recently metamorphosed larvae, but previously we have looked to echinoderms such as *Odontaster* (Dayton et al., 1974) or *Ophionotus* (Fratt and Dearborn, 1984) or crustaceans such as phoxocephalid amphipods (Oliver and Slattery, 1985b). But because the foraminifera occur in such high densities, their effects could be more important.

E. Life Below Antarctic Ice Shelves

The existence of large ice shelves has stimulated questions about marine life below the ice. Scott's expeditions to McMurdo Sound found benthic organisms on the surface around Black and White islands and in other areas far removed from the edge of the ice shelf, and Littlepage and Pearse (1962) collected benthic animals through a crack in southwestern McMurdo Sound. Dayton and Oliver (1977) report dense growths of most components of the McMurdo sponge community at White Island. Clearly, these organisms are feeding on material advected from open water. However, Heywood and Light (1975) and Lipps et al. (1977) report fish and benthic microbiota under the ice shelf over George VI Sound. This ice shelf varies from 100 to 500 m thick and the collection site was more than 100 km from the nearest open water; however, the site was under a proglacial lake, which conceivably allowed enough light to permit a local biological oasis that was not general for George VI Sound. Local productivity was impossible under the Ross Ice Shelf, which was studied by the Ross Ice Shelf Project 430 km from the open Ross Sea. A hole was drilled through 420 m of ice which overlaid a 273-m water column (Lipps *et al.,* 1979). A variety of limited techniques demonstrated an errant crustacean fauna of possible mysids and/or euphausiids (detected by a television camera), three species of amphipods (thousands of individuals of *Orchomene;* Stockton, 1982), and a 7-cm isopod *Serolis trilobitoides.* The abundance of the predatory and scavenging *Orchomene* certainly hints at an autocthonous food supply despite the fact that Azam *et al.* (1979) estimated very low amounts of living particulate matter in the water column. Bruchhausen *et al.* (1979) caught a fish 80 km from the ice edge, again emphasizing the importance of food ultimately advected from open water. Foraminifera, bivalve gastropod polychaetes, and ostracod skeletons were observed in the mud. Bacterial densities and organic carbon were equivalent to deep-sea values (Azam *et al.,* 1979), but no living infauna was collected despite the fact that a similar deep-sea collection would yield many infaunal individuals. It is not known whether this is a sampling artifact or a real difference, but the density of the vagile epifauna certainly demonstrates that life can exist under the ice shelves.

VIII. Antarctic Benthic Macroalgae

As is true for the fauna and for the Arctic, Antarctic macroalgae were sampled by some of the earliest expeditions and have been studied by some of the real giants such as Hooker, Skottsberg, Kylin, Foslie, Neushul, and Papenfuss. As Dell (1972) noted, it is ironic that the great Swedish phycologist C. Skottsberg towers over the 20th-century research despite the fact that his own apparently remarkable collection on the Swedish South Polar Expedition was lost. Skottsberg (1964) was always convinced that the Antarctic Convergence separates subantarctic from Antarctic flora, a point well supported in the comprehensive review by Neushul (1968), who outlined five biogeographical regions. The recent review by Heywood and Whitaker (1984) reports some 700 species of macroalgae in the Southern Ocean. Most of these species are subantarctic, and most of these can be found on other cold southern hemisphere shores. Regarding truly Antarctic vegetation, they also summarize data from Papenfuss's (1964) review showing that endemism increases with latitude.

Skottsberg (1964) introduced an interesting problem regarding the effects of Pleistocene glaciation on obligate shallow benthic organisms such as macroalgae. He speculated, as did Newman and Ross (1971) for barnacles, that the macroalgae must have been destroyed on the continent by the glaciers that spread over the shelf and that algae must have had a refuge on the subantarctic islands (in contrast, the shallow-water acorn barnacles disappeared). This hypothesis does not explain the presently observed clean separation between subantarctic and Antarctic flora and the relatively high amount of Antarctic endemism. Perhaps Pleistocene coastal polynyas were present and served as refugia for the algae, especially in the Antarctic Peninsula. When more of their biogeography is known, the algae may form the basis of a future Antarctic synthesis, such as the one of Nesis (1983) for the Arctic. This synthesis will almost certainly have to include the role of latitudinal effects of solar declination.

Dell (1972) summarizes patterns of zonation and ecology. In some areas the ice recedes in the summer and there is a lush, if ephemeral, growth of algae in the intertidal region, where a rough zonation can be noted (Hedgpeth, 1971; Stockton, 1973). Some of these intertidal species are annual (e.g., species of *Urospora, Ulothrix, Porphyra, Monostroma,* and *Adenocystis*), but others such as the lichens *(Verrucaria serpuloides)* and encrusting red algae such as *Hildenbrandia, Lithophyllum,* and *Lithothamnion* are perennials, which simply withstand freezing and ice abrasion or must live in crevices or other refugia.

Below the ice zone along the Antarctic Peninsula there is extensive cover of coralline algae and a rich association of species which may continue down to 20–30 m (Moe and DeLaca, 1976). Some of these species are annual but probably most are perennial. There are many species, especially in more shallow depths, which perennate (Dixon, 1965) or are "pseudoperennial" (Heywood and Whitaker, 1984), i.e., species which behave as annuals in that the foliage grows quickly and is lost at the end of the summer but maintains a permanent holdfast which sprouts new foliage the next season. The deeper waters from 10 to as much as 50 m tend to be dominated by brown algae, *Macrocystis pyrifera,* in the subantarctic, whereas various Desmarestiales *(Desmarestia* spp. and especially *Himantothallus grandifolium)* dominate in the low Antarctic. Considering how much overlap there is between the subantarctic islands and the Antarctic Peninsula (Neushul, 1968; Heywood and Whitaker, 1984), it seems remarkable that *Macrocystis* does not penetrate below the convergence. *Macrocystis* disperses effectively via drifting plants (Dayton, 1985), and it seems likely that propagules have penetrated to the Antarctic Peninsula but have failed to colonize for physiological or ecological reasons. Considering its success in other habitats (Dayton, 1985), these barriers must be strong and deserve study. The most obvious and overlooked biogeographic factor which merits consideration is the latitudinal effect of seasonal solar declinations on large perennial plants. The growth and life history of *Himantothalus,* a remarkable Desmarestiales, have been studied at Signy Island by Hastings (reviewed in Heywood and Whitaker, 1984) and Anvers Island (Moe and Silva, 1977, 1981).

Not much phycological research has been published from the high Antarctic. Zaneveld (1968) summarizes the patterns from the few collections. We have observed coralline crusts on most rocky outcrops on the southern end of Hut Point Peninsula on Ross Island; we have also found a few fronds of *Phyllophora antarctica* growing from encrusting systems at McMurdo Station. It seems likely that the conditions at McMurdo Station represent the southern limits of the distribution of foliose algae. Within 5 km north the *Phyllophora* are common, and at nearby Cape Evans there are thick mats of *Phyllophora, Iridaea obovata,* and occasionally a green alga, probably *Monostroma hariotii.* One interesting enigma is the *"Laminaria"* Hodgson (1907) reported at Hut Point. Although it probably was a specimen of *Himantothallus,* extensive collecting and diving in the McMurdo Sound area has not confirmed this or any other brown alga since.

The foliose algae at Cape Evans grow well to at least 60 m. There are many records of deep-water algae from the Antarctic, and, indeed, *Ballia callitricha* has not been recorded more shallow than 37 m (reviewed by Heywood and Whitaker, 1984). Certainly, *Himantothallus* grows below 50 m, and Heywood and Whitaker list many species which have been collected below 100 m and even at depths in excess of 300 m! Some of these records are in error or represent material uplifted, possibly by anchor ice (Dayton et al., 1969), and transported to deep water. Nonetheless, many of these records are legitimate and the combination of extremely clear water and cold temperatures probably deepens the compensation point. Indeed, Wagner and Zaneveld (1988) detail many collections from depths well over 300 m, including *M. hariotii* from 348 m off Possession Island in the Ross Sea. Clearly, the evaluation of the mechanisms of deep-water growth of macroalgae is a fertile research topic.

Heywood and Whitaker (1984) offer a comprehensive review of the relative importance of environmental limiting factors. In general, the low temperature reduces the compensation depth because the initial stage of photosynthesis is photochemical and thus light rather than temperature dependent; the primary effect of low temperature thus is to reduce respiration and lower the compensation point. This would seem to hold true for benthic algae, suggesting that their growth and distribution are light limited. In most temperate oceans, sea urchin grazing has the potential to eliminate most kelp (reviewed by Dayton, 1985). However, there are high densities of the sea urchin *Sterechinus neumayeri* at Cape Evans, Ross Island, where they do not graze the dense macrophytes. Although it appeared from a 1975 survey that *Sterechinus* maintained a coralline crust "barren" at Granite

Harbor, this was not experimentally tested and other hypotheses, such as seasonal ice or sedimentation, are equally likely. W. Stockton (personal communication) reports lush *Iridea* growth at nearby Dunlap Island in the presence of dense populations of *Sterechinus*. In most areas the productivity driving the benthic system probably comes from phytoplankton and microalgae, but in areas with large standing stocks of macroalgae the trophic role of detrital food webs may be significant. Thus, research on the fates of stable carbon isotopes such as that of Dunton and Schell (1987) in the Arctic might provide new insights into Antarctic food webs.

IX. Discussion

Marine ecologists familiar with polar oceans are well aware that there are many differences between the two, both in the water column and in the benthos. Birds and mammals aside, the northern polar seas seem dominated by fishes while the Southern Ocean is characterized by invertebrates. This is true despite the remarkable radiation of Nototheniiforme fish, which are still stereotyped compared to the diverse fish faunas of the northern seas. In addition to their diversity, the northern fish fauna supports many sustainable fisheries with very large yields while the Southern Ocean appears to be dominated by species of krill.

Less well publicized but perhaps more fundamental are the benthic differences between the two areas. The Antarctic shelf is much smaller and deeper and is also characterized by a profoundly different and in many ways simpler fauna. That is, it has a high diversity in some taxa but it lacks the crabs, sharks, most of the benthic fishes, the many species of balanomorph barnacles, many types of snails and polychaetes, and the large productive populations of clams and ampeliscid amphipods so characteristic of northern faunas. Although there are many bivalves in the Antarctic and some, especially *Adamussium* and *Laternula,* can be locally abundant, they do not dominate a habitat and are never a "key industry" species such as they are for walrus and possibly crabs and the asteroids in the Arctic. Finally, and perhaps most important, the Antarctic benthos in general is not subject to most of the common and important types of biotic disturbances which characterize the Arctic (Table 12.2). Gallardo (1987) reviews many types of disturbances to the Antarctic benthos, and in some areas scour by drifting ice and subsequent predation is a very important source of disturbance (e.g., Zamorano *et al.,* 1986). But in general the Antarctic bottoms are deeper and lack the persistent overwhelming disturbances characterizing the Arctic benthos. For example, in addition to the physical stresses, the Arctic has many more surface-burrowing species such as echiuroids, polychaetes, echinoderms, and especially crustacea. This is interesting because one might consider the extent of the scour and expect the opposite. Moreover, in many areas of the Arctic there are large and often very abundant crabs which are voracious predators. The king crabs, for example, can be extremely large (ca. 100 kg), are highly motile, and are generalized carnivores capable of eating almost any epifaunal animal. Probably most important, disturbance analogs to the bottom-feeding fishes (especially flatfishes and rays), gray whales, and walrus are lacking in the Antarctic. To emphasize the importance of these larger disturbances, Nelson and Johnson (1987) estimate that the gray whales resuspend 172 million metric tons of sediment every summer in the restricted Chirikov Basin of the northeastern Bering Sea; this is three times the suspended sediment deposited by the Yukon River each year.They also calculate that the walrus disturbs about 100 million metric tons of sediment each year in the same area. Oliver and Slattery (1985b) speculate that this lack of physical disturbance is one of the most important factors maintaining the dense assemblage they studied at McMurdo Sound.

The rate processes may also be different in a very general sense. Grebmeier *et al.* (1988) and Gallardo (1987) review the clear relationships between primary production and the benthic communities in the Arctic and Antarctic, respectively. However, the Arctic productivity seems to be efficiently transferred within the food web and populations often have relatively high turnover rates, but the Antarctic seas seem to exhibit the paradox Hedgpeth (1977) discussed; i.e., the water column seems to be productive whereas the benthic communities are characterized by low growth rates and secondary production (White, 1984; Picken, 1985). Furthermore, Dunbar *et al.* (1989) argue that there are areas which receive a tremendous amount of carbon influx, yet the carbon does not seem to be accumulating despite the fact that bacterial growth rates may be low (White *et al.*, 1984, 1985). The difference might be made up by protozoans and meiofaunal production, but such data have not been collected. Certainly, this discrepancy between high water-column productivity and apparently low benthic utilization represents a fertile topic for future research.

Another related comparison between the two systems is the very different patterns of decomposition. Large Arctic coastal areas receive more carbon from terrestrial peat than from marine primary productivity, yet, with a single exception, marine metazoans do not utilize the terrestrial carbon (Schell, 1983). Nevertheless, the carbon is not accumulating and efficient microbial processes must be at work (Griffiths and Morita, 1981). Presumably this is also an important source of nutrients for marine flora. While decomposition *per se* is little studied in either system, this Arctic situation seems to contrast strongly with the picture of large amounts of marine carbon reaching the Antarctic sea floor, where there is a complete absence of peat and where the microbial activity may be relatively slow.

The two polar benthic biotas represent profoundly different evolutionary situations. The shallow-water benthic habitat in the Arctic is heavily disturbed physically and biologically and the fauna exhibits a low endemism and strong ties to the Atlantic and the Pacific. Although it is said to be a young fauna, the habitat has been available for colonization by a large number of Atlantic and Pacific species since the last ice age. In contrast, the Antarctic system is sometimes thought to be an old system which may have been isolated since the Jurassic (Hedgpeth, 1977). Although it exhibits a great deal of endemism, most of the actual radiation is restricted to relatively few groups (White, 1984) and many forms common in the Arctic (crabs, sharks, flatfishes, shallow balanomorph barnacles, etc.) are absent from the Antarctic. In some cases the missing groups were last abundant in the Eocene (e.g., crab: Feldman, 1984; Feldman and Zinsmeister, 1984), but in other cases were recently present in the Antarctic (shallow-water balanomorphs: Newman and Ross, 1971) and may have become extinct during a recent period of intense ice scour. Nevertheless, the lack of adaptive radiation in the Antarctic to parallel the life forms common in the Arctic suggests that the genetic template of the original founder populations is not very plastic or that the selective pressures are less diverse. That is, a fascinating evolutionary question is whether this pattern results from old genetic founder groups persisting relatively unchanged in a simplified or less stressful environment or whether the relatively few radiating groups are themselves so ecologically dominant that they cause the patterns. This discussion emphasizes the fact that all evolutionary, biogeographical, and ecological research absolutely depends on competent systematic research. Considering this, it is appropriate that Picken's (1985) review lists active systematics. Ecologists must unite to reverse the decline in support of systematic research.

Because they present novel problems with novel suites of species, polar ecosystems may be useful areas in which to study the interplay between ecosystem and evolutionary ecology. Ecosystem ecologists study fluxes and mass balances with the intent of elucidating large-scale patterns (Parsons *et al.*, 1977; Mann, 1982; Smith, 1984), whereas evolutionary ecologists focus on population-level processes and ask such questions as: What limits the distribution and abundance of a given population? What factors select for the evolution of particular reproductive patterns or other life history phenomena? What limits the number of species in a given habitat? It seems self-evident that the latter evolutionary questions are influenced by factors such as mass flux, nutrient regeneration, and patterns of productivity, yet the two approaches are not often combined.The stark contrasts between the benthic systems in the two polar areas (the historic and evolutionary differences between the sources of the species, habitat differences, degree of biotic disturbance, etc.) are intertwined with similarities (pulsed primary production, cold temperatures, reduction in planktotrophic larvae, etc.) inviting many types of comparative research. That is, given that both the genetic material and the environment determine the evolutionary patterns, the evaluation of these differences between the Arctic and the Antarctic could lead to fertile comparative research projects on subjects such as the following:

1. Nutrients, primary production, and growth/reproduction rates.
2. Benthic–pelagic coupling, especially as it affects larval settlement and nutrient transfer.
3. Patterns of succession and persistence stability (see complete discussion in Gallardo, 1987).
4. Consequences of different predator–prey relationships.
5. Ecosystem consequences of perturbations caused by humans, especially through fishing.

Such questions, together with more emphasis on autoecological research (Picken, 1985), could help define the selective pressures maintaining different evolutionary patterns in larval development and dispersal, development and growth of individuals, and their various reproductive patterns. I emphasize with White (1984) that such progress depends on more emphasis on basic natural history and systematic information, most of which must come from difficult field work which often explicitly includes continuous observations.

Acknowledgments

I dedicate this review to Curly Wohlschlag and George Llano, who got me into the business! I thank Walker Smith for soliciting it, having the patience to see it finished, and careful editorial help. I am grateful to Gordon Robilliard for three decades of wonderful companionship and for sharing a box of literature for this review. Jim Barry and John Oliver have shared many years of polar experiences with me. The review was also significantly enhanced by Ken Dunton, who shared copies of obscure but important Arctic references, and with S. Schonberg and J. Grebmeier, who made many editorial improvements. Robert Wilce contributed many corrections and observations to the section on Arctic algae. Other colleagues who have helped include F. Bacon, J. Bernhard, M. Boudrais, E. Buch, A. Clarke, J. Dearborn, S. France, J. Gutt, G. Hempel, L. Lowry, K. Miller, B. Mordida, J. Reeve, S. Thrush, and W. Stockton. Friends who have particularly expanded my understanding of polar systems include T. Berg, J. Dearborn, A. DeVries, L. Giddings, J. Hedgpeth, R. Hofman, G. Llano, L. Lowry, J. Macdonald, G. Robilliard, I. Stirling, and J. Twiss. L. Jackson and E. Crissman have helped with typing and editorial assistance. The Division of Polar Programs of the National Science Foundation has generously supported my research.

References

Alexander, S. P. & T.E. DeLaca. 1987. Feeding adaptations of the foraminiferan *Cibicides refulgens* living epizoically and pasasitically on the Antarctic scallop *Adamussium colbecki*. *Biol. Bull (Woods Hole, Mass.)* **173**: 136–159.

Alongi, D. M. 1985. Effect of physical disturbance on population dynamics and trophic interactions among microbes and meiofauna. *J. Mar. Res.* **43**: 351–364.

Andriashev, A. P. 1965. A general review of the Antarctic fish fauna. Biogeography and ecology in Antarctica. *Monogr. Biol.* **15**: 491–550.

Azam, F., J. R. Beers, L. Campbell, A.F. Carlucci, O. Holm-Hansen, F. M. H. Reid & D. M. Karl. 1979. Occurrence and metabolic activity of organisms under the Ross Ice Shelf, Antarctica, at Station J9. *Science* **203**: 451–453.

Barnes, P. W., E. Reimnitz & D. Fox. 1982. Ice rafting of fine grained sediment, a sorting and transport mechanism, Beaufort Sea, Alaska. *J. Sediment. Petrol.* **42**: 493–502.

Barnes, P. W., D. M. Rearic & E. Reimnitz. 1984. Ice gouging characteristics and process. *In* "The Alaskan Beaufort Sea: Ecosystems and Environments" (P. W. Barnes, D. M. Schell & E. Reimnitz, eds.), pp. 185–212. Academic Press, Orlando, Florida.

Barry, J. P. 1988. Hydrographic patterns in McMurdo Sound, Antarctica and their relationship to local benthic communities. *Polar Biol.* **8**: 377–391.

Barry, J. P. & P. K. Dayton. 1988. Current patterns in McMurdo Sound, Antarctica and their relationship to biological production of local benthic communities. *Polar Biol.* **8**: 367–376.

Bernhard, J. M. 1987. Foraminiferal biotopes in Explorers Cove, McMurdo Sound, Antarctica. *J. Foraminiferal Res.* **17**: 286–297.

Blacker, R. W. 1957. Benthic animals as indicators of hydrographic change in Svalbard waters. *G. B. Minist. Agric., Fish. Food Fish. Invest., Ser. 2* **20**: 1–49.

Bowser, S. S., T. E. DeLaca & C. L. Rieder. 1986. Novel extracellular matrix and microtubule cables associated with pseudopodia of *Astrammina rara*, a carnivorous Antarctic foraminifer. *J. Ultrastruct. Mol. Struct. Res.* **94**: 149–160.

Brotzky, V. A. 1930. Materials for the quantitative evaluation of the Storfjord (East Spitzbergen). *Ber. Wiss. Biol.* **4**: 47–61.

Bruchhausen, P. M., J. A. Raymond, S. S. Jacobs, A. L. De Vries, E. M. Thorndike & H. H. DeWitt. 1979. Fish, crustaceans, and the sea floor under the Ross Ice Shelf. *Science* **203**: 449–451.

Buck, P. 1958. "Vikings of the Sunrise." Whitcombe & Tombs, Ltd., London.

Bullivant, J. S. 1969. Bryozoa. *Antarct. Map Folio Ser.* **11**: 22–23.

Bullivant, J. S. & J. H. Dearborn. 1967. The fauna of the Ross Sea. 5. General accounts, station lists and benthic ecology. *N.Z. Oceanogr. Inst. Mem.* **32**: 1–77.

Bunt, J. S. 1963. Diatoms of Antarctic sea-ice as agents of primary production. *Nature (London)* **199**: 1255–1257.

———. 1964. Primary productivity under sea ice in Antarctic waters. *Antarct. Res. Ser.* **1**: 13–31.

Bunt, J. S. & C. C. Lee. 1970. Seasonal primary production in Antarctic sea ice at McMurdo Sound in 1967. *J. Mar. Res.* **28**: 304–320.

Carey, A. G., Jr. 1987. Particle flux beneath fast ice in the shallow southwestern Beaufort Sea, Arctic Ocean. *Mar. Ecol.: Prog. Ser.* **40**: 247–257.

Carey, A. G., Jr. & R. E. Ruff. 1977. Ecological studies of the benthos in the western Beaufort Sea with special reference to bivalve molluscs. *In* "Polar Oceans" (M. J. Dunbar, ed.), pp. 505–530. Arctic Inst. North Am., Calgary.

Carey, A. G., Jr., R. E. Ruff, J. G. Castillo & J. J. Dickinson. 1974. Benthic ecology of the

western Beaufort Sea continental margin. *In* "The Coast and Shelf of the Beaufort Sea" (J. C. Reed & J. E. Sater, eds.), pp. 665–680. Arctic Inst. North Am., Arlington, Virginia.
Carey, A. G., Jr., M. A. Boudrias, J. C. Kern & R. E. Ruff. 1984a. "Selected Ecological Studies on Continental Shelf Benthos and Sea Ice Fauna in the Southwestern Beaufort Sea," Final Rep. 23. NOAA-OCSEAP, Washington, D.C.
Carey, A. G., Jr., P. H. Scott & K. R. Walters. 1984b. Distributional ecology of shallow SW Beaufort Sea (Alaska) bivalve Mollusca. *Mar. Ecol.: Prog Ser.* **17**: 125–134.
Carter, L., J. S. Mitchell & N. J. Day. 1981. Suspended sediment beneath permanent and seasonal ice, Ross Ice Shelf, Antarctica. *N. Z. J. Geol. Geophys.* **24**: 249–262.
Chapman, A. R. O. & J. E. Lindley. 1980. Seasonal growth of *Laminaria solidungula* in the Canadian high Arctic in relation to irradiance and dissolved nutrient concentrations. *Mar. Biol. (Berlin)* **57**: 1–5.
Collen, J. D. 1979. Marine invertebrates from the Ross Ice Shelf, Antarctica. *Search* **10**: 274–275.
Curtis, M. A. 1975. The marine benthos of Arctic and subarctic continental shelves. *Polar Rec.* **17**: 595–626.
Dall, W. H. 1875. Arctic marine vegetation. *Nature (London)* **12**: 166.
Dayton, P. K. 1979. Observations of growth, dispersal, and population dynamics of some sponges in McMurdo Sound, Antarctica. *Colloq. Int. C. N. R. S.* **291**: 273–282.
———. 1985. Ecology of kelp communities. *Annu. Rev. Ecol. Syst.* **16**: 215–245.
———. 1989. Interdecadal variation in an Antarctic sponge and its predators resulting from oceanographic climate shifts. *Science* **245**: 1484–1486.
Dayton, P. K. & J. S. Oliver. 1977. Antarctic soft-bottom benthos in oligotrophic and eutrophic environments. *Science* **197**: 55–58.
Dayton, P. K., G. A. Robilliard & A. L. DeVries. 1969. Anchor ice formation in McMurdo Sound, Antarctica, and its biological effects. *Science* **163**: 273–274.
Dayton, P. K., G. A. Robilliard & R. T. Paine. 1970. Benthic faunal zonation as a result of anchor ice at McMurdo Sound, Antarctica. *In* "Antarctic Ecology" (M. W. Holdgate, ed.), Vol. 1, pp. 244–258. Academic Press, London.
Dayton, P. K., W. A. Newman, R. T. Paine & L. B. Dayton. 1974. Ecological accommodation in the benthic community at McMurdo Sound, Antarctica. *Ecol. Monogr.* **44**: 105–128.
Dayton, P. K., W. A. Newman & J. S. Oliver. 1982. The vertical zonation of the deep-sea Antarctic acorn barnacle, *Bathylasma corolliforme* (Hoek): Experimental transplants from the shelf into shallow water. *J. Biogeogr.* **9**: 95–109.
Dayton, P. K., D. Watson, A. Palmisano, J. P. Barry, J. S. Oliver & D. Rivera. 1986. Distribution patterns of benthic standing stock at McMurdo Sound, Antarctica. *Polar Biol.* **6**: 207–213.
Dearborn, J. H. 1965. Ecological and faunistic investigations of the marine benthos at McMurdo Sound, Antarctica. Ph.D. Dissertation, Stanford Univ., Stanford, California.
———. 1977. Foods and feeding characteristics of Antarctic asteroids and ophiuroids. *In* "Adaptations Within Antarctic Ecosystems" (G. A. Llano, ed.), pp. 293–326. Gulf Publ. Co., Houston, Texas.
Dearborn, J. H., F. D. Ferrari & K. C. Edwards. 1986. Can pelagic aggregations cause benthic satiation? Feeding biology of the Antarctic brittle star *Astrotoma agassizii* (Echinodermata: Ophiuroidea). *Antarct. Res. Ser.* **44**: 1–28.
DeLaca, T. E. 1986. The morphology and ecology of *Astrammina rara*. *J. Foraminiferal Res.* **16**: 216–223.
DeLaca, T. E. & J. H. Lipps. 1976. Shallow-water marine associations, Antarctic Peninsula. *Antarct. J. U.S.* **11**: 12–19.
DeLaca, T. E., J. H. Lipps, & R.R. Hessler. 1980. The morphology and ecology of a new large

agglutinated Antarctic foraminifer (Textulariina: Notodendrodidae nov.). *J. Linn. Soc. London, Zool.* **69**: 205–224.

Dell, R. K. 1972. Antarctic benthos. *Adv. Mar. Biol.* **10**: 1–216.

Denton, G. H. & T. J. Hughes. 1981. "The Last Great Ice Sheets." Wiley, New York.

DeWitt, H. H. 1971. "Coastal and Deep-Water Benthic Fishes of the Antarctic," Antarct Map Folio Ser., Folio 15. Am. Geogr. Soc., New York.

Dixon, P. S. 1965. Perennation, vegetative propagation and algal life histories, with special reference to *Asparagopsis* and other Rhodophyta. *Proc. Mar. Biol. Symp., 5th,* pp. 67–74.

Dunbar, R. B. & A. R. Leventer. 1986. Opal and carbon fluxes beneath ice-covered regions of McMurdo Sound. *Antarct. J. U.S.* **21**: 132–133.

Dunbar, R. B., J. B. Anderson & E. W. Domack. 1985. Oceanographic influences on sedimentation along the Antarctic continental shelf. *Antarct. Res. Ser.* **43**: 291–312.

Dunbar, R. B., J. B. Anderson & W. L. Stockton. 1989. Biogenic sedimentation in McMurdo Sound, Antarctica. *Mar. Geol.* **85**: 155–179.

Dunton, K. & D. M. Schell. 1986. Seasonal carbon budget and growth of *Laminaria solidungula* in the Alaskan high Arctic. *Mar. Ecol.: Prog. Ser.* **31**: 57–66.

———. 1987. Dependence of consumers on marcoalgal *(Laminaria solidungula)* carbon in an Arctic kelp community: ^{13}C evidence. *Mar. Biol. (Berlin)* **93**: 615–625.

Dunton, K., E. Reimnitz & S. Schonberg. 1982. An Arctic kelp community in the Alaskan Beaufort Sea. *Arctic* **35**: 465–484.

Ekman, S. 1953. "Zoogeography of the Sea." Sidgwick & Jackson, London.

Ellis, D. V. 1960. Marine infaunal benthos in Arctic North America. *Tech. Pap.—Arct. Inst. North Am.* **5**.

Ellis, D. V. & R. T. Wilce. 1961. Arctic and subarctic examples of intertidal zonation. *Arctic* **35**: 224–235.

Ellsberg, E. 1938. "Hell on Ice. The Saga of the 'Jeannett'." Dodd, Mead, New York.

Fay, F. H. 1981. Marine mammals of the eastern Bering Sea Shelf: An overview. *In* "The Eastern Bearing Sea Shelf: Oceanography and Resources" (D. W. Hood & J. A. Calder, eds.), Vol. 2, pp. 807–811. Univ. of Washington Press, Seattle.

Feder, H. M. & S. C. Jewett. 1981. Feeding interactions in the eastern Bering Sea with emphasis on the benthos. *In* "The Eastern Bering Sea Shelf: Oceanography and Resources" (D. W. Hood & J.A. Calder, eds.), Vol. 2, pp. 1229–1261. Univ. of Washington Press, Seattle.

Federle, T.W., R. J. Livingston, L. E. Wolfe & D. C. White. 1986. A quantitative comparison of microbial community structure of estuarine sediments from microcosms and the field. *Can. J. Microbiol.* **32**: 319–325.

Feldman, R. M. 1984. Decapod crustaceans from the late Cretaceous and the Eocene of Seymour Island, Antarctic Peninsula. *Antarct. J. U.S.* **19**: 4–5.

Feldman, R. M. & W. J. Zinsmeister. 1984. New fossil crabs (Decapoda: Brachyura) from the La Meseta Formation (Eocene) of Antarctica: Paleographic and biogeographic implications. *J. Paleontol.* **58**: 1046–1601.

Francis, J. E. 1988. A 50-million year old fossil forest from Strathcona Fiord, Ellesmere Island, Arctic Canada: Evidence for a warm polar climate. *Arctic* **41**: 314–318.

Fratt, D. B. & J. H. Dearborn. 1984.Feeding biology of the Antarctic brittle star *Ophionotus victoriae* (Echinodermata: Ophiuroidea). *Polar Biol.* **3**: 127–139.

Frost, K. J. & L. F. Lowry. 1984. Trophic relationships of vertebrate consumers in the Alaskan Beaufort Sea. *In* "The Alaskan Beaufort Sea: Ecosystems and Environments" (P. W. Barnes, D. M. Schell & E. Reimnitz, eds.), pp. 381–401. Academic Press, Orlando, Florida.

Fry, W. G. 1964. The pycnogonid fauna of the Antarctic continental shelf. *In* "Biologie Antarctique" (R. Carrick, M. W. Holgate & J. Prevost, eds.), pp. 263–269. Hermann, Paris.

Fuhrman, J. A. & F. Azam. 1980. Bacterioplankton secondary production estimates for coastal waters of British Columbia, Antarctica and California. *Appl. Environ. Microbiol.* **39**: 1085–1095.
Gallardo, V. A. 1987. The sublittoral macrofaunal benthos of the Antarctic shelf. *Environ. Int.* **13**: 71–81.
Gallardo, V. A., J. C. Castillo, M. A. Retamal, A. Yanez, H. I. Moyano & J. G. Hermosilla. 1977. Quantitative studies on the soft-bottom macrobenthic animal communities of shallow Antarctic bays. *In* "Adaptations Within Antarctic Ecosystems" (G. A. Llano, ed.), pp. 361–387. Gulf Publ. Co., Houston, Texas.
Golikov, A. N. & A. G. Averincev. 1977. Distribution patterns of benthic ice and ice biocenoses in the high latitudes of the polar basin and their part in the biological structure of the world ocean. *In* "Polar Oceans" (M. J. Dunbar, ed.), pp. 331–364. Arctic Inst. North Am., Calgary.
Gray, J. S. 1981. "The Ecology of Marine Sediments." Cambridge Univ. Press, Cambridge.
Grebmeier, J. M. 1987. The ecology of benthic carbon cycling in the northern Bering and Chukchi seas. Ph.D. Dissertation, Univ. of Alaska, Fairbanks.
Grebmeier, J. M., C. P. McRoy & H. M. Feder. 1988. Pelagic–benthic coupling on the shelf of the northern Bering and Chukchi Seas. I. Food supply source and benthic biomass. *Mar. Ecol.: Prog. Ser.* **48**: 57–67.
Grebmeier, J. M., H. M. Feder & C. P. McRoy. 1989. Pelagic–benthic coupling on the shelf of the northern Bering and Chukchi seas. II. Benthic community structure. *Mar. Ecol.: Prog. Ser.* **51**: 253–268.
Grebmeier, J. M. & C. P. McRoy. 1989. Pelagic-benthic coupling on the shelf of the northern Bering and Chukchi seas. II. Benthic food supply and carbon cycling. *Mar. Ecol.: Prog. Ser.* **53**: 79–91.
Griffiths, R. P. & R. Y. Morita. 1981. Study of microbial activity and crude oil–microbial interactions in the waters and sediments of Cook Inlet and the Beaufort Sea. *In* "Environmental Assessment of the Alaskan Continental Shelf," Final Rep., Res. Unit 190. NOAA-OCSEAP, Rockville, Maryland.
Gruzov, E. N. 1977. Seasonal alterations in coastal communities in the Davis Sea. *In* "Adaptations Within Antarctic Ecosystems" (G. A. Llano, ed.), pp. 263–279. Gulf Publ. Co., Houston, Texas.
Gulliksen, B. 1974. Marine investigations in Jan Mayen in 1972. *Misc.—K. Nor. Vidensk. Selsk., Mus.* **19**: 1–46.
———. 1979. Shallow water benthic fauna from Bear Island. *Astarte* **12**: 5–12.
———. 1988. Marinbiologiske forhold i Svalbards Territoriale Farvann. *Nor. Polarinst. Rapp. Ser.* **42**: 1–43.
Gulliksen, B., T. Haug & O. K. Sandnes. 1980. Benthic macrofauna on new and old lava grounds at Jan Mayen. *Sarsia* **65**: 137–148.
Gulliksen, B., B. Holte & K.-J. Jakola. 1985. The soft bottom fauna in Van Mijenfjord and Raudfjord, Svalbard. *In* "Marine Biology of Polar Regions and Effects on Marine Organisms" (J. S. Gray and M. E. Christiansen, eds.), pp. 199–215. Wiley, Chichester.
Hansen, J. A., D. M. Alongi, D. J. W. Moriarity & P. L. Pollard. 1987. The dynamics of benthic microbial communities in winter at Davies Reef, Great Barrier Reef, Australia. *Coral Reefs* **6**: 63–70.
Hedgpeth, J. W. 1969. Introduction to Antarctic zoogeography. *Antarct. Map Folio Ser.* **11**: 1–9.
———. 1971. Perspective of benthic ecology in Antarctica. *In* "Research in the Antarctic" (L. O. Quam, ed.), Publ. No. 93, pp. 93–136. Am. Assoc. Adv. Sci., Washington, D. C.

———. 1977. The Antarctic marine ecosystem. *In* "Adaptations Within Antarctic Ecosystems" (G. A. Llano, ed.), pp. 3–10. Gulf Publ. Co., Houston, Texas.
Hempel, G. 1985. Antarctic marine food webs. *In* "Antarctic Nutrient Cycles and Food Webs" (W. R. Siegfried, P. R.Condy & R. M. Laws, eds.), pp. 266–270. Springer-Verlag, Berlin.
Herman, Y. & D. M. Hopkins. 1980. Arctic oceanic climate in late Cenozoic time. *Science* **209**: 557–562.
Heywood, R. B. & J. J. Light. 1975. First direct evidence of life under Antarctic shelf ice. *Nature (London)* **254**: 591–592.
Heywood, R. B. & T. M. Whitaker. 1984. The Antarctic marine flora. *In* "Antarctic Ecology" (R. M. Laws, ed.) Vol. 2, pp. 373–410. Academic Press, Orlando, Florida.
Hodgson, T. V. 1907. On collecting in Antarctic seas. *Natl. Antarct. Exped. 1901–1904 (Zool.-Bot.)*.
Hodson, R. E., F. Azam, A. F. Carlucci, J. A. Fuhrman, D. M. Karl & O. Holm-Hansen. 1981. Microbial uptake of dissolved organic matter in McMurdo Sound, Antarctica. *Mar. Biol. (Berlin)* **61**: 89–94.
Hood, D. W. & J. A. Calder. 1981. Consideration of environmental risks and research opportunities on the eastern Bering Sea shelf. *In* "The Eastern Bering Sea Shelf: Oceanography and Resources" (D. W. Hood & J. A. Calder, eds.), Vol. 2. pp. 1299–1322. Univ. of Washington Press, Seattle.
Horner, R. 1984. Phytoplankton abundance, chlorophyll *a*, and primary productivity in the western Beaufort Sea. *In* "The Alaskan Beaufort Sea" (P. W. Barnes, D.W. Schell & E. Reimnitz, eds.), pp. 295–310. Academic Press, Orlando, Florida.
Horner, R. & G. C. Schrader. 1982. Relative contributions of ice algae, phytoplankton, and benthic microalgae to primary production in nearshore regions of the Beaufort Sea. *Arctic* **35**: 485–503.
Idelson, M. S. 1930. A preliminary quantitative evaluation of the bottom fauna of the Spitzbergen bank. *Ber. Wiss. Merresunters.* **4**: 25–46.
Jewett, S. C. & H. M. Feder. 1980. Autumn food of adult starry flounder *Patichthys stellatus* from the NE Bering Sea and SE Chukchi Sea. *J. Cons., Cons. Int. Explor. Mer* **39**: 71–4.
———. 1981. Epifaunal invertebrates of the continental shelf of the eastern Bering and Chukchi seas. *In* "The Eastern Bering Sea Shelf: Oceanography and Resources" (D. W. Hood & J. A. Calder, eds.), pp. 1131–1155. Univ. of Washington Press, Seattle.
Johnson, K. R. & C. H. Nelson. 1984. Side-scan sonar assessment of gray whale feeding in the Bering Sea. *Science* **225**: 1150–1152.
Kaufman, Z. S. 1974. Sexual cycles and gametogenesis of invertebrates from the White Sea. *Issled. Fauny Morei* **13**: 191–271.
Kellogg, D. E., T. B. Kellogg, J. H. Dearborn, K. C. Edwards & D. B. Fratt. 1983. Diatoms from brittle star stomach contents: Implications for sediment reworking. *Antarct. J. U.S.* **17**: 167–169.
Kelly, B. P. 1988. Bearded seal, *Erignathus barbartus*. *In* "Selected Marine Mammals of Alaska" (J. W. Lentfer, ed.), pp. 77–94. Marine Mammal Commission, Washington, D.C.
Kennett, J. P. 1978. The development of planktonic biogeography in the Southern Ocean during the Cenozoic. *Mar. Micropaleontol.* **3**: 301–345
Kirwan, L. P. 1960. "A History of Polar Exploration." Norton, New York.
Knox, G. A. & J. K. Lowry. 1977. A comparison between the benthos of the Southern Ocean and the north polar ocean with special reference to the Amphipoda and the polychaetes. *In* "Polar Oceans" (M. J. Dunbar, ed.), pp. 423–462. Arctic Inst. North Am., Calgary.
Kottmeier, S. T., S. M. Grossi & C. W. Sullivan. 1987. Sea ice microbial communities. VIII. Bacterial production in annual sea ice of McMurdo Sound, Antarctica. *Mar. Ecol.: Prog. Ser.* **35**: 175–186.

Kussakin, O. G. 1967. Fauna of Isopoda and Tanaidacea in the coastal zones of the Antarctic and subantarctic waters. *Issled. Fauny Morei* **4**: 12.
Lamb, I. M. & M. H. Zimmerman. 1977. Benthic marine algae of the Antarctic Peninsula. *Antarct. Res. Ser.* **23**: 130–229.
Lawrence, J. M. 1975. On the relationship between plants and sea urchins. *Oceanogr. Mar. Biol.* **13**: 213–286.
Lee, R. K. S. 1973. General ecology of the Canadian Arctic benthic marine algae. *Arctic* **26**: 32–43.
———. 1980. A catalogue of the marine algae of the Canadian Arctic. *Natl. Mus. Can. Publ. Bot.* **9**: 2–82.
Lentfer, J. W. (ed.). 1988. "Selected Marine Mammals of Alaska species accounts with research and management recommendations." Marine Mammal Commission, Washington, D.C.
Leventer, A. & R. B. Dunbar. 1987. Diatom flux in McMurdo Sound, Antarctica. *Mar. Micropaleontol.* **12**: 49–64.
Lewbel, G. S. (ed.). 1983. "Bering Sea Biology: An Evaluation of the Environmental Data Base Related to Bering Sea Oil and Gas Exploration and Development." LGL Alaska Research Associates, Inc., Anchorage.
Lipps, J. H. & C. S. Hickman. 1982. Origin, age and evolution of Antarctic and deep-sea faunas. *In* "Environment of the Deep Sea" (W. G. Ernst & J. G. Morin, eds.), Vol. 2, pp. 324–356. Prentice-Hall, Englewood Cliffs, New Jersey.
Lipps, J. H., W. N. Krebs & N. K. Temnikov. 1977. Microbiota under Antarctic ice shelves. *Nature* **265**: 232–233.
Lipps, J. H., T. E. Ronan & T. E. DeLaca. 1979. Life below the Ross Ice Shelf, Antarctica. *Science* **203**: 447–449.
Littlepage, J. L. 1965. Oceanographic investigations in McMurdo Sound, Antarctica. *Antarct. Res. Ser.* **5**: 1–37.
Littlepage, J. L. & J. S. Pearse. 1962. Biological and oceanographic observations under an Antarctic ice shelf. *Science* **137**: 679–681.
Lopez, B. 1986. "Arctic Dreams." Picador, London.
Lowry, J. K. 1975. Soft bottom macrobenthic community of Arthur Harbour, Antarctica. *Antarct. Res. Ser.* **23**: 1–19.
Lowry, L. F. and F. H. Fay. 1984. Seal eating by walruses in the Bering and Chukchi seas. *Polar Biol.* **3**: 1–18.
MacGinitie, G. 1955. Distribution and ecology of the marine invertebrates of Point Barrow, Alaska. *Smithson. Misc. Collect.* **128**: 1–201.
MacIntosh, R. A. & D. A. Somerton. 1981. Large marine gastropods of the eastern Bering Sea. *In* "The Eastern Bearing Sea Shelf: Oceanography and Resources" (D. W. Hood & J. A. Calder, eds.), Vol. 2, pp. 1215–1228. Univ. of Washington Press, Seattle.
Malveyeva, T. A. 1974. Ecology and life cycles of mass species of gastropods in the Barents and White seas. *Issled. Fauny Morei* **13**: 65–190.
Mann, K. H. 1982. "Ecology of Coastal Waters: A Systems Approach." Univ. of California Press, Berkeley.
McKenna, M. C. 1980. Eocene paleolatitude, climate, and mammals of Ellesmere Island. *Palaeogeogr., Palaeoclimatol., Palaeoecol.* **30**: 349–362.
Menzies, R. J., R. Y. George & G. T. Rowe. 1973. "Abyssal Environment and Ecology of the World Oceans." Wiley, New York.
Mileykovsky, S. A. 1970. Seasonal and daily dynamics in pelagic larvae of marine shelf bottom invertebrates in nearshore waters of Kandalaksha Bay (White Sea). *Mar. Biol (Berlin)* **5**: 180–194.
———. 1971. Types of larval development in marine bottom invertebrates, their distribution and ecological significance: A re-evaluation. *Mar. Biol (Berlin)* **10**: 193–213.

——. 1972. The "pelagic larvation" and its role in the biology of the world ocean, with special reference to pelagic larvae of marine bottom invertebrates. *Mar. Biol. (Berlin)* **16**: 13–21.

Mills, E. L. 1975. Benthic organisms and the structure of marine ecosystems. *J. Fish. Res. Board Can.* **32**: 1657–1663.

Moe, R. L. & T. E. DeLaca. 1976. Occurrence of macroscopic algae along the Antarctic Peninsula. *Antarct. J. U.S.* **11**: 20–24.

Moe, R. L. & P. C. Silva. 1977. Antarctic marine flora: Uniquely devoid of kelp. *Science* **196**: 1206–1208.

——. 1981. Morphology and taxonomy of *Himantothallus* (including *Phaeoglossum* and *Phyllogigas*), an Antarctic member of the Desmaresiales (Phaeophyceae). *J. Phycol.* **17**: 15–29.

Mohr, J. L., N. J. Wilimousky & E. Y. Dawson. 1957. An Arctic Alaskan kelp bed. *Arctic* **10**: 45–52.

Nakajima, Y., K. Watanabe & Y. Naito. 1982. Diving observations of the marine benthos at Syowa Station, Antarctica. *Proc. Symp. Antarct. Biol., 5th,* pp. 44–54.

Neiman, A. A. 1963. Quantitative distribution of benthos onthe shelf and upper continental slope in the eastern part of the Bering Sea. *Sov. Fish. Invest. Northeast Pac., Part I*, pp. 143–217.

Nelson, C. H. & K. R. Johnson. 1987. Whales and walruses as tillers of the sea floor. *Sci.Am.* **255**: 112–117.

Nelson, C. H., R. W. Rowland, S. W. Stoker & B. R. Lawson. 1981. Interplay of physical and biological sedimentary structures of the Bering continental shelf. *In* "The Eastern Bering Sea Shelf: Oceanography and Resources" (D. W. Hood & J. A. Calder, eds.), Vol. 2, pp. 1265–1296. Univ. of Washington Press, Seattle.

Nesis, K. N. 1959. The distribution of boreal benthic fauna off the coast of west Spitzbergen. *Dokl. Akad. Nauk SSSR* **127**: 677–680.

——. 1960. Variations in the bottom fauna of the Barents Sea under the influence of fluctuation in the hydrological regime (on the section of the Kola meridian). *In* "Soviet Fisheries Investigations in North European Seas" (Y. Marti, ed.), pp. 129–138. Rybnoye Khozyaistvo, Moscow.

——. 1983. A hypothesis on the origin of western and eastern Arctic distribution areas of marine bottom animals. *Sov. J. Mar. Biol.* **9**: 235–243.

Neushul, M. 1965. Diving operations of sub-tidal Antarctic marine vegetation. *Bot. Mar.* **8**: 234–243.

——. 1968. Benthic marine algae. *Antarct. Map Folio Ser.* **10**: 9–10.

Newbury, T. K. 1983. Under landfast ice. *Arctic* **36**: 328–340.

Newman, W. A. & A. Ross. 1971. Antarctic Cirripedia. *Antarct. Res. Ser.* **14**: 1–257.

Oliver, J. S. & P. N. Slattery. 1985a. Destruction and opportunity on the sea floor: Effects of gray whale feeding. *Ecology* **66**: 1965–1975.

——. 1985b. Effects of crustacean predators on species composition and population structure of soft-bodied infauna from McMurdo Sound, Antarctica. *Ophelia* **24**: 155–175.

Oliver, J. S., R. G. Kvitek & P. N. Slattery. 1985. Walrus feeding disturbance: Scavenging habits and recolonization of the Bering Sea benthos. *J. Exp. Mar. Biol. Ecol.* **91**: 233–246.

Otto, R. S. 1986. Management and assessment of eastern Bering Sea king crab stocks. *Can. J. Fish. Aquat. Sci.* **92**: 83–106.

Palmisano, A. C. & C. W. Sullivan. 1983. Sea ice microbial communities (SIMCO). I. Distribution, abundance and primary production of ice microalgae in McMurdo Sound, Antarctica in 1980. *Polar Biol.* **2**: 171–177.

Papenfuss, G. F. 1964. Catalogue and bibliography of Antarctic and sub-Antarctic benthic marine algae. *Antarct. Res. Ser.* **1**: 1–76.

Parsons, T. R., M. Takahashi & B. Hargrave. 1977. "Biological Oceanographic Processes." Pergamon, Oxford.
Paul, A. Z. & R. J. Menzies. 1974. Benthic ecology of the high Arctic deep sea. *Mar. Biol. (Berlin)* **27**: 251–262.
Pawson, D. L. 1969a. Holothuroidea. *Antarct. Map Folio Ser.* **11**: 36–38.
———. 1969b. Echinoidea. *Antarct. Map Folio Ser.* **11**: 38–41.
Peckham, V. 1964. Year-round scuba diving in the Antarctic. *Polar Rec.* **12**: 143–146.
Petersen, G. H. 1964. The hydrography, primary production, bathymetry and "tagsaq" of Disko Bugt, West Greenland. *Medd. Groenl.* **159**: 1–45.
Picken, G. B. 1980. Reproductive adaptations of Antarctic benthic invertebrates. *Biol. J. Linn. Soc.* **14**: 67–75.
———. 1985. Benthic research in Antarctica: Past, present and future. *In* "Marine Biology of Polar Regions and Effects of Stress on Marine Organisms" (J. S. Gray & M.E. Christiansen, eds.), pp. 167–183. Wiley Chichester.
Platt, H. M. 1979. Free-living marine nematodes of Antarctica. A current appraisal. *Ann. Soc. R. Zool. Belg.* **108**: 93–101.
———. 1980. Ecology of King Edward Cove, South Georgia: Macrobenthos and the benthic environment. *Br. Antarct. Surv. Bull.* **49**: 231–238.
Reimnitz, F., E.W. Kempana & P. W. Barnes. 1986. Anchor ice and bottom-freezing in high-latitude marine sedimentary environments: Observations from the Alaskan Beaufort Sea. *U.S. Geol. Surv. Rep.* **86–298**.
———. 1987. Anchor ice, seabed freezing, and sediment dynamics in shallow Arctic seas. *J. Geophys. Res.* **92**: 14,671–14,678.
Rhoads, D. C. 1974. Organism–sediment relations on the muddy seafloor. *Oceanogr. Mar. Biol.* **12**: 263–300.
Richardson, M. G. 1979. The distribution of Antarctic macro-algae related to depth and substrate. *Br. Antarct. Surv. Bull.* **49**: 1–13.
Richardson, M. D. & J. W. Hedgpeth. 1977. Antarctic soft-bottom macrobenthic community adaptations to a cold, stable, highly productive, glacially affected environment. *In* "Adaptations Within Antarctic Ecosystems" (G. A. Llano, ed.), pp. 181–196. Gulf Publ. Co., Houston, Texas.
Sambrotto, R. N., J. J. Goering & C. P. McRoy. 1984. Large yearly production of phytoplankton in the western Bering Strait. *Science* **225**: 1147–1150.
Schell, D. M. 1983. Carbon-13 and carbon-14 abundances in Alaskan aquatic organisms: Delayed production from peat in Arctic food webs. *Science* **219**: 1068–1071.
Sease, J. L. & D. O. Chapman. 1988. Pacific walrus, *Odobenus rosmarus divergens*. *In* "Selected Marine Mammals of Alaska" (J. W. Lentfer, ed.), pp. 17–38. Marine Mammal Commission, Washington, D.C.
Sigurdsson, A. 1972. The benthic coastal fauna of Surtsey in 1969. *Surtsey Res. Prog. Rep.* **6**: 91–96.
Skjaeveland, S. H. 1973. Echinoderms of Jan Mayen Island. *Astarte* **6**: 69–74.
Skottsberg, C. J. F. 1964. Antarctic phycology. *In* "Biologie Antarctique" (R. Carrick, M. Holdgate & J. Prevost, eds.), pp. 147–154. Hermann, Paris.
Smith, G. A., P. D. Nichols & D. C. White. 1986. Fatty acid composition and microbial activity of benthic marine sediments from McMurdo Sound, Antarctica. *FEMS Microb. Ecol.* **38**: 219–231.
Smith, G. A., P. D. Nichols, A. C. Palmisano & D. C. White. 1990. Benthic near-shore microbial communities of McMurdo Sound and Arthur Harbor, Antarctica: Microbial biomass, community structure and metabolic activity. *SCAR Symp., 5th* (in press).
Smith, S. V. 1984. Phosphorus versus nitrogen limitation in the marine environment. *Limnol. Oceanogr.* **29**: 1149–1160.

Sparks, A. K. and W. T. Pereyra. 1966. Benthic invertebrates of the southeastern Chukchi Sea. *In* "Environment of the Cape Thompson Region, Alaska" (N. J. Wilimovsky & J. N. Wolfe, eds.), pp. 817–838. U. S. At. Energy Comm., Oak Ridge, Tennessee.

Stockton, W. L. 1973. An intertidal assemblage at Palmer Station. *Antarct. J. U.S.* **8:** 305–307.

———. 1982. Scavenging amphipods from under the Ross Ice shelf. *Deep-Sea Res.* **29:** 819–835.

Stoker, S. W. 1973. Winter studies of under-ice benthos on the continental shelf of the northeastern Bering Sea. M. S. Thesis, Univ. of Alaska, Fairbanks.

———. 1978. Benthic invertebrate macrofauna of the eastern continental shelf of the Bering and Chukchi seas. Ph.D. Dissertation, Univ. of Alaska, Fairbanks.

———. 1981. Benthic invertebrate macrofauna of the eastern Bering/Chukchi continental shelf. *In* "The Eastern Bering Sea Shelf: Oceanography and Resources" (D. W. Hood & J. A. Calder, eds.), Vol. 2, pp. 1069–1090. Univ. of Washington Press, Seattle.

Svendsen, P. 1959. The algal vegetation of Spitsbergen. A survey of the marine algal flora of the outer part of Isfjorden. *Skr. M.* **116:** 4–51.

Taylor, W. R. 1954. Algae: Non-planktonic. *Bot. Rev.* **20:** 363–399.

Thomson, D. H. 1982. Marine benthos in the eastern Canadian high Arctic: Multivariate analysis of standing crop and community structure. *Arctic* **35:** 61–74.

Thorson, G. 1934. Contributions to the animal ecology of the Scoresby Sound fjord complex (east Greenland). *Medd. Groenl.* **100**(3): 1–67.

———.1936. The larval development, growth, and metabolism of Arctic marine bottom invertebrates. *Medd. Groenl.* **100**(6): 1–155.

——— 1957. Bottom communities (sublittoral or shallow shelf). *Mem.—Geol. Soc. Am.* **67:** 461–534.

———. 1966. Some factors influencing the recruitment and establishment of marine benthic communities. *Neth. J. Sea Res.* **3:** 267–293.

Thorsteinson, L. K. 1987. Invertebrates. *In* "The Diapir Field Environmental and Possible Consequences of Planned Offshore Oil and Gas Development" (P. R. Becker, ed.), Outer Cont. Shelf Environ. Assess. Program, pp. 77–109. NOAA, Anchorage, Alaska.

Tucker, M. J. & H. R. Burton. 1987. A survey of the marine fauna in shallow coastal waters of the Vestfold Hills and Rauer Islands, Antarctica. *ANAR Res. Notes* **55:** 1–24.

Van Blaricom, G. R. 1982. Experimental analyses of structural regulation in a marine sand community exposed to oceanic swell. *Ecol. Monogr.* **52:** 283–305.

Vibe, C. 1939. Preliminary investigations on shallow water animal communities in the Upernavik and Thule districts (northwest Greenland). *Medd. Groenl.* **124:** 1–42.

von Bodungen, B., V. S. Smetacek, M. M. Tilzer & B. Zeitzschel. 1986. Primary production and sedimentation during spring in the Antarctic Peninsula region. *Deep-Sea Res.* **33:** 177–194.

Wagner, H. P. & J. S. Zaneveld. 1988. Xanthophyceae and Chlorophyceae of the western Ross Sea, Victoria Land, Antarctica and Macquarie Island collected under the direction of Prof. Dr. J. S. Zaneveld (1963–1967). *Blumea* **33:** 141–180.

Walsh, J. M., C. P. McRoy, L. K. Coachman, J. J. Goering, J. J. Nihoul, T. E. Whitledge, T. H. Blackburn, P. L. Parker, C. D. Wirick, P. G. Shuert, J. M. Grebmeier, A. M. Springer, R. D. Tripp, D. Hansell, S. Djenidi, E. Deleersnijder, K. Henriksen, B. A. Lund, P. Andersen, F. E. Muller-Karger & K. Dean. 1989. Carbon and nitrogen cycling within the Bering/Chukchi seas: Source regions for organic matter affecting AOU demands of the Arctic Ocean. *Prog. Oceanogr.* (in press).

Ward, B. L., P. J. Barrett & P. Vella. 1987. Distribution and ecology of benthic foraminifera in McMurdo Sound, Antarctica. *Palaeogeogr., Palaeoclimatol., Palaeoecol.* **58:** 139–153.

Warnke, D. A., J. Richter & C. H. Oppenheimer. 1973. Characteristics of the nearshore environment off the south coast of Anvers Island, Antarctic Peninsula. *Limnol. Oceanogr.* **18**: 131–142.
White, D. C., G. A. Smith & G. R. Stanton. 1984. Biomass, community structure and metabolic activity of the microbiota in benthic marine sediments and sponge spicule mats. *Antarct. J. U.S.* **19**: 125–156.
White, D. C., G. A. Smith, P. D. Nichols, G. R. Stanton & A. C. Palmisano. 1985. The lipid composition and microbial activity of selected recent Antarctic benthic marine sediments and organisms: A mechanism for monitoring and comparing microbial populations. *Antarct. J. U.S.* **19**: 130–132.
White, M. G. 1984. Marine benthos. *In* "Antarctic Ecology" (R. M. Laws, ed.), Vol. 2, pp. 421–462. Academic Press, London.
Wilce, R.T. 1959. The marine algae of the Labrador Peninsula and northwest Newfoundland (ecology and distribution). *Bull.—Natl. Mus. Can.* **158**: 1–103.
Wilimovsky, N. J. 1966. Synopsis of previous scientific expeditions. *In* "Environment of the Cape Thompson Region, Alaska" (N. J. Wilimovsky & J. N. Wolfe, eds.) pp. 1–5. U.S. At. Energy Comm., Oak Ridge, Tennessee.
Zamorano, J. H., W. E. Duarte & C. A. Moreno. 1986. Predation upon *Laternula elliptica* (Bivalvia, Anatinidae): A field manipulation in South Bay, Antarctica. *Polar Biol.* **6**: 139–143.
Zaneveld, J. S. 1968. Benthic marine algae, Ross Island to Balleny Islands. *Antarct. Map Folio Ser.* **10**: 10–12.
Zenkevitch, L. 1963. "Biology of the Seas of the U.S.S.R." Unwin, London.

13 Particle Fluxes and Modern Sedimentation in the Polar Oceans

Susumu Honjo
Woods Hole Oceanographic Institution
Woods Hole, Massachusetts

I. Introduction   688
II. Ocean Particles and Mechanisms of Settling   691
III. Methods of Particle Flux Studies in the Polar Oceans: Field Experiments   694
IV. Arctic Oceans and Their Marginal Seas: Pelagic Particle Fluxes   697
 A. The Nordic Seas   697
 B. The Norwegian and Greenland Current Areas: A Contrast   699
 C. Fluxes of Ice-Rafted Sediment   706
V. Antarctic Oceans: Pelagic Particle Fluxes   707
 A. A Flux Experiment in the North-Central Weddell Sea   708
 B. Ice Edge Development and the Maximum Flux Period   710
 C. Modes of Particle Sedimentation in the Pelagic Weddell Sea   711
 D. Carbonate Fluxes in the Pelagic Weddell Sea   713
 E. Questions on Apparently Very Small Organic Carbon Flux in the Southern Ocean   713
VI. Comparison of Pelagic Fluxes in the Arctic and Antarctic Oceans   715
 A. Biogeochemical Contrasts in Settling Particles   716
 B. Coccolithophorid Distribution in the Polar Oceans   717
 C. Differences in Physical and Biogeochemical Settings in the Polar Oceans   718
VII. Processes of Neritic Sedimentation in the Polar Oceans   722
 A. Particle Fluxes in the Neritic Antarctic Ocean   722
 B. Krill-Mediated Ecosystem and Enhanced Silica Flux in the Ross Sea   724
 C. Neritic Sedimentation in the Arctic Ocean   724
 D. Winter Sedimentation in the Barents Sea   725
 E. A Large Carbon Sink in Arctic Shelf Fronts   730
VIII. Summary and Conclusions   732
References   734

I. Introduction

The finding of living animals on the abyssal ocean floor by one of the first oceanographic expeditions (the H.M.S *Lightning* and *Porcupine* by C. W. Thompson in 1868 and 1870, in Mills, 1983) provoked vigorous discussions on the quality and quantity of energy resources and nutrients which must be supplied to a habitat existing miles below the productive ocean surface. The mechanism which allows communities living at the surface to share nutrients with communities in the deep ocean also was debated for many years (e.g., Menzies, 1962; Vinogradov, 1962). Meanwhile, the mass accumulation rate of skeletal, preserved, and refractory particles on the ocean floor is the key to understanding the geological evolution of the oceans and the earth itself (e.g., Lizitsin, 1978). These major questions in oceanography have been revised more recently in view of the urgent need to understand the biogeochemical cycle of carbon in relation to the recent anthropogenic increase of atmospheric CO_2 (Brewer *et al.*, 1986).

One way to answer these questions is to measure the rate of sedimentation, usually called "particle flux," and examine the sources and origins of particles which settle toward the deep-sea floor. A number of open-ocean experiments have begun to assess particle fluxes and to explain them in relation to other oceanographic criteria. However, the number of observations is limited in time and space; thus, a major effort is required to reach a plausible global model.

The production, transport, and settling of oceanic particles in the polar oceans involve intriguing problems which are different from those in lower-latitude ocean environments. Polar oceans are strongly influenced by the extent of ice coverage, particularly of dynamic marginal ice zones (MIZ; e.g., Smith and Nelson, 1986), where occurrences of short-lived blooms of phytoplankton supply a significant amount of the annual production of biogenic ocean particles. In general, because of the insolation pattern, the seasonal variations in primary production are very large; thus, biogenic fluxes are highly variable. As explained in this chapter, the principal components of polar ocean fluxes are controlled by the biological interactions in the surface layers, although the sedimentation of coarse ice-rafted rock detritus involves only physical processes. One essential requirement for studying oceanic fluxes in the polar oceans is to sample on appropriate space and time scales so as to document this intense seasonal variability. For example, a flux measurement taken only during the open-water season or under ice-fast conditions cannot be used to measure the annual flux; the former results in a serious overestimation and vice versa. Therefore, flux measurements and collection of settling particles must be done in all seasons continuously. In the interior of polar oceans, where winter access is usually impossible and the sedimentary environment changes quickly, deployment of an automated long-term, high-time-resolution sampler appears to be the only way to collect these data at present.

Systematic particle flux experiments using sediment traps started only recently in both polar regions; these began with relatively short deployment durations during the austral summer of 1981–1982 in the Drake Passage, Antarctica (Wefer *et al.*, 1982), and the Bransfield Strait and Ross Sea areas (Dunbar, 1984). The systematic deployment of fully automated, four-season, time-series sediment traps began in the Fram Strait and northern Nordic Seas in 1983, funded by the U.S. Office of Naval Research with West German and Icelandic ship support. By the summer of 1987, 17 sediment traps had been deployed, each of which collected samples for about 1 year at approximately 1-month intervals (Honjo *et al.*, 1987a) and this program is continuing today. A multiyear, time-series sediment trap experiment in the pelagic Weddell Sea, Antarctica, has been carried out since 1985 under

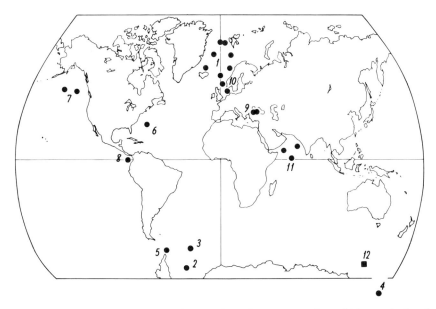

Figure 13.1 Approximate positions of time-series sediment trap stations which were deployed at the interior of the deep ocean and are relevant to the text. Much more data may be available from other stations not shown in this figure. 1, Nordic Seas (Honjo *et al.*, 1987a). 2, Atlantic Southern Ocean (Fischer *et al.*, 1988). 3, South Georgia Basin (unpublished data from the British Antarctic Survey and WHOI). 4, Ross Sea area (Dunbar *et al.*, 1985; Dunbar and Leventer, 1987). 5, Antarctic Peninsula–Drake Passage area (Wefer *et al.*, 1982, 1988). 6, Bermuda time-series station (e.g., Deuser *et al.*, 1981). 7, Ocean Station P (e.g., Wong and Honjo, 1984). 8, Panama Basin station (Honjo, 1982). 9, Black Sea stations (e.g., Honjo *et al.*, 1987c). 10, Southern North Sea (Kempe *et al.*, 1987). 11, Gulf of Arabia (Nair *et al.*, 1989). 12, Pacific Southern Ocean, short-term deployment (Noriki *et al.*, 1985).

international cooperation with R/V *Polarstern* (Fischer et al., 1988). Time-series particle flux measurement programs are also being conducted on the Ross Sea shelves and slopes (Dunbar and Leventer, 1987) and the coastal Beaufort Bay area (Dr. K. Iseki, personal communication, 1988) (Figs. 13.1 and 13.2).

This chapter concerns the formation of recent polar sediments inferred from the particle fluxes at the deep interior of the ocean. However, attempts have been made to understand the origin and quantity of particles, particularly in relation to the plankton ecosystem in the upper layers. Since our traps were deployed in low-energy, deep-environment oceans and flux data from shallow layers were not always available, there are limitations when one

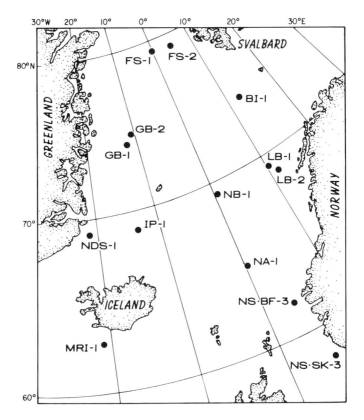

Figure 13.2 Distribution of time-series sediment trap stations in the Nordic Seas. At each station, one or two (GB stations) time-series sediment traps were deployed continuously for at least one full year. Traps were deployed approximately 400 m above the seafloor (381–622 m) at deep ocean stations; at the shelf stations, NS and NDS, traps were deployed at 100 m above the seafloor (Table 1 of Honjo et al., 1987a).

tries to relate deep-ocean particulate organic carbon (POC) flux data to surface productivity. In many instances, they are decoupled (for example, Fig. 13.5).

Considering the vast environmental variability of the polar oceans, our accomplishments in gathering reliable data on particle fluxes are still inadequate for drawing a more complete picture of the biogeochemical role of oceanic particles. Particle flux data from the central Arctic Ocean, Bering Sea, the Sea of Okhotsk, and the Indian Ocean sectors of the Southern Ocean are scarce or nonexistent. However, if one relates particle flux data to the recent rapid advance of polar oceanography, a preliminary comprehension may be possible. Prominent examples of more recent advances which have contributed to explanations of the variabilities of particle fluxes in polar oceans in this chapter are (1) a better understanding of the marginal ice zone (MIZ) in both Arctic and Antarctic regions with regard to productivity and ecosystem sequences in time and space (e.g., Smith and Nelson, 1986; Chapter 9, this volume), (2) findings in Arctic biogeochemistry (e.g., Anderson and Dyrssen, 1981; Östlund and Hut, 1984; Chapter 8, this volume) and new information on the controls of ocean fertility (e.g., Martin and Gordon, 1988), and (3) studies of the neritic sedimentation in Antarctic shelves and slopes (e.g., Anderson et al., 1984; Dunbar et al., 1985). It is clear that more multidisciplinary research is warranted, particularly in the areas of primary productivity, ecodynamics, physical oceanography, and bottom sediments in ice-covered and open environments to gain a better explanation and more useful models of particle fluxes in the polar oceans.

One conclusion drawn from this study is that the processes and fate of oceanic particles in the polar oceans are strongly controlled by the variability of primary production, just as is seen in low-latitudinal oceans, with the exception of the production and transportation of coarse glacial detritus. The temporal variability of biogenic particle production responds sharply to changes of polar ecosystems with the evolution of marginal ice zones. Fine and light particles, organic as well as clay, need to be aggregated by biological mediators in order to settle through the water column to the ocean interior and seafloor. Thus, fluxes of biogenic and fine lithogenic matter are strongly coupled in polar oceans.

II. Ocean Particles and Mechanisms of Settling

Oceanic particle studies are relatively new, and many concepts, hypotheses, and interpretations have emerged recently. Some of these ideas are still controversial and many others are unconventional, when compared with more established oceanographic dogma. The following is a brief explanation of the basic understanding of ocean particles and their fluxes which forms the basis for further discussions in this chapter.

Two major sources supply particles to the ocean layers. Biogenic (ocean) particles are formed in the upper ocean as a result of planktonic metabolism. Lithogenic (ocean) particles (mostly clay and rock detritus), in contrast, are transported from continents by rivers, coastal erosion, and wind. Also, a large volume of lithogenic particles is transferred within the ocean interior as a result of resuspension of sediment from all sides of the oceans. Particles, whether produced *in situ* or transported from other locations, would not settle toward the ocean floor unless they had a higher specific gravity than the surrounding seawater. Particles which are large and heavy enough to settle vertically remove solid matter from the upper and deeper layers of the ocean and transport it to the ocean interior. These settling particles are capable of supplying benthic animals with food. Lithogenic particles and skeletal biogenic material (calcium carbonate shells, tests and shields, silica skeletons and frustules) unaffected by chemical dissolution at the seafloor surface accumulate as ocean sediments. Particle flux, which is the rate of sedimentation of solid matter per unit of time and area at a given depth, is therefore applicable only to particles which settle with a velocity (a negative value is possible). The term "ocean flux" is applicable only to settling particles and not to suspended particles which are not involved with vertical settling (Honjo, 1984, 1987).

The majority of particles provided to the water column, whether produced by biological activity in the upper oceans or as lithogenic particles originating at the periphery of oceans, are either too small or too light to settle by themselves in seawater without the aid of sinking-speed acceleration. For example, according to straight Stokesian reasoning, a coccolith of a few to several micrometers would take thousands of years to settle through the long, calcium carbonate–undersaturated water column, i.e., no coccolith should reach the abyssal ocean floor (Honjo, 1976). Nevertheless, coccoliths indeed are found in deep-sea sediments. Another example is oil and wax droplets, which are substantially lighter than seawater but are commonly found in deep-ocean sediments.

A number of mechanisms are known which accelerate the sinking speeds of fine particles. One is via incorporation into fecal pellets of metazoan zooplankton. These animals graze, by filter feeding, mostly on phytoplankton, detritus, and lithogenic particles and eject membrane-covered pellets into the water column. Organic matter is usually only partially digested by the host; thus a pellet itself can provide relatively high food value. Pellets settle at relatively high speed; e.g., a typical oceanic *Calanus* pellet descends at the rate of about 200 m day^{-1} (Small *et al.*, 1979), taking only a few to several weeks to reach the deep-ocean floor.

The other form of biogeochemical aggregate which accelerates the flux of fine and light individual ocean particles is marine snow. "Marine snow" is an ambiguous term used to define aggregates of millimeter size made up of numerous fine particles that are not usually seen in normal reflective light but will show up as a light-scattered image under certain conditions of illumination. It is almost impossible to see marine snow in the water from above. The amorphous detritus of organic matter, often produced from the decomposition processes of macroplankton such as jellyfish, remains of larvacean houses, and entangled diatoms, serves as a matrix for the aggregate (Silver and Alldredge, 1981). Although it is speculative, we believe that independent suspended particles are caught by this hydrated, adhesive mass. The collection of fine but dense particles, such as clay minerals and calcite (coccoliths), increases the total specific gravity of an aggregate, and as an aggregate descends through the upper water column it agglutinates more independent particles and increases its speed.

Analogous to fecal pellets which have lost their surface cover, an aggregate also will lose particles while settling through the deep water column and supply fine particles to the water column. Suspended particles, sloughed off from either fecal pellets or marine snow aggregates, can be scavenged by another settling particle and again gain settling speed. This switching between suspended and settling status may happen many times while an ocean particle settles through a long water column (Honjo, 1975). Studies of the efficiency of particle trapping using a radiochemical tracer showed that nearly 100% of the transuranium elements generated in the water column above were found in a sediment trap (Bacon et al., 1985). Therefore, we suspect that all suspended particles eventually settle down with large particles after changing back and forth from suspended to settling. A suspended particle can also be injected into any depth after being sloughed off from an aggregate; therefore, no time-stratigraphic relationship will be found among suspended particles distributed throughout a deep water column. The predicted random vertical relation of suspended particles was confirmed by tracing the grade of dissolution on calcite coccoliths in the deep Pacific Ocean, where seawater is strongly undersaturated with calcium carbonate (Honjo, 1975).

The fast descent of settling particles has been observed at a number of stations. For example, at Ocean Station P, Gulf of Alaska, the arrival of settling particles at 3800 m was delayed for about 2 weeks after their arrival at 1000 m, as shown in samples from a pair of synchronized time-series sediment traps. This residence time applies to all particles smaller than 1 mm and is independent of origin, indicating that the bulk settling speed of particles in the bathypelagic layer is about 200 m day^{-1} (Wong and Honjo, 1984; Honjo, 1984) and is equivalent to that of *Calanus* fecal pellets (Small *et al.*, 1979). Similar settling speeds were estimated in the Black Sea (Honjo *et al.*, 1987c) and the Gulf of Arabia (Nair *et al.*, 1989, in review).

The quality and quantity of biogenic particles which arrive at the ocean interior depend largely on a succession of interactions between phyto- and zooplankton in the upper ocean. Hypothetically, if phytoplankton growth and zooplankton growth are in phase (i.e., occur at the same time), biogenic particles will be transported toward the ocean interior mostly as fecal pellets. The pattern of biogenic particle fluxes in the pelagic Weddell Sea suggests that this is the case (see Section V). If the growth of phytoplankton and that of zooplankton are offset, a large quantity of nongrazed plant tissue may descend through the water column in amorphous form during and after a bloom. Carbon-rich sediments on the shelves of the northern Barents Sea were probably formed by this process (see Section VII). It is commonly observed in the ocean that zooplankton populations grow in parallel with phytoplankton, but the former are somewhat temporally delayed. In this case, sedimentation begins with amorphous settling particles such as marine snow, and fecal pellet flux increases as the grazer population increases. The maximum fecal pellet flux should occur when the two curves, of declining phytoplankton and of increasing grazer populations, intersect.

Organic flux measured in the ocean interior should be expressed as primary production minus the mass utilized by zooplankton activity and microbial metabolism in the ocean layers above, just below the bottom of the euphotic layer, plus the mass of settling dead zooplankton. The long time-integrated fluxes of nondigestible mineralized tissues, such as diatom frustules and coccoliths, produced by a primary producer should be equal to the production rate in the ocean surface layer, because such tissues are chemically unaffected while passing through a metazoan gut, even during coprophagy (Honjo and Roman, 1978). Thus, frustules and coccoliths are valuable tracers in understanding the relationship between particle flux in the ocean interior and "new-production" carbon, the primary production based on allochthonous nutrients (Eppley and Peterson, 1979).

III. Methods of Particle Flux Studies in Polar Oceans: Field Experiments

The most important logistical task in particle flux studies in polar oceans is that of maintaining a sediment trap experiment throughout all four seasons of the year with time fractionations to cover the expected strong variability of fluxes, which reflects the large amplitude of biological and physical variables in polar environments. Because of the importance of short but prominent events, such as the passing of the MIZ, the resolution of time fractionation

13 Particle Fluxes and Modern Sedimentation in the Polar Oceans

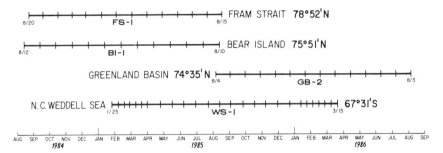

Figure 13.3 Comparison of time-series sampling schedules at four sediment trap stations in boreal versus austral regions deployed from August 1984 to August 1986. Samples from Atlantic Southern Ocean station WS-1 can be directly compared with those from a number of Nordic Seas stations with respect to seasonality, which overlapped for about a half-year. Detailed sampling schedules can be referred to in Honjo et al. (1987a).

should be as short as possible and sample collection continuous. We have developed a time-series sediment trap with 12 or 13 time fractionations which can be deployed continuously for over a year and which is reliable at freezing-seawater temperatures at any depth. For the Weddell Sea, where particularly short bloom events were expected, we deployed sediment traps with 24 time fractionations of uneven intervals. The particle flux occurring at a given time period in the Arctic and Antarctic can be directly compared using this method (Fig. 13.3). When the unit collection period is short, a larger physical collection area is required; thus we used traps with a 1.2-m² opening (Mark 5) in the Weddell Sea experiment. Mark 6 traps with a 0.5-m² opening were deployed, with few exceptions, at the rest of the stations in the Nordic Seas. A microprocessor-based power controller is used to define the sequential collecting periods and to monitor program execution (Honjo and Doherty, 1987).

In polar oceans where there is no ice cover at any time, such as at stations in the Norwegian Sea, conventional bottom-tethered mooring arrays with time-series trap(s) can be applied. In areas which are partially covered by seasonal sea ice, such as the Weddell Sea and the Greenland Sea, we also applied conventional arrays, but the mooring sites were chosen with particular attention to the fact that the sea could be expected to be free of ice during the planned recovery period. At this time, we have no means of safely recovering a large sediment trap when it is caught under an ice sheet or floe. Sediment traps were deployed in the Nordic Seas at 400 to 600 m above the seafloor to avoid the possible influence of resuspended particles from the bottom (Fig. 13.4). One array in the central Greenland Sea had an additional sediment trap tethered about 2000 m above the deeper trap. A continuously recording transmissometer was deployed with a Fram Strait mooring for 1 year.

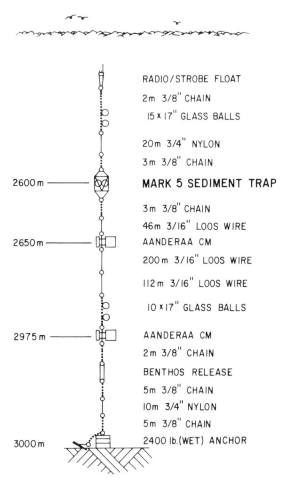

Figure 13.4 Typical mooring array deployed in the deep Nordic Seas stations. At most of the stations, Mark 6 sediment traps were used (0.5-m^2 aperture; Honjo and Doherty, 1987); at the LB, GB, and BI stations, Mark 5 traps were used.

In our earlier experiments the sample collectors on sediment traps contained 0.4% $HgCl_2$ as a preservative; however, the effectiveness of this poison in the deep polar ocean environment was not clear. An experiment to test the effectiveness of this poison has not shown any significant difference between poisoned and nonpoisoned samples. No significant effect on possible metazoan visitors or recolonization ("swimmers"; Knauer et al., 1979) was noted during our experiments. With few exceptions, current meters were deployed with all sediment trap arrays. All sediment trap arrays were deployed in deep, low-energy areas with current speeds of less than 5 cm s^{-1}, except for the Fram Strait, where the current reached 27 cm s^{-1} during January and March (Honjo et al., 1987b). No significant relationship between current vectors and mass fluxes was found in our studies. The seasonal variability of fluxes overwhelmed the deviation of current speed in the deep ocean; thus, no statistical relationship was found between those parameters. The radiochemical trapping efficiency of the Mark 6 at a subtropical North Atlantic station was estimated to be nearly 100%, using a ^{230}Th and ^{231}Pa ratio (Bacon et al., 1985), and similar results were obtained from deep trap samples from the Lofoten basin (Dr. M. Bacon, personal communication).

IV. Arctic Oceans and Their Marginal Seas: Pelagic Particle Fluxes

A. The Nordic Seas

Particle fluxes in the Arctic Ocean *per se* have not yet been studied by direct collection of samples and rate measurements. However, field experiments using time-series sediment traps have begun in the major marginal seas of the Arctic Ocean: the Nordic Seas. Detailed research on particle fluxes in the Nordic Seas, based on the basinwide distribution of sediment trap stations (Fig. 13.2), is beginning to provide useful information and hypotheses leading to a better understanding of oceanic sedimentation in the Arctic Ocean.

The Nordic Seas include the eastern Greenland Sea, Norwegian Sea, and western edge of the Barents Sea, with the Faeroe Islands as the southern boundary (Hurdle, 1986). The Nordic Seas form the largest marginal sea of the Arctic Ocean, with an area roughly equivalent to the European Mediterranean, and bridge the Arctic Ocean with the North Atlantic. The Nordic Seas are characterized by a prominent oceanic environmental contrast between the eastern and western halves. To the west, the Transpolar Drift flows southward along the east coast of Greenland, transporting sea ice and ice floes from the high Arctic Ocean and grading into marginal ice zones along its eastern boundary. The area dominated by this southward-flowing Transpolar Drift in the western Nordic Seas is called the Greenland Current area. The long front roughly parallels the east coast of Greenland and longitudinally oscillates over many miles; during winter, the area covered by water at 0°C in the Greenland Current area is twice as large as during the summer months. The southern extension of the front remains off southeast Greenland and does not oscillate in latitude very much with the seasons (Wadhams, 1986).

In contrast, the eastern half of the Nordic Seas is covered by the warm,

Table 13.1 Annual Fluxes in the Nordic Seas and Comparison with Other Oceans[a]

Station[b]	Year	Flux (g m^{-2} yr^{-1})				Ca/Si	
		Total	Ca	Si[c]	C[d]	Weight ratio	Mole ratio
Bear Island (BI-1)	1984/85	28.3	2.6	0.9	2.9	2.9	2.0
Lofoten Basin (LB-1)	1983/84	22.8	4.6	0.5	1.4	9.2	6.4
Aegir Ridge (NA-1)	1985/86	17.4	3.7	0.8	0.6	4.6	3.2
East Jan Mayan (NB-1)	1985/86	16.8	3.6	0.7	0.5	5.1	3.6
Fram Strait (FS-1)	1984/85	6.1	0.6	0.3	0.4	2.0	1.4
Greenland Basin (GB-2, 3)	1985/86	10.2	1.3	1.2	0.4	1.1	0.8
Denmark Strait (NDS)	1986/87	12.5	1.2	0.7	0.5	1.7	1.2
Ocean Station Papa[e]	1982/83	56.8	7.8	13.6	2.4	0.6	0.4
	1983/84	16.5	3.4	2.4	0.8	1.4	1.1
	1984/85	22.5	4.4	2.9	0.7	1.5	1.1
Black Sea[f]	1982/83	36.1	4.3	1.5	5.2	2.9	2.0
	1983/84	37.3	4.3	1.5	4.3	2.9	2.0
	1984/85	36.1	4.3	1.4	4.3	3.1	2.2
Arabian Sea							
West[g]	1986/87	41.5	9.8	4.5	1.8	2.2	1.5
Central	1986/87	32.0	8.5	1.7	1.7	5.0	3.5
East	1986/87	27.1	5.5	1.7	1.8	3.2	2.2
Panama Basin[h]	1979/80	93.3	17.5	5.1	3.7	3.4	2.4
Weddell Sea[i]	1985/86	0.4	0.005	0.14	0.02	0.04	0.03
South Georgia Island[j]	1987/88	1.29	0.10	0.41	0.04	0.24	0.17

[a] Biogenic Ca and Si fluxes in sediment trapped at bathypelagic depths from the Nordic Seas and other selected pelagic ocean stations (except the Denmark Strait station, which was neritic).
[b] Station locations in the Nordic Seas are shown in Figure 13.2.
[c] Silica, in the form of biogenic opal only.
[d] Carbonate-carbon free.
[e] See, e.g., Wong and Honjo (1984), Honjo (1984).
[f] See, e.g., Honjo et al. (1987).
[g] Nair et al. (1989).
[h] Honjo (1982).
[i] Fischer et al. (1988).
[j] Unpublished data, P. Barker, C. Pudsey (British Antarctic Survey), and S. Honjo.

northward-flowing Norwegian Current (or Norwegian-Atlantic Current). Thus, the surface temperature in a wide zone along Norway is maintained at 8-11°C during summer and 5-7°C during winter (Dietrich, 1969; Johannessen, 1986). Scoresby Sound, Greenland, and offshore Lofoten Island, Norway, for example, are both located at about 70°N latitude and are only approximately 1300 km apart, yet the water temperature difference exceeds 10°C during the summer and 6°C during the winter.

B. The Norwegian and Greenland Current Areas: A Contrast

In order to understand the relationship between the particle fluxes and oceanographic variabilities in the Nordic Seas, we deployed, from 1983 to 1986, time-series sediment traps at 10 widely separated open-ocean stations in the deep interior of the Nordic Seas (Fig. 13.2). The sediment traps operated throughout the year, collecting settling particles continuously in 12 or 24 increments (2-week or 1-month periods) at approximately 400 m above the seafloor (Honjo et al., 1987a). The annual particle fluxes in the Greenland Current area ranged from 7.2 to 10.2 g m^{-2} (averaging about 8.5 g m^{-2}), while the annual fluxes in the Norwegian Current area ranged from 16.6 to 28.4 g m^{-2} (averaging 21.3 g m^{-2}) during 1983 to 1986 (Tables 13.1, 13.2, 13.3). Peinert et al. (1987) reported similar values from their summer experiments conducted at a station set about 2000 m deep, southeast of Station LB-1. Particle fluxes were relatively large in all size and compositional categories; in particular, carbonate and organic carbon fluxes were several times higher in the Norwegian Current area than in the Greenland Current area.

The annual total organic carbon flux in the Norwegian Current area ranged from 0.5 g C m^{-2} (at a far offshore station) to 2.8 g C m^{-2} (at a near-shelf station south of Spitsbergen, where the influence of shelf-to-basin export is strong. A year-round observation of primary production is rarely possible in the Nordic Seas. The annual primary carbon production measured in 1979 near one of our sediment trap sites, LB-1 (Fig. 13.4), was 90 g C m^{-2} (Rey, 1981a). Taking into account a large year-to-year difference in primary production and flux, the portion of annually produced organic carbon which arrives at the interior of the Lofoten basin (1.4 g C m^{-2}; Table 13.2) is roughly 1.6% and the proportion is smaller than the values found at some temperate ocean stations (Honjo, 1984).

A spring bloom occurs in the April-July period with year-to-year consistency, as in other high-latitude North Atlantic regions (Rey, 1981a,b). A large increase of primary production occurring in May-June in the Lofoten area may be represented by a large peak of carbon flux appearing in July-August (Fig. 13.5), which suggests that the surface bloom is delayed by at least 1

Table 13.2 Mass Fluxes of Major Sedimentological Constituents for Five Selected Stations in the Northern Nordic Seas, 1983–1986

Area	Norwegian–Atlantic Current			East Greenland/Fram Strait		
Trap station	LB-1	BI-1	NA-1	NB-1	FS-1	GB-21[a]	GB-23[b]
Latitude	69°30'N	75°51'N	65°31'N	70°00'N	78°52'N	74°35'N	75°35'N
Longitude	10°00'E	11°28'E	00°64'E	01°58'W	01°22'E	06°43'W	06°43'W
Trap depth (m)	2760	1700	2630	2749	2440	1966	2871
Flux (g m^{-2} yr^{-1})							
Total	22.80	28.30	17.36	16.78	6.61	8.79	10.21
Carbonate	11.40	6.61	9.18	8.93	1.40	2.59	3.28
Noncombustible	8.07	16.31	5.94	6.24	4.26	3.69	5.73
Combustible	3.37	5.38	2.31	1.90	0.92	2.50	1.23
Biogenic opal	1.12	1.96	1.68	1.44	0.60	—[c]	2.61
Lithogenic	6.95	14.35	4.26	4.65	4.00	—	3.12
Organic carbon	1.37	2.85	0.59	0.53	0.41	0.94	0.40
Nitrogen	0.18	0.30	0.08	0.08	0.06	0.16	0.06

[a] Sediment trap deployed 881 m deep.
[b] Sediment trap deployed 2823 m deep.
[c] No data available.

Table 13.3 Comparison of Average Total Mass Fluxes and Average Fluxes in Norwegian Current Area and Transpolar Drift Area, 1983–1986

	Norwegian Current area	Nordic Sea Transpolar Drift area
Moorings	LB-1, BI-1, NA-1, NB-1	FS-1 and GB
Flux (g m^{-2} yr^{-1})		
Total	21.3 (5.4)^a	8.5 (1.8)
Carbonate	9.0 (2.0)	2.4 (0.9)
Noncombustible	9.1 (4.9)	4.6 (1.0)
Combustible	3.2 (1.5)	1.6 (0.8)
Biogenic	1.6 (0.4)	2.0 (1.4)
Lithogenic	7.6 (4.7)	3.6 (0.7)
Organic carbon	1.3 (1.0)	0.6 (0.3)
Nitrogen	0.2 (0.1)	0.1 (0.1)

^{a} Standard deviations are given in parentheses.

month and that the residence time of bloom-generated settling particles is longer in the Nordic Seas than in temperate waters. This is not conclusive because the years when carbon appeared in the ocean interior did not coincide with time of high surface productivity.

The mode of variability of fluxes of biogenic and lithogenic particles in the Norwegian Current area differs significantly in summer and winter (see Fig. 13.5). The standard deviation in monthly mass fluxes from the annual average was as high as 90% in the deep Lofoten basin station. The annual maximum particle flux occurs during the autumn (in September and October) in the Norwegian Current area. This "late-autumn flux maximum" is an unusual phenomenon which has been observed only in the Norwegian Current area and, to a lesser extent, in the deep Fram Strait (Fig. 13.6) (Honjo et al., 1987a, 1988) and is characterized by a large flux of zooplankton remains. About half of the carbonate flux in September and October consisted of pteropod shells and fully grown planktonic foraminifer tests, in contrast to the particles which settled during July and August, which were mostly coccolithophorids and diatom frustules. Primary production decreases dramatically in the fall when daylight becomes progressively shorter after September (Rey, 1981a, b) (Fig. 13.5, lower right), and the late-autumn flux maximum seems unrelated to high primary production. The reason why the organic flux maximum occurs in late autumn in the Norwegian Current area is not readily apparent, but it appears that the phytoplankton are efficiently consumed within the upper layers during the spring bloom and are not immediately delivered as a pulse. In the late summer/autumn, the zooplankton remains settle following the period when primary production is drastically reduced or has ceased.

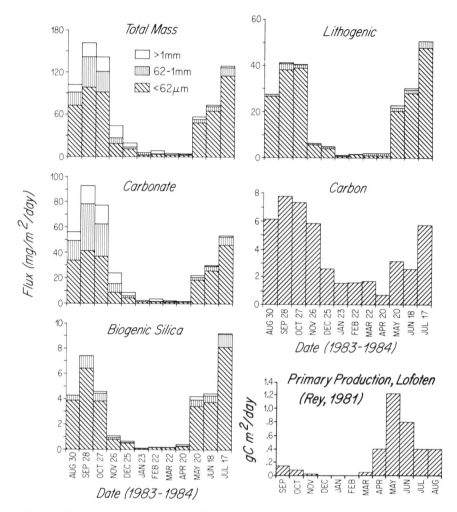

Figure 13.5 Particle flux measured off Lofoten Island during 1983 and 1984. Annual total mass and major sedimentological constituents of trap-collected sediments at the LB-1 station (Honjo et al., 1987a) (see Fig. 13.2) compared with monthly primary production observed at a nearby station in 1975 and 1976 (Rey, 1981a).

The annual flux in the Greenland Current or Transpolar Drift area was, in general, less than half that from the Norwegian Current area (Table 13.3). In contrast to the Norwegian Current area, the flux in the Greenland Current area is much less varied throughout a given year. No late-fall flux maximum was observed in this area (Figs. 13.5, 13.6, and 13.7). The standard deviation in the average of monthly flux was about 40% at Greenland Current stations, compared with 90% in the Norwegian Current area. The sources of settling

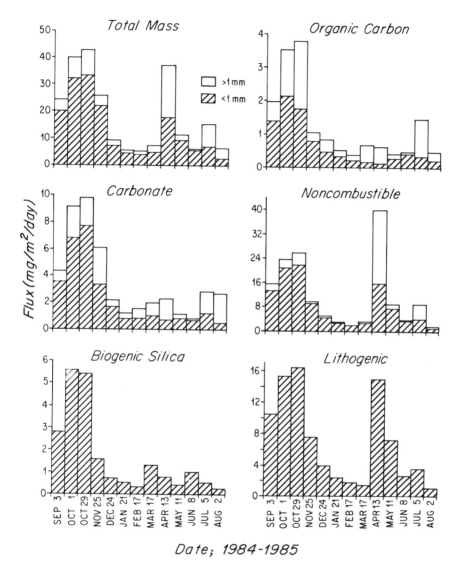

Figure 13.6 Particle flux from Fram Strait during 1984 and 1985. Annual total mass and major sedimentological constituents of trap-collected sediments at the FS-1 station. Biogenic silica and lithogenic particles were analyzed in samples from which all particle fractions were combined (S. Honjo and G. Wefer, unpublished).

particles in the Norwegian Current area were similar to those in the Greenland Current area, except that more lithogenic particles, supplied from sea ice, were contained in the latter (Table 13.2). The flux of pteropod shells and planktonic foraminifer tests was far less in the Greenland area. Surprisingly,

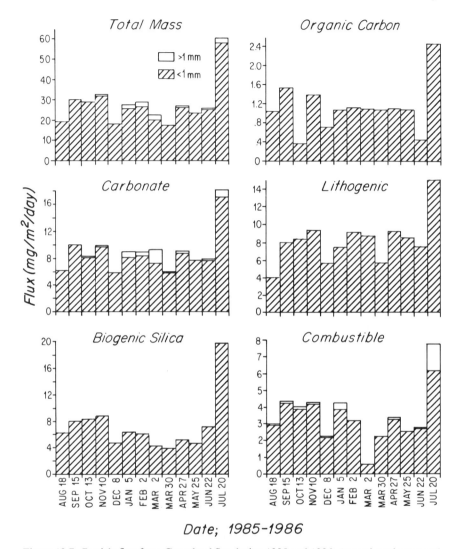

Figure 13.7 Particle flux from Greenland Sea during 1985 and 1986. Annual total mass and major sedimentological constituents of trap-collected sediments from the GB-2 station (Honjo et al., 1987a). Note that compared to stations in the Norwegian Current area, the seasonality in particle sedimentation was smaller except in the sequence of biogenic silica flux.

the annual total coccolith flux was about equal in both areas; however, the flux of planktonic foraminifera was less than in the Norwegian Current area, making the total carbonate flux smaller in the Greenland Current area (see Section VI).

The Greenland basin stations (e.g., Station GB-2; Fig. 13.7) were located in a region which was a marginal ice zone during the summer and covered by sea ice during winter. The surface water temperature at these stations was about 0°C throughout the year and occasionally reached 4°C during July and August (Norsk Meteorological Institute, Oslo). The relationship between the ice extent and the flux at the GB stations seemed to be more complex than in the north-central Weddell Sea, where the development of the ice edge and the variability of flux are clearly related (see Section V). This complexity may be attributed to the fact that at these Greenland Current area stations the MIZ during the boreal summer was not well defined, and there was frequent coverage by ice floes even during midsummer (information from the Norsk Meteorological Institute, Oslo; Wadhams, 1986). It also may be due to differences in growth patterns of phytoplankton in the Greenland Current areas (see Section VI).

In the Norwegian Current area, the organic and lithogenic particle fluxes were well correlated (Fig. 13.5). However, their relationship was less clear in the Greenland Current or Transpolar Drift area (Fig. 13.7). Lithogenic particles settling in the former area were dominated by fine clay, and the majority probably originated by resuspension from the shelf and continental slopes off Norway (Honjo et al., 1988). By themselves, these particles do not settle through a deep-ocean water column on a practical time scale unless accelerated by concentration into larger, heavier aggregates.

Settling metazoan fecal pellets played a smaller role as an agent in the transfer of fine lithogenic particles through the water column in the Norwegian Current area. Less than 10% of organic carbon flux was due to sinking fecal pellets in this area. Although no direct evidence, such as direct observation by a "marine snow" camera, has been obtained, it seems that the major mechanism of fine-particle sedimentation in this area involves settling marine snow aggregates, as in other temperate pelagic oceans (Pilskaln and Honjo, 1987). The close relationship of fine-particle sedimentation with biogenic flux can thus be explained: during a bloom period, more organic matrices are produced and these form aggregates, such as marine snow, which more efficiently scavenge suspended particles including clay minerals and transport them to the ocean interior more rapidly than would be possible by independent settling. This means that the variability of the primary supply of fine lithogenic particles from land (coastline erosion and riverine input) or via resuspension from the shelf and continental slope may not necessarily be the primary influence on variability in lithogenic flux. Such productivity-dependent vertical flux of fine lithogenic matter also has been observed at temperate stations (Honjo et al., 1982).

In contrast, the relationship between lithogenic and biogenic fluxes was less clear in the Greenland Current area (Table 13.2; Figs. 13.5-13.7). At a Fram Strait station (FS-1), a period of large organic flux during the fall was associated with a large lithogenic flux (Fig. 13.6) as in the Norwegian Current area; however, large pulses of lithogenic flux also occurred independently of the biogenic flux.

C. Fluxes of Ice-Rafted Sediment

The Transpolar Drift transports ice which has resided in the Arctic Ocean for several years (Östlund and Hut, 1984), and passes it through Fram Strait at a rate of $0.75-1.3 \times 10^6$ km^{-2} yr^{-1} (Vinje, 1976). Multiyear sea ice in the southern Transpolar Drift is characterized by "dirty" ice which is rich in ice-rafted particles of silt to sand size. Northern Greenland, Svalbard, and other Arctic alpine areas often have exposures of rugged rock formation, and low precipitation and high winds result in a dense covering of airborne particles over the nearby ice fields. If fine sand particles are lifted to an initial height of 6500 m, mean North Atlantic wind velocities are capable of providing up to 600 km of horizontal transport, about equal to the width of the Fram Strait (Shaw et al., 1974).

Sea ice drifts with the Transpolar Drift from the Arctic Ocean to the Greenland Current area of the Nordic Seas and eventually melts in the MIZ of the Greenland Sea. Ice-rafted debris in the sea ice is sloughed off ice floes from melting sea ice, supplying fine to coarse sand sediment to the bottom. It seems that bottom sediments in the south-central Fram Strait area receive a large flux of sand-sized ice raft rock fragments. For example, a peak of lithogenic flux was observed at a station in central Fram Strait during April 1985, when organic production and flux virtually ceased (Fig. 13.6). Almost all of the material collected in a sediment trap during this April maximum period was sand-sized rock fragments, about 60% of which were larger than 1 mm; the total flux of the rock fragments during this period was 1.1 g m^{-2}. A tansmissometer deployed horizontally 50 m above the seafloor on the sediment trap mooring at this station recorded the event when rock fragments rushed to the bottom. Sudden changes of transmissivity suggested that this ice-rafted sediment supply to the water column lasted for about 12 days (S. Honjo, unpublished data). It is hypothesized that the warm Western Spitsbergen Current penetrates under the Transpolar Drift at the central Fram Strait, accelerating ice melt and enhancing the flux of ice-rafted particles at this station.

Icebergs supply a large amount of detritus to the neritic and pelagic ocean. Davis Strait and the Labrador Sea are characterized by a dense distribution of giant icebergs which calve from western Greenland coastal glaciers. The annual production fluctuates, but on average $20-30 \times 10^3$ icebergs are discharged from this area every year (Murray, 1968). On average, newly produced icebergs along the western Greenland coast are estimated to be 300 m^3

in size and 23×10^6 tons in weight; they then decrease to 10-15% of their original weight by the time they reach Labrador (Fillon et al., 1981). About 2500 icebergs pass south of 60°N, but only about 400 travel south of 48°N (an iceberg has been recorded at 28°N, 48°W, 350 km southeast of Bermuda). Most of the ice-rafted sediment from giant icebergs is discharged in the area south of 58°N when icebergs meet water at 4-9°C. Production of giant icebergs is limited along the east side of Greenland, compared to the west coast, except for glaciers from King Christian IX. Land such as Kangerdlugssaug fjord, where glaciers are backed by a large supply of ice from the continental ice cap. "Black icebergs," which appear dark-colored from a distance, have been spotted off southeastern Greenland (J. Milliman, personal communication, 1988).

Three types of glacial sediments are included in a typical iceberg: (1) tillite, up to boulder size; (2) unsorted diamicton; and (3) relative small amounts of assorted sand and fine clay-sized particles. Although the density of glacial debris in icebergs is highly variable, icebergs are a significant source of open-ocean detrital material. The flux of ice-rafted detritus may differ a great deal according to location and year. Only an average taken over a long period of time, such as is found in bottom core samples, may be useful in understanding long-term temporal patterns (e.g., Fillon et al., 1981). Fillon and Duplessy (1985) determined that the flux of sand in the northwestern Labrador Sea mostly delivered by icebergs was 1100 mg cm^{-2} (1000 yr)$^{-1}$, equivalent to approximately 30.4 mg m^{-2} day^{-1}.

V. Antarctic Oceans: Pelagic Particle Fluxes

Although particle fluxes in the neritic Ross Sea and Antarctic Peninsula have begun to be more clearly understood (e.g., Dunbar et al., 1985; Wefer et al., 1982), studies of particle fluxes in the pelagic southern oceans are still very limited. Only a few year-round sediment trap deployments have been made, and these are insufficient for drawing general conclusions about pelagic fluxes in southern oceans. Two recent long-term time-series experiments which were conducted in the north-central Weddell Sea by R/V *Polarstern* (Fischer et al., 1988) and in the South Georgia basin (unpublished WHOI and British Antarctic Survey data, 1988, S. Honjo, C. Pudsey and P. Barker) were used as a basis for a preliminary explanation of findings on pelagic flux in this region.

Although the Weddell Sea is as seasonally covered by sea ice as the Greenland Current area, there are a number of fundamental differences in the modes and processes of sedimentation between these two apparently similar regions. Sediment traps deployed at seasonal MIZ stations in summer open-ice areas, and under winter pack ice coverage in the southern oceans showed a characteristic mode of sedimentation, which can be summarized as follows: (1) organic carbon and calcium carbonate fluxes were far smaller than in other oceans; (2) the majority of particles were diatom frustules packaged into fecal pellets; (3) a strong seasonal pulse of vertical flux was related to the regression of the ice edge; (4) virtually no particle fluxes occurred during the ice-covered winter; and (5) no coccolithophorid flux occurred throughout the year.

A. A Flux Experiment in the North-Central Weddell Sea

The annual particle flux measured by a time-series sediment trap set at 863 m (water depth, 3880 m) in the northern Weddell Sea (62°26.5'S, 34°45.5'W) in 1985 (January 15, 1985 to January 23, 1986) was 0.37 mg m^{-2} (Fischer *et al.*, 1988). The biogenic constituents occupied over 99% of the annual flux and lithogenic particles were scarce (less than 1% of the total flux). Of the total flux, 79% consisted of biogenic opal and 17% was combustible matter. The carbonate particle flux was only 3% of the total flux, compared to 50–80% in other lower-latitude stations (Honjo, 1984) and 20–50% in the Nordic Seas (Honjo *et al.*, 1987a) (Table 13.1). The sea ice began opening during the first week of January over the sediment trap station. The ice edge rapidly retreated southward and in mid-February, the time of minimum sea ice coverage for the year, was located about 800 km south of the trap site (Antarctic Ice Chart, Naval Polar Oceanographic Center, 1985–1986).

As soon as the waters over the sediment trap became ice free, the flux of biogenic particles rapidly increased (periods 1–4; Fig. 13.8) and reached a maximum in mid-March (9.2 mg m^{-2} day^{-1}; period 5). Just after this period the particle flux decreased abruptly to about half of the maximum (period 6) and further decreased to less than 0.2 mg m^{-2} day^{-1} (periods 7–10). The advance of sea ice started in early April, but at lower speeds than the retreat. During the periods when the particle flux decreased rapidly (i.e., April and May, periods 8, 9, and part of 10), the ice edge was far south of the trap site. Although some of the samples which contained the austral winter materials (July–November) were lost during recovery, we concluded that very few particles were collected during that period. In 1986, the ocean above the sediment trap opened about 10 days later than during the previous year, but the maximum distance from the trap site to the ice edge was almost the same as in the previous year (ca. 800 km in early March). The trap site was again covered by sea ice in late May or early June. The sediment trap array was recovered in early March, 1986 (period 25), and thus the total flux during the 1986 summer period remains unknown.

13 Particle Fluxes and Modern Sedimentation in the Polar Oceans

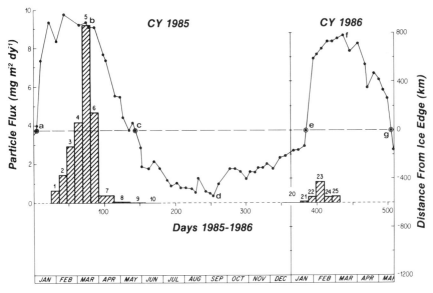

Figure 13.8 Total flux for each period (bar graph, shadowed) at the north-central Weddell Sea station during 1985 and 1986. The periods are at the top of the bars. Superimposed is the closest distance from the approximate ice edge to the sediment trap site 3 weeks before, during, and after the experiment, based on weekly Antarctic Ice Charts (NPOC) compiled from the NOAA polar orbiter, NASA Nimbus-7 Scanning Multichannel Microwave Radiometer (SMMR), GEOSAT altimeter, and visual data using NOPC ice coverage scales. (a) Ice-edge passage over the trap site during the 1985 regression. (b) The maximum opening of ice lasted about 60 days. The ice edge rapidly moved northward after early April 1985. (c) Ice edge passage over the trap site during the 1985 transgression. (d) 1985 maximum ice extension. The ice edge was 550 km north of the trap site. (e) Ice edge passage over the trap site during the 1986 regression. (f) Maximum ice opening during 1986. (g) Ice edge passage over the trap site during the 1986 transgression. From Fischer et al. (1988).

The sharp increase of biogenic flux following the ice retreat, the termination of particle sedimentation while the ocean above the trap was ice free, and the marked flux maximum early in the austral summer were primarily related to the spring thaw process and not to the availability of a larger open-sea area. Although the breakup of pack ice began in late November of the previous calendar year, the flux started to increase only after the sea ice had almost completely melted (Fig. 13.8). The maximum biogenic flux occurred about 75 days after the ice opened at this site, and the particle flux virtually ceased about 100–110 days after the ice opening. This observation matches well with the finding that the bloom period is initiated in the marginal ice zone as the ice retreats, lasting typically for about 60 days (Smith and Nelson, 1986). The arrival of the bloom signal at the 800-m-deep layer may be offset for at least a week, considering that small copepod fecal pellets sink at a rate of 100–150 m day^{-1} (Small et al., 1979). Primary production declines after the bloom period (e.g., Sakshaug and Holm-Hansen, 1984) and by fall the system approaches oligotrophy (Hopkins and Torres, in press), reflecting the virtual cessation of particle flux. In any case, the major bloom in the surface water is reflected by a flux peak within a short period of time; the particle sedimentation rate must be accelerated by at least two orders of magnitude more than the passive sinking rate of diatoms, which is 0.2–2.5 m day^{-1} as measured from Weddell Sea samples (Johnson and Smith, 1986).

B. Ice Edge Development and the Maximum Flux Period

The close relationship between the ice opening and the spring thaw in the open Weddell Sea warrants further explanation. Trapped sediment, representing the very early stage of ice opening (periods 1 and 2 in Fig. 13.8), included relatively large amounts of dark-colored algae or bacterial mat-like material up to a few millimeters in diameter. These were probably not produced in the water column but were released from the melting ice. Diatom frustules also were abundant in the sediment during the earlier periods. Some of the frustules collected during the early periods of the spring flux grew within the water, but probably the majority were an ice flora which were released from the melting ice contemporaneously with the mat-like fragments mentioned above.

The particle flux in the ocean interior must be coupled with algal blooms in the water column, and this phenomenon warrants an investigation of the relationship between production in the upper ocean and the organic carbon flux in the interior (Eppley and Petterson, 1979). Primary productivity measured at the trap site during deployment was 350 mg C m^{-2} day^{-1} (Fischer et al., 1988), roughly equivalent to the entire amount of in-ice organic carbon made available by the spring thaw. If this value is applied to the 60-day open-water period, the available organic carbon at this station is estimated to be 45 g m^{-2}. However, the rapid decrease of flux after its maximum during the time of ice-free conditions indicates that such a high rate of primary production may not be maintained throughout the ice-free period.

The general standing stock of organic carbon contained in the Weddell Sea ice column is not well known at present, but estimates from the published literature range from 46 to 530 mg C m^{-2} during the summer. These values are the sum of (a) the organic carbon in ice-bound diatoms [estimated to be about 30 mg C m^{-2} (Ackley et al., 1979); the chlorophyl *a* value was converted to carbon using Whiteker's (1977) conversion factor], (b) the ice bacterial biomass, estimated to be 6–400 mg C m^{-2} (Miller et al., 1984), and (c) the organic carbon from ice microheterotrophs, estimated to be 10–

13 Particle Fluxes and Modern Sedimentation in the Polar Oceans 711

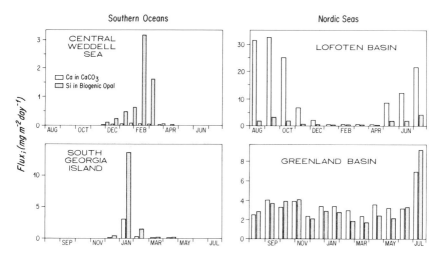

Figure 13.9 Comparison of biogenic calcium (Ca content in calcium carbonate in coccoliths, shells, and tests of planktonic foraminifera and pteropods) and biogenic silica (Si content in opal from diatom frustules, skeletons, and shells from silicoflagellates and radiolaria) from the Atlantic Southern Ocean (off South Georgia Island, S. Honjo, C. Pudsey and P. Barker unpublished results) and from the Nordic Seas. Note the different scale of fluxes.

400 mg C m^{-2} (Garrison *et al.*, 1984). In the Weddell Sea the major source of organic carbon is found in the core of sea ice (Ackley *et al.*, 1979; Miller *et al.*, 1984). Therefore, complete melting is required to release the large mass of ice-related particulate organic carbon. Despite the breakup of pack ice which began in late November of the previous calendar year (Figs. 13.8 and 13.9), a large amount of POC was released into the water only after more complete melting of sea ice.

At present, the estimate of the sum of organic carbon available from the water column community during the ice-free period, the POC released from ice by thawing, and primary production in the open water is far too general to compare with the particle flux. However, it can be concluded that the ratio of annual organic carbon flux (20 mg C m^{-2}) at the ocean interior to the carbon produced annually (ca. a few grams per square meter) was less than 1% and was extremely small compared with the "new production" of other oceans (Eppley and Peterson, 1979).

C. Modes of Particle Sedimentation in the Pelagic Weddell Sea

A strong characteristic of the mode of particle sedimentation in the Atlantic Southern Ocean stations is the prominent role of fecal pellets, including all particle categories except planktonic foraminifera tests and relatively rare occurrences of radiolarian shells. The only time fecal pellets did not predominate was at the very beginning of the open-sea period. Two morphologically different types of pellets were involved in the north-central Weddell Sea station: one with a peritrophic membrane and the other without a membrane and with exposed contents. Both types of pellets were about 0.5 mm in diameter, were slightly elongate, and appeared to be produced by small metazoan grazers such as copepods. These pellets are filled with diatom frustules and occasionally with silicoflagellate skeletons and other small skeletal remains.

In the period immediately after the ice opening (periods 1 and 2 in Fig. 13.8), the particles were included in amorphous aggregates and the fecal pellet flux was less. The contribution of fecal pellets to total flux increased toward the maximum flux occurrence, when the total flux made up of fecal pellets reached 80–85%. During the ice-covered period the settling particles consisted entirely of fecal pellets and foraminifera tests. These observations led to the following explanation of the succession of zooplankton behavior with reference to the MIZ: Immediately after the spring thaw, the zooplankton population did not utilize all of the POC which had been released from the ice (perhaps in a short period); some of the POC settled out of, or was lost to, the surface water in a form similar to marine snow aggregates (Honjo *et al.*, 1984). The zooplankton activity intensified about a month later, and the particles from ice as well as primary production in the open water were more efficiently utilized, resulting in a large production of fecal pellets. The zooplankton migrated toward the south, perhaps pursuing ice edge conditions, and the site was abandoned by the majority of zooplankton.

Throughout the austral winter there was a quantitatively minor flux of fecal pellets, a few units per square meter per day. This suggested that a small population of zooplankton were maintained under the ice throughout the austral winter and, most likely, were grazing on ice diatoms. The diatom species assemblages in the fecal pellets collected from under the pack ice were not significantly different from the assemblages produced, presumably in open-ice waters, during the flux maximum in the austral summer. Also, the shape and size statistics of fecal pellets collected during the mid-austral winter were identical to those of pellets collected during the austral summer. This implies that the phytoplankton species and their grazers were consistent throughout the seasons regardless of ice coverage, as suspected by Hopkins and Torres (1989).

The observation of fecal pellets at MIZ stations agrees with results of zooplankton gut content studies in the same general area. The contents of fecal pellets were composed, regardless of the kind, of small diatoms, *Nitzichia curta* (approximately 35%), *Thalassiosira gracilis* (approximately 30%), and the rest mostly *Nitzichia* spp. (Fischer *et al.*, 1988). Hopkins and Torres (1989, observed that during the bloom period most food was derived by grazing on small particles, mainly by small diatoms. The zooplankton species responsible for producing the fecal pellets collected in the sediment traps have not been identified. It is puzzling that, despite the relatively high species diversity of grazers at the Weddell Sea MIZ area (for example, Hopkins and Torres, 1989, listed 16 copepod species), apparently only two forms of fecal pellets were identified in the material flux. No pellets from krill, *Euphausia superba,* were found at those pelagic stations. At the neritic Antarctic station in Bransfield Strait, Wefer, *et al.* (1988) found that virtually the entire particle flux was a result of pellets produced by krill which swarmed, feeding on phytoplankton (see Section VI).

D. Carbonate Fluxes in the Pelagic Weddell Sea

One conspicuous characteristic of the particle sedimentation scheme in the pelagic Weddell Sea is its small carbonate fluxes (Table 13.1). In both the north-central Weddell Sea and the South Georgia basin stations, no carbonate hard tissues produced by plants (i.e., coccoliths) have been collected by sediment traps. The relatively small carbonate flux in those stations consisted exclusively of dwarfed planktonic foraminifera, mostly *Neogloboquadrina pachyderma.* The flux of planktonic foraminifera was discontinuous, particularly during the ice-covered winter. In the north-central Weddell Sea an average of 4 individual tests m^{-2} day^{-1} were caught in the ice-free summer and 6 individual tests m^{-2} day^{-1} were caught under pack ice throughout the winter. A large number of foraminifera, 21–26 tests, fell into the sediment trap soon after the end of the flux peak in 1986.

The origin of planktonic foraminifera which settled during the ice-covered winter is not clear, but a combination of two mechanisms is suggested. First, planktonic foraminifera may live under the ice throughout the winter, feeding on the ice-bottom community. It seems that more tests were settling during winter than during the open-ocean periods. Second, new frazil ice scavenges suspended particles, such as planktonic foraminifera tests (Gow *et al.,* 1982; Cox and Weeks, 1988), and thus some of the living tests were incorporated into the pack ice. The bottom of the pack ice melts slowly even during the winter; thus, tests would have been released constantly from the pack ice into the water column.

E. Questions on Apparently Very Small Organic Carbon Flux in the Southern Ocean

The majority of biogenic particles collected in the north-central Weddell Sea were from metazoan fecal pellets, except for those collected during the early stages of opening ice, and the period of settling particle sedimentation was short. Thus, one may speculate that primary production and zooplankton growth were in phase and the organic material was efficiently utilized by grazers. However, the fluxes of opal (from skeletal remains of primary producers) were also very small compared to their Nordic seas counterparts. Most opal frustules would remain intact even though ingested by grazing metazoans, although they may be mechanically fragmented. The production of opal and organic carbon in diatoms are well correlated Lisitzin, 1972). Therefore, the small organic flux at the interior of the Atlantic Southern Ocean suggests small new production in the upper ocean.

This observation does not agree with the general assumption that productivity in the Southern Ocean, including pelagic environments, is high. Some direct measurements yielded relatively small primary productivity, but such results were explained by the large spatial and temporal variability in the Southern Ocean (e.g., El-Sayed and Turner, 1977). However, a model used by Jennings *et al.* (1984) gives spatially and temporally averaged results; based on the difference in nutrient profiles between the summer (GEOSECS) and the winter (WEPOLEX), they estimated that the productivity in the Southern Ocean was 220–420 mg m^{-2} day^{-1}. Chen (1986) estimated an even larger value based on the summer versus winter difference in carbonate profiles. Fogg (1977) reported 10 g C yr^{-1} m^{-2} from the Antarctic.

One should note that the carbon and biogenic silica fluxes observed at the Atlantic Southern Ocean trap stations were almost exclusively made up of relatively fast-sinking, more protected fecal pellets.

Opaline SiO_2 from diatom frustules represented 93% of the total flux, and the annual flux at the north-central Weddell Sea station was only 293 mg m^{-2} for 1984 (Fischer *et al.*, 1988). Assuming the mass of frustules is preserved through (multiple) coprophagy and coprohexy, the original carbon produced with the production of opal can be estimated after the carbon and opal are decoupled in the upper-middle water column (Lisitzin, 1972). The organic carbon which was associated with these frustules thus was estimated at 150 mg, and this was much smaller than the estimated annual primary production, as mentioned above. Vigorous dissolution of diatom frustules before and after they reach a sediment trap does not seem reasonable because of the low temperatures (Hurd, 1983), and microscopic studies of fecal pellets favor preservation. Another explanation could involve the relatively large production of organic carbon by primary producers other than hard-tissue-producing plants, although no evidence has been found.

Noriki *et al.* (1985) observed an organic carbon flux (about 60 mg m^{-2} day^{-1}) at a southern Pacific station (3580 m deep) which is far larger than the organic carbon flux peak at the South Georgia Island station for an equivalent season (about 1 mg m^{-2} day^{-1}). The station occupied by Noriki *et al.*

(1985) could have been north of the Antarctic Convergence zone. However, this difference suggests that the generalization about extremely small new carbon production in all southern oceans may not be correct.

VI. Comparison of Pelagic Fluxes in the Arctic and Antarctic Oceans

Comparable year-round sediment trap experiments revealed a number of basic differences between the modes of particle fluxes in the pelagic northwestern Nordic Seas (Fram Strait and Greenland Current area) and the Atlantic Southern Ocean, even though the upper-ocean oceanography of both areas is largely driven by processes related to the seasonal MIZ. The major differences are as follows:

1. The annual organic carbon flux in the pelagic Nordic Seas is relatively large, almost equivalent to those found in temperate and tropical oceans. However, organic carbon flux in the Atlantic Southern Ocean, particularly in the pelagic Weddell Sea, is far smaller.
2. The Nordic Seas form a "carbonate ocean" and the Atlantic Southern Ocean forms a "silica ocean." The predominant biogenic particle flux in the Nordic Seas consists of coccoliths, whereas diatom frustules are the major constituent in the Southern Ocean beyond the Antarctic Convergence.
3. Particle fluxes in the Atlantic Southern Ocean are strongly related to the melting of pack ice; during maximum melting the flux reaches its maximum, and then it decreases virtually to zero after the phytoplankton bloom. In the Nordic Seas a relatively large biogenic flux, including coccoliths, continues throughout the year regardless of ice conditions.
4. Fecal pellets are the principal form of aggregates known which accelerate sedimentation of fine particles in the Atlantic Southern Ocean; in the Nordic Seas, a medium other than fecal pellets (possibly amorphous aggregates) is the major mechanism by which fine particles settle.
5. The lithogenic flux in the Weddell Sea of the Southern Ocean is several orders of magnitudes smaller than the flux in the Transpolar Drift area of the Nordic Seas.
6. The contrast in mass flux between neritic and pelagic Antarctic waters was not seen in the Arctic (see Section VII).

A. Biogeochemical Contrasts in Settling Particles

The annual organic carbon flux in the Norwegian Current area is roughly 1.4 g C m^{-2} and is comparable to that in the temperate to tropical oceans of the world (Fig. 13.1). The Greenland Current area receives less organic carbon than the Norwegian Current area, where there is no ice cover in any season, but it still ranges from 0.5 g C m^{-2} (Fram Strait) to 1.1 g C m^{-2} (central Greenland basin at 881 m) annually; these carbon fluxes are comparable to those at a near-Bermuda station in the Sargasso Sea (Deuser *et al.*, 1981). In contrast, the annual carbon flux in the north-central Weddell Sea station in 1985 was only about 20 mg C m^{-2}, and it appeared to be even smaller during 1986. In the South Georgia basin station the annual organic carbon flux was 40 mg C m^{-2}. Furthermore, almost the entire annual organic carbon flux in the Atlantic Southern Ocean was produced during one short period at the beginning of the austral summer and virtually no particles settled during the rest of the year (Fig. 13.8), whereas organic carbon seemed to settle continuously in the Greenland Current area (Fig. 13.7).

Ocean particles derived from phytoplankton are fundamentally different in the northern and southern extremes of the Atlantic Ocean. The "hallmark" mineralized plant tissue settling in the Nordic Seas is coccoliths from coccolithophorids, whereas in the Southern Ocean it is frustules from diatoms. Our studies found no coccolithophorids settling to the interior of the ocean north of the Antarctic Convergence zone in the Weddell Sea area, and very probably they are not produced in the surface waters there. Other carbonate tissues from planktonic foraminifera and pteropods (usually limited to the Norwegian Current area) also contribute to the carbonate fluxes in the Nordic Seas.

The annual carbonate carbon fluxes in the Norwegian Current and Greenland Current areas were 1.4 and 0.4 g C m^{-2}, respectively. In the central Weddell Sea the annual carbonate carbon flux is only 2 mg C m^{-2}. The entire carbonate flux in the South Georgia basin station was due to planktonic foraminifera tests and the annual carbonate carbon flux was 40 mg C m^{-2}. The ratio of organic carbon to carbonate carbon in the Norwegian Current area was 1.4 and for the Greenland Basin was 2.4. In the Atlantic Southern Ocean this ratio was always greater than 1. In fact, there was 10 times more organic carbon than carbonate carbon in the annual flux at the Weddell Sea station. In comparison, the ratio was often less than 1 at temperate and tropical stations (Honjo *et al.*, 1982).

Why, then, are the Nordic Seas relatively fertile (with a larger new carbon input) and characterized as a carbonate ocean, whereas the Atlantic Southern Ocean is a silica ocean with a far smaller annual organic carbon flux? Also, how do other Arctic marginal seas, such as the Bering Sea, Barents Sea, or the Sea of Okhotsk, compare with the Nordic Seas? No definite explanation is available, but a few speculations can be put forth.

B. Coccolithophorid Distribution in the Polar Oceans

The temperature limit tolerated for growth of coccolithophorids is thought to be a few degrees Celsius. Thus, the presence of coccoliths in pelagic sediment has been generally regarded as indicating that the sediment was formed outside a polar front. This notion is applicable to the Southern Ocean environment; however, our sediment trap experiments in the Nordic seas indicated that coccoliths and coccolithophores settle to the deep ocean throughout the year in the Norwegian Current area (including the West Spitsbergen Current area) and in the Greenland basin and Fram Strait. The dominating coccolith species collected in sediment traps at pelagic stations throughout all annual seasons in both the Greenland and Norwegian Current areas was *Coccolithus pelagicus,* and *E. huxleyi* was a minor component. Furthermore, fully grown individuals of *C. pelagicus* were found in abundance under water as far north as 86°N, 20°E in the summer of 1987 (S. Pfirman and S. Honjo, unpublished). A coccosphere of *C. pelagicus* includes 20–50 times more calcium carbonate than does *E. huxleyi. Coccolithus pelagicus* is reported as the most stenothermal coccolithophorid species and is adapted to water temperatures 0 to 15°C (Okada and McIntyre, 1979).

Coccolithophorids are photosynthetic, and it was therefore surprising that a large flux of *C. pelagicus* was found in samples collected during the dark winter and early spring at this high latitude, particularly at the Greenland Current area stations, which were covered by pack ice during the boreal winter and which had seawater temperatures below zero (-1.7 to -1.8°C throughout the winter below 20 m; S. Honjo, unpublished). A supply of fresh coccospheres to the interior of the Greenland Sea by advection is not feasible, because there is no source area of *C. pelagicus* in the Nordic Seas, considering the southerly direction of the Transpolar Drift ice and current. If coccoliths and coccospheres were supplied from melting ice, they should appear in large quantities in the ice located north of Station GB; this is not observed. Electron microscope studies of ice residue from the Fram Strait did not reveal large numbers of coccoliths (S. Honjo, unpublished). Heterotrophic growth during the boreal winter is thus suspected for this coccolithophorid species in the Nordic Seas (Paasche, 1968; Okada and Honjo, 1973). If this hypothesis is right, why does *C. pelagicus* produce in a heterotropic mode in the Nordic seas and not in the Southern Ocean, where insolation, temperature, and ice conditions are similar?

The Ca/Si ratio in biogenic fluxes in the Atlantic Southern Ocean was far smaller than that in the Nordic Seas (Table 13.1). The largest Ca/Si ratio was observed at the Lofoten Basin station, and the annual Ca flux was nearly an order of magnitude more than the Si flux (by weight). On the other hand, in the Weddell Sea the annual Si flux was about 30 times more than that of Ca (Fig. 13.9). The Ca flux comprised dwarfed foraminifera in Southern Ocean stations and no coccolithophorid remains were found. But the major Ca flux in the Nordic Seas, in both the Greenland basin and Norwegian Current areas, was made up of coccolithophorids *(C. pelagicus),* with a minor proportion of foraminifera. The Ca/Si ratio in the Greenland basin (Stations GB-1 and GB-2) was larger during wintertime and decreased to less than 1 during the spring season (Fig. 13.9). Both Ca and opal were more consistent throughout the year in the Greenland basin station, which was covered by Transpolar Drift ice sheet most of the year. In contrast, the flux in the non-ice-covered Norwegian Current area was highly seasonal, and particle fluxes at Southern Ocean stations were largely controlled by ice edge development (e.g., Fig. 13.8).

C. Differences in Physical and Biogeochemical Settings in the Polar Oceans

The relatively high fertility and relative enrichment of carbonate in the Nordic Seas do not seem to be coupled to (major) nutrient enrichment, such as is seen in upwelling areas of other oceans. Rather, the NO_3 concentrations in the Fram Strait and Greenland Sea area are relatively low (11 to 12 μmol kg^{-1} at about 50-m depth; Chapter 8). The salinity in the upper, very-surface layer is diluted by ice melt and nutrients are less concentrated. In contrast, the nutrient concentration in the pelagic Weddell Sea is far higher: more than 25 μmol kg^{-1} (Chapter 8). More solar radiation is available at the Weddell Sea station (62°S), for example, than at the Greenland basin stations (74°N).

A number of hypotheses can be postulated to explain the contrast in particle fluxes between the Nordic Seas and the Atlantic Southern Ocean. One such explanation may include consideration of (1) differences in surface water circulation and (2) possible differences in growth-controlling micronutrient levels in the two regions. The Norwegian Current, together with the East Iceland Current, is an extension of the Gulf Stream; thus, it maintains relatively high temperatures and salinities for the region. The Norwegian Current probably mixes most vigorously with the Greenland Current (Transpolar Drift) at the Fram Strait and forms a large gyre in the Greenland basin. Annual particle fluxes under the Norwegian Current are relatively large (Tables 13.1 and 13.3). The carbonate content made up primarily of coccolithophores and seasonal pteropod shell flux, is as large as those observed in the low-latitude Atlantic stations (Table 13.2). It is conceivable that the biogeochemical setting, based on the circulation patterns of the Norwegian Current, may strongly influence the quality and quantity of particle flux in the Greenland Current area. The Weddell Sea, which shows very small biogenic flux, is isolated from such temperate current systems.

What keeps biogenic particle fluxes in the Greenland Sea at constantly high levels throughout the year? The average carbonate fluxes during midwinter (January–April) at Lofoten (LB-1) and Greenland Sea (GB-2) stations were compared and were 1.1 and 8.1 mg m^{-2} day^{-1}, respectively (Figs. 13.5 and 13.7). Another factor which strongly characterizes the Greenland Current area may be the supply of micronutrients carried by the Transpolar Drift; this phenomenon may influence the biogeochemical setting in the northern Norwegian Current area and be in sharp contrast to the biogeochemical regime in the Atlantic Southern Ocean.

One of the basic differences between the surface waters of the Arctic Ocean and those of the Southern Ocean may be accessibility to dissolved and particulate matter supplied by large continental rivers to the northern oceans. All of the large Siberian rivers discharge into the Arctic Ocean (except for the Amur River, which discharges mainly into Tartan Strait). The Mackenzie River also has a significant discharge into the basin. Although the Yukon River opens to the Bering Sea, its major discharge flows into the Arctic Ocean through the Bering Strait. Therefore, the Yukon can be regarded as an Arctic Ocean river in a sense similar to Siberian rivers (Fig. 13.10). The ratio of the estimated annual discharge of rivers into the Arctic basin (2.4 × 10^6 km^3; Alexander, 1986) to the area of the Arctic Ocean (14.1 × 10^6 km^2; Bialek, 1966) is unusually large on a global scale.

Geochemical evidence has been found which indicates that the Arctic ice and upper water (holocline water) contain characteristics of river water, particularly in terms of their excess specific alkalinity (Anderson and Dyrssen, 1981; Tan *et al.*, 1983; Chapter 8, this volume). The Transpolar Drift brings Arctic sea ice from the East Siberian Sea to the Greenland Sea within a few to several years (Östlund and Hut, 1984) and thus preserves its original river water characteristics as ice far better than other, low-latitude river–ocean interfaces.

The Transpolar Drift flows through the Eurasian Arctic Ocean and eventually melts in the MIZ of Fram Strait and the Greenland Sea. Therefore, one could postulate that Fram Strait is the actual estuary of all the great Siberian rivers combined. This "Fram Strait estuarine" hypothesis may help to explain the relatively high fertility of the Nordic seas.

Although much still remains to be learned, micronutrients work as a complex of trace elements and organic compounds, including vitamins. However, it is well known that "garden-soil extract" is essential in many laboratory cultures, although the specific elements and compounds which play a role in phytoplankton growth are unclear. Although it is speculative,

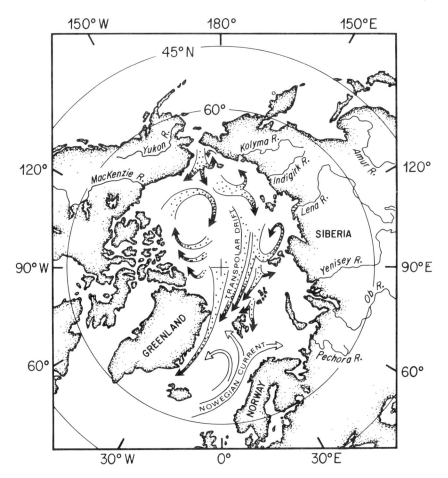

Figure 13.10 Major river input to the Arctic Ocean. Siberian rivers discharge more than 1885 km^3 yr^{-1} and the rivers from the North American plate discharge about 472 km^3 yr^{-1} (Alexander, 1986), for a total of approximately 2.4 million km^3 into a Mediterranean-type ocean of approximately 14 million km^2 (Bialek, 1966). Part of the river water turns into sea ice and drifts toward the south as the Transpolar Drift, reaching as far south as the Denmark Strait. Vigorous melting of Transpolar Drift ice is observed in the western Nordic Seas, particularly where it meets the warm southerly Norwegian Current (flow patterns in this figure are conceptual). Thus, the Fram Strait can be regarded as a collective estuary for all of these rivers.

in the low temperatures and generally low metabolic activities of the ice-packed Arctic Ocean, micronutrients could be enriched and transported to the MIZ in Fram Strait and the Greenland Sea via the Transpolar Drift (Fig. 13.10).

Among the essential trace metals, the role of dissolved iron in ocean primary production has begun to be investigated and applied to global oceanography in time and space (Martin and Gordon, 1988). River waters contain high concentrations of dissolved iron in stable forms which are chelated with dissolved organic carbon compounds and transported further into the pelagic ocean. Danielsson and Westerlund (1983), for example, found very high concentrations of dissolved iron (10-20 nmol Fe kg^{-1}) in surface waters with 1.8 nmol Zn kg^{-1} (estimated from Fig. 4, p. 91, Danielsson and Westerlund, 1983) at Eurasian station. Indeed, this iron concentration was higher than the values reported from California's continental margins (9 nmol Fe kg^{-1}) by Martin and Gordon (1988), where the annual productivity was estimated as 200-500 g C m^{-2} (Berger et al., 1987). If we apply the Morel and Hudson (1985) optimal phytoplankton growth ratio of trace iron (106 C to 0.01 Fe), then Danielsson and Westerlund's Eurasian station waters should support up to 24 g C m^{-2} yr^{-1}, supposing that the euphotic layer is 20 m thick. Subba Rao and Platt (1984) reported an estimated annual primary production rate in Arctic waters which ranged from 12 to 98 g C m^{-2} yr^{-1}, based on a 120-day growth season. However, reliable trace-metal data collected using noncontaminating sampling techniques from the polar regions are scarce, and the single example of Fe concentrations mentioned above cannot be generalized.

Martin and Gordon (1988) emphasized the role of lithogenic aerosols as a primary source of dissolved Fe to the open oceans. Duce (1986) postulated that 16-96% of growth-controlling trace metals in the pelagic Pacific and Atlantic are supplied by aerosol fallout of lithogenic dust from the Asian drylands, whose particles reach the Arctic Ocean as "Arctic haze" (Rahn, 1981). In large-scale transport, the Canadian coastal region and Lincoln Sea areas are bordered by vast exposures of rock formations. Due to strong prevailing winds and low precipitation, this Arctic desert supplies mineral particles, including iron resources, to the Arctic, and these are eventually transported to the Nordic seas. The volume of airborne lithogenic particles reaching the interior of the Arctic Ocean, however, is unknown.

In contrast to the Arctic region, the supply of mineral resources to the Southern Ocean is limited to neritic areas by coastal erosion. The Antarctic continent is covered by a thick ice cap with no rivers comparable to those of Siberia. The atmospheric circulation in the southern hemisphere is much more latitudinally restrained than in the northern hemisphere; thus, airborne particles seldom reach beyond the Antarctic Convergence zone. However, recent ice core studies have shown that the lithogenic dust flux at Bostok Camp during the last glaciation was orders of magnitude higher than the present-day values (DeAngelis et al., 1987). The annual fluxes of lithogenic particles were extremely small in the north-central Weddell Sea, where the advective transport of shelf-originated particles was negligible (Fischer et al., 1988). Such conditions may result in a deficiency of micronutrients in the Southern Ocean, in sharp contrast to the Nordic Seas.

The Bering Sea is comparable in size to the Nordic seas and is a major marginal sea of the Arctic Ocean. The essential circulation features of Bering Sea surface waters are (1) a strong inflow of North Pacific water through the west-central Aleutian Archipelago, with a narrow outflow at the East Kamchatka Current (Hughes et al., 1974), and (2) a well-defined Alaskan Coastal Current to the north through the Bering Strait, which carries the major portion of the Yukon outflow to the Chukchi Sea (Arsen'ev, 1967; Reynolds and Pease, 1984; Dr. N. Maynard, personal communication). The deep layers of the Aleutian basin are characterized by high levels of dissolved silica and deep-water alkalinity at near-bottom depths (Tsunogai et al., 1979; Craig et al., 1981). The Bering Sea has been regarded as one of the most productive pelagic oceans and supports vigorous offshore fishing activities. Berger et al. (1987) estimated an average annual primary production of 100–200 g C m^{-2}. The dissolved organic carbon (DOC) is thought to be equivalent to that in the North Pacific (Hood and Reeburgh, 1974). Unfortunately, no reliable data on growth-controlling trace metal concentrations are available for the Bering Sea. Martin and Gordon (1988) showed that Fe concentrations in the Gulf of Alaska (Ocean Station P) were extremely low (<0.1 nmol kg^{-1} in surface water). Although no time-series flux measurement experiment has been attempted, there is much evidence to support the idea that the surface Bering Sea may be a "silicate ocean" environment, somewhat similar to the Southern Ocean. Coccolithophorids are very rarely encountered in the Bering Sea; in contrast, over 70 diatom species were found (Motoda and Minoda, 1974).

VII. Processes of Neritic Sedimentation in the Polar Oceans

A. Particle Fluxes in the Neritic Antarctic Ocean

Sediment trap experiments in the Ross Sea and Antarctic Peninsula indicate several strong characteristics of particle flux in the Antarctic nearshore environment, some of which have global implications. In those areas the annual particle flux, both biogenic and lithogenic, was orders of magnitude larger than in their pelagic counterparts. The annual flux was confined to the short period of the austral summer similar to that in offshore counterparts (Fischer et al., 1988); for example, in the time-series sediment trap experiment at Bransfield Strait, over 95% of the annual flux occurred during December and January (Wefer et al., 1988). However, the annual fluxes in those areas were far larger than in their offshore counterparts.

Wefer et al. (1988) reported that the annual flux at Bransfield Strait in 1984 was 107 g m^{-2}, of which 50% consisted of biogenic particles (Fig. 13.1). This was one of the largest annual fluxes observed in neritic deep stations.

Virtually all of the annual flux at this station occurred during December and January. Dunbar (1984) reported 50–600 mg m^{-2} day^{-1} at a neritic Ross Sea station; if the spring bloom persists for about 2 months, as at other locations such as the Bransfield Strait station, the estimated annual flux at this station could reach 3–36 g m^{-2}. Matsuda et al. (1987) reported similar concentrations of flux in the austral summer, measured by traps deployed 5 and 25 m deep at fast-ice–covered stations a few hundred meters offshore in eastern Lutzow–Holm Bay, western Indian Ocean sector of the Antarctic, which had an estimated annual flux of about 120 g m^{-2} (estimated from Fig. 2, p. 26, Matsuda et al., 1987) and an organic carbon flux of 1.5–136 mg C m^{-2} day^{-1}.

The rate of silica accumulation in the southwestern Ross Sea by settling biogenic opal produced by diatoms may be among the highest in the world's oceans. Annual accumulation at the ocean floor of the neritic Ross Sea was estimated, using ^{210}Pb chronology in bottom sediments, to be as high as 130 g SiO$_2$ m^{-2} (Ledford-Hoffman et al., 1986), whereas the world's average is estimated as 1 g SiO$_2$ m^{-2} (Calvert, 1983). Indeed, Ledford-Hoffman et al. (1986) estimated that a significant amount of silica input from the world's rivers to the oceans is balanced by deposition in the neritic Ross Sea. Compared to surface productivity measurements (e.g., El-Sayed, 1970; Holm-Hansen et al., 1977), one must expect that all of the surface production of opal is transported to the seafloor. On the other hand, the organic carbon content of the neritic Ross Sea bottom sediment is relatively low, less than 1.5% (e.g., Lisitzin, 1972), and a large portion of the primarily produced carbon appears to be demineralized before reaching the bottom. Thus, two categories of primary production, organic carbon and biogenic silica, are characteristically decoupled in the neritic Ross Sea environment (Dunbar et al., 1985).

The efficient uptake of primary production and quick delivery to the ocean bottom of frustules in well-packaged metazoan fecal pellets in the cold neritic Antarctic water leave little opportunity for the dissolution of opal particles in the upper water column. Leventer and Dunbar (1987) observed a significant decrease in frustule flux at the McMurdo Sound station from the surface to the 100-m and 200-m traps. Jousé et al. (1971) and Kozlova (1971) concluded that diatom frustules had undergone extensive dissolution in the upper water layers in the Pacific and Indian sections of the Antarctic. Diatoms produce less silicified, fragile, and dissolution-susceptible frustules during the bloom period. However, Leventer and Dunbar (1987) did not find a clear sign of dissolution at stations nearby, and there are fundamental difficulties in interpretations of the suspended-particle distribution. It is possible that the major dissolution of opal occurs after it arrives at the neritic seafloor.

B. Krill-Mediated Ecosystem and Enhanced Silica Flux in the Ross Sea

The mode of sedimentation in the Antarctic Peninsula area is characterized by the fact that nearly 100% of the particles are transported by *Euphausia superba* fecal pellets (Wefer *et al.*, 1982, 1988; Dunbar, 1984). This contrasts sharply with pelagic stations in the north-central Weddell Sea or South Georgia basin, where the fecal pellets were from far smaller metazoans. In the neritic Antarctic, krill swarms consume the majority of the primary production and provide transport of biogenic particles in large fast-sinking pellets to the seafloor. A "superswarm," which is an extraordinarily dense swarm of krill, has been observed in the neritic environment (Shulenberger, 1983), and such densities of grazers may exhaust the phytoplankton within a short period. Krill continuously produce fecal pellets as large as 150 μm in diameter and usually several hundred micrometers in length (Marchant and Nash, 1986). The sinking speed of these pellets is an order of magnitude faster than that of pellets from other metazoan species such as *Calanus* (Fowler and Small, 1972; Small *et al.*, 1979). The residence time of krill fecal pellets in the neritic water is estimated to be a few days.

The cylinder-shaped surface of krill fecal pellets is covered by a peritrophic membrane (Gauld, 1957). Laboratory experiments on *Calanus* pellets showed that the preservation of the peritrophic membrane is temperature dependent (Honjo and Roman, 1978). In the laboratory a peritrophic membrane was degraded within a few hours at 20–25°C by microbial growth (Honjo and Roman, 1978). When the peritrophic membrane is lost, a pellet becomes susceptible to disintegration while settling in the water column as well as after arriving at the seafloor. On the other hand, Honjo and Roman (1978) found that the peritrophic membranes of *Calanus* pellets were intact for at least 20 days at 5°C. This experiment suggests that krill pellets can arrive at the bottom intact, virtually unaffected by microbial degradation during their descent through the shallow and cold water column of Antarctic neritic oceans. Such a mode of mass transport may explain the extraordinarily efficient sedimentation of diatom frustules in the Antarctic neritic environment.

C. Neritic Sedimentation in the Arctic Ocean

A wealth of information is available on the origin, cycles, and fate of recent marine sediments in time and space in the Arctic coastal and neritic environments (e.g., volumes edited by Tolmachev, 1970; Andrews, 1985). Whereas the Antarctic continent is isolated from other high-latitude terrains and bordered by glacial coastlines with fast ice, seasonal rock exposure, and essentially no vegetation, the sedimentological background along the Arctic

Ocean is more closely related to high-latitude terrains with diversified coastline physiography. Most of the Eurasian Arctic consists of vegetated lowlands where large rivers discharge to the Arctic Ocean. On the other hand, the North American Arctic, including northern Greenland, which roughly coincides with the North American plate in the Arctic, is covered by dryland with more exposed basal rock formations and virtually no consistently discharging rivers. In these areas, there is a short- to medium-distance airborne transport of particles to the neritic water through fast-ice melting. The Beaufort coastal area is adjacent to orogenic topography and involves high-energy erosional input to the coastal area. More than half of the Arctic Ocean is composed of continental shelf, such as the East Siberian Sea and Barents Sea. The interactions among three components (coastline, shelf, and deep basin) in the Arctic warrant further study.

As pointed out previously, Antarctic neritic sedimentation is strongly characterized by the ingestion of diatoms by the Antarctic krill *E. superba,* which transports surface production to the neritic bottom via fecal pellets with extraordinary efficiency. But Arctic neritic seas lack *E. superba* (Chapter 10), and this seems to make its sedimentary environment significantly different from that in the Antarctic. However, the neritic sedimentation in the Arctic is also largely mediated by the plankton ecosystem at large; an example is mentioned below.

D. Winter Sedimentation in the Barents Sea

Some of the sediment trap experiments deployed at Nordic Seas stations, relatively close to the shelves and slopes, have provided information on how phytoplankton on the Arctic shelves is transported to the outer slopes during the midwinter interaction with specific physical conditions of the regions. As a result, a large organic carbon sink is formed along the shelf breaks of this Arctic region (Honjo et al., 1988). Settling particles were collected continuously southwest of Spitsbergen for a year in 1984 and 1985 by a time-series sediment trap (Station BI-1). The trap collecting increment was 12 equal time periods, each of which lasted for 1 month. The total annual particle flux measured during those years was 28.3 g m^{-2} (Table 13.1). Fifty-one percent of the mass flux was lithogenic and 49% was composed of biogenic particles (Table 13.2).

The quality and quantity of particle sedimentation at this station showed three phases (Fig. 13.11): phase 1, May–July; phase 2, August–November; and phase 3, December–April. The distinction and timing of these sedimentary phases in the year were roughly identical to those of the three phases observed in the Lofoten basin (Fig. 13.5) and some similarity was found to Fram Strait (Fig. 13.6) (S. Honjo, unpublished). Phase 1 sediments, which

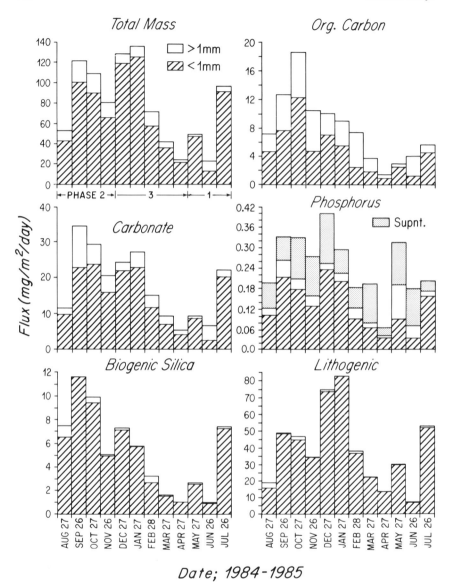

Figure 13.11 Particle flux from Bear Island during 1984 and 1985. Annual total mass and major sedimentological constituents of trap-collected sediments from the BI-1 station (Honjo et al., 1988). Note the high lithogenic particle flux during midwinter (December and January). The phosphorus flux is the total value of acid-soluble and organic flux. The dotted boxes indicate phosphors dissolved from the sample while stored (in excess of ambient content) (Honjo et al., 1988). The organic carbon content in the trapped sediment was 10% during December, 9% during January, 8% during February, and 4% in March and April.

contributed 18% of the annual flux, consisted of finer particles and contained relatively more biogenic opal and fewer carbonate particles than the sediments in other phases. Phase 2 sediments contributed 38% of the annual total flux and were dominated by biogenic particles. Phase 3, differing from the other phases by its increased flux of fine lithogenic particles, contributed as much as 44% of the annual flux.

The annual organic carbon flux of 2.9 mol C m^{-2} measured at this northernmost Norwegian Current area seems comparatively large; it was several times larger than that in the Sargasso Sea (0.4–0.7 mol C m^{-2}; Deuser et al., 1981) or the North Pacific (0.7–2.4 mol C m^{-2}; Honjo, 1984) and roughly equivalent to that measured in hemipelagic stations in the southern Black Sea (Honjo et al., 1987c) and Panama basin (Honjo et al., 1982). However, most conspicuous and puzzling is the fact that about 25% of the annual carbon flux occurred during phase 3 in midwinter (Fig. 13.11). During January 1984, the lithogenic flux reached 83 mg m^{-2} day^{-1} at station BI-1 (Figs. 13.11 and 13.12). Such a large flux of lithogenic fine particles is rare at any open-ocean station (Honjo et al., 1987a).

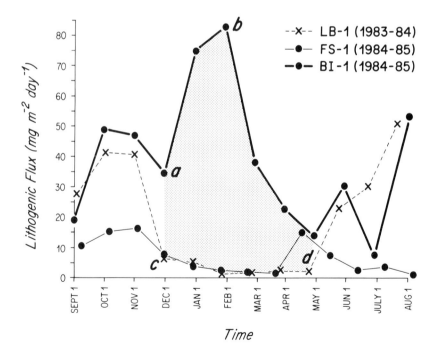

Figure 13.12 Comparison of lithogenic fluxes at three stations in the Norwegian Current area. LB, Lofoten basin; FS, Fram Strait; BI, Bear Island (for positions, see Fig. 13.2). The integral of the area a, b, c, and d is estimated as the "outburst" of lithogenic flux (Honjo et al., 1988).

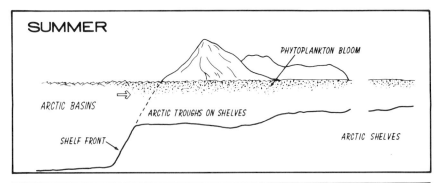
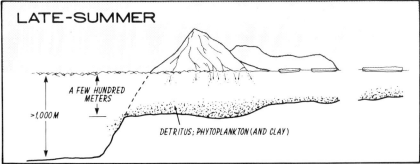
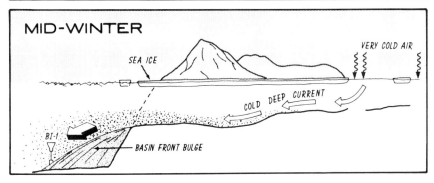

Figure 13.3 Hypothesis for the mechanism of the organic, carbon-rich "winter outburst" sedimentation (Honjo et al., 1988) in the northwestern Barents Sea. During the boreal summer a phytoplankton bloom takes place (upper diagram). A significant portion of the production settles at the bottom of the shelves, particularly in depressions such as south of Storfjord, because the grazing population is not able to consume all that is produced in the upper layers. This detritus is mixed with lithogenic particles supplied from glacial erosion and advective transport (middle diagram). When the very dense deep water cascades down the shelf front, the fresh detrital layer is also drawn down into the deeper basin and forms organically rich basin front layers (lower diagram).

The total lithogenic flux throughout phase 3 was 7.0 g m^{-2}, whereas at other Norwegian Current stations the phase 3 wintertime fluxes of lithogenic particles were less than 0.4 g m^{-2} (Fig. 13.13; Honjo et al., 1987a). Assuming that such a small lithogenic flux is normal during winter periods in the pelagic Norwegian current area, an "excess" of lithogenic particles of 6.6 g m^{-2} was supplied to the BI-1 station. Other sedimentary components also increased during this phase, although not as dramatically as the lithogenic particles (Fig. 13.12). Excess carbonate, opal, and organic carbon fluxes during phase 3 were 1.9, 0.5, and 0.8 g m^{-2}, respectively. The total sediment above "normal" which settled during phase 3 (10.7 g m^{-2}) was enriched by lithogenic particles, particularly quartz particles, and contained less carbonate and other biogenic components than the sediments during other phases.

It is not plausible to expect ice-rafted detritus to supply such large amounts of lithogenic sediment because this station was free from sea ice during the period of the experiment. Contents of the sediments, in particular the high organic carbon content (Table 13.3), as well as scanning electron microscopy observations, do not support an ice-rafted origin for the particles. Storfjord and the southern coast of Spitsbergen started to freeze in late January and were ice free in late May during 1984 (Norsk Meteorological Institute, Oslo). Therefore, freezing of the nearby coast also seems to have no direct relation to intensified lithogenic sedimentation.

One hypothesis to explain the large winter flux of organic-rich sediments (Fig. 13.13) involves the interaction between the specific sequence of primary production on Arctic shelves and the formation of dense water which intensifies its flow through the bottom topography in midwinter. On the Barents shelf, particularly where the warmer West Spitsbergen Current flows into Storfjord and southern Spitsbergen, for example, phytoplankton grow rapidly during the spring and summer. Wassmann (1989) reported that the flux at a shallow layer of the Berents Sea was 250 mg C m^{-2} day^{-1} during the spring bloom and 110 mg C m^{-2} day^{-1} in the summer. More recently, it was found that spring bloom flux reaches as high as 950 mg C m^{-2} day^{-1} (Wassmann, personal communication, 1990). With the arrival of fall, zooplankton become inactive; thus, a large portion of organic particles remains in those areas. The unutilized particulate matter sinks down to the relatively shallow shelves and is deposited as new and unconsolidated sediment. In the meantime, fine lithogenic particles supplied from glaciers with the summer thaw are also deposited along with the organic detritus.

Quadfasel et al. (1988) observed an outflow of dense water from Storfjord into the Norwegian basin and then into Fram Strait during summer. According to the authors, a 20–40-m-thick bottom layer with temperatures close to the freezing point and unusually high salinities lay under warmer, fresher waters produced by summer melting. Atlantic water formed an intermediate layer. The salinities near the bottom were above 35.4 per mil, higher than those of any water mass in the area. This dense water flows as a gravity current along the continental slope of western Spitsbergen. Quadfasel *et al.* (1988) speculated that the spreading Storfjord water contributes 5–10% of the newly formed deep waters of the Arctic Ocean. Blindheim (1987) reported on the Storfjord bottom water with a salinity of 34.9–35.1 per mil and its contribution to the Norwegian Sea by cascading along the slope to intermediate depths.

A higher rate of production of dense bottom water can be expected during the maximum winter months when more sea ice forms (Pfirman, 1985; Swift, 1986). Although it is speculative, the downflow of dense bottom water at the steep shelf-break topography may trigger an avalanche of a large-scale bottom flow over a wider area on the shelf, which in turn may flush out newly deposited unconsolidated sediment from the floor of the fjords and shelves, causing a winter maximum of sediment transport. If our assumption of advective transport of particles is correct, however, sediment trap measurements of vertical flux may involve large errors caused by hydrodynamic complexities. The velocity and volume of bottom water which could cause such sediment transport are unknown at this time. Fresh sediments are unconsolidated and can be resuspended with much less velocity, such as the velocity known to resuspend fine particles in the absence of bedforms at 1 m (U_{1m}, 42 cm s^{-1}; Butman and Moody, 1983) or the minimum velocity of bottom water measured *in situ* which causes seabed erosion (U_{1m}, 7.5–22 cm s^{-1}; Young and Southard, 1978).

E. A Large Carbon Sink in Arctic Shelf Fronts

A winter flux maximum is a significant phenomenon which may help us to understand the oceanographic interface between the Arctic shelf and deep basin frontage. For example, we do not know the origins of a succession of large bulges which have developed along the boundary between the Barents Sea shelf and the northern Norwegian basin, such as the frontages of Storfjord Trough and Bear Island Trough (Fig. 13.14). Seismic profiles indicate that these bulges did not exist at the end of the last glaciation in the Barents Sea (Elverhoi and Solheim, 1983). One can hypothesize that winter sediments produced that topography during the last postglaciation period. If the above explanation of sediment export is feasible, these bulges could be an efficient sink of organic carbon, and if the correct geothermal conditions are met, organic matter in a bulge can be an important future source of fuel hydrocarbons. In fact, at a number of locations in the Arctic, e.g., parts of the Kara Sea and Laptev Sea, a significant winter maximum may have occurred.

13 Particle Fluxes and Modern Sedimentation in the Polar Oceans 731

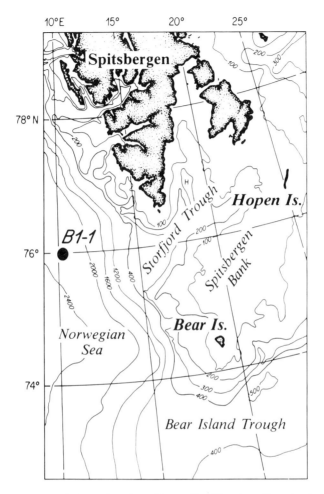

Figure 13.14 Bathymetric chart of southwest Barents Sea. Note the sedimentary bulges in front of the Storfjord and Bear Island troughs, indicative of submarine "deltas." From Perry *et al.*, (1980), simplified.

Romankevich (1984) noted particularly high contents of organic carbon in the upper sedimentary layer at several locations in the eastern Arctic Ocean. The export of organic particulate matter to the Arctic deep basin's frontage and subsequent (permanent) sink is particularly important with regard to the global carbon cycle. This unusual process involving biological and physical oceanography clearly warrants further research.

VIII. Summary and Conclusions

A limited number of particle flux measurements and collection of settling particles by time-series sediment traps deployed in polar oceans have revealed a number of new findings concerning the biogeochemical cycling of biogenically and lithogenically produced and transported particles. Particle sedimentation in the neritic or pelagic environments in polar oceans is strongly mediated by the upper-ocean biology, which is related to ice conditions and development of phytoplankton blooms at the ice edge. As in low-latitude oceans, fine particles in the polar ocean water columns are packaged in relatively large aggregates and delivered to the ocean interior and seafloor at relatively high speeds of descent. Thus, high seasonality in the polar oceans is well preserved in the annual sequences of particle fluxes. The contrast between the winter minimum and the summer maximum is strong in the Norwegian Current area of the Nordic Seas and at Antarctic stations in both neritic and pelagic environments. The flux of fine lithogenic particles, such as clay minerals, was controlled by the availability of a settling medium and not by the variability in density of lithogenic particles in the water column.

There are wide contrasts in particle fluxes between Arctic and Antarctic environments. The annual biogenic fluxes in the pelagic Nordic Seas are relatively large, almost equivalent to those found in low-latitude stations of other oceans. However, our measurements of annual biogenic carbon fluxes in the Atlantic Southern Ocean revealed far smaller amounts, suggesting that the pelagic Weddell Sea at least has an extremely small annual production of new carbon. In contrast, the Nordic Seas are far more fertile, despite the fact that there is a smaller nutrient supply to this area than in the Southern Ocean. The ratio between biogenic calcium carbonate and biogenic silica flux on both sides of the Nordic Seas was far larger than those in the Atlantic Southern Ocean, implying that the former is a "carbonate ocean" and the latter a "silica ocean." Coccolithophores were produced throughout the winter in the Greenland Sea under pack ice; no new production was observed in the Antarctic austral winter. The major settling particles in pelagic and neritic environments in the Southern Ocean were metazoan fecal pellets, whereas in the Nordic Seas amorphous aggregates seemed to be the dominant form of settling of fine particles.

The differences in the quantity and mode of particle sedimentation may be due to the hydrographic setting. A mixture of North Atlantic elements with those of the Greenland Current occurs in the northern Nordic Seas, whereas the Weddell Sea is isolated from exchange with the rest of the Atlantic basin. Another hypothesis is that the Nordic Seas receive Siberian and Alaskan river components transported by the Transpolar Drift, and thus

Fram Strait can be regarded as a biogeochemical estuary of all these large rivers and the recipient of high levels of micronutrients. On the other hand, the Antarctic continent provides virtually no river water and the supply of aerosols to the pelagic ocean is severely limited. Thus, the availability of micronutrients may be far less in Antarctic water than in its northern counterpart, and this may cause low fertility despite a higher level of macronutrients.

The neritic Antarctic is characterized by efficient vertical transport of diatom frustules, mediated by the production of fecal pellets by dense populations of *E. superba* during the austral summer at the MIZ. Such metazoan activity appears to be responsible for the globally significant flux of silicon to Antarctic shelves/slopes. Because populations of *E. superba* are absent or because the growth of phytoplankton and that of zooplankton are out of phase, fecal pellet-mediated particle sedimentation was not observed in Arctic neritic environments. In the Arctic, high summer productivity results in a significantly underutilized primary production which descends to the bottom of some Arctic shelves. When the generation of very cold and highly saline bottom water on these shelves reaches a maximum in midwinter, the dense water flows toward the deep basins and delivers a significant volume of organic matter. Such "winter outbursts" of sediment may be significant in providing a permanent sink of organic carbon along Arctic basin frontages.

To answer critical questions such as the global role of polar oceans in terms of the biogeochemical cycle and sedimentation of carbon, more data on the quality and quantity of settling particles are required. Is the extremely small organic carbon flux observed at pelagic Atlantic Southern Ocean stations an isolated case? Annual time-series flux measurements must be extended to all sectors of the pelagic Southern Ocean. Also, the Bering Sea, which is an important Arctic marginal sea, is predicted to be a "silica ocean" with respect to new production. Flux samples will have to be collected to test this prediction. In order to understand the entire system of particle flux in relation to primary production in the polar ocean, one needs to develop technology to measure the flux at the bottom of the euphotic layer as well. Indeed, particle flux studies in polar oceans have just begun and much remains to be learned.

Acknowledgments

This chapter is a review, but much published and nonpublished data from our joint programs with colleagues from Germany, Great Britain, and Iceland who study particle fluxes in polar oceans have been incorporated. First, I thank G. Wefer, University of Bremen, for his partnership in carrying out many field experiments discussed in this chapter. Without the generous ship time granted to Dr. Wefer and us on the R/V *Polarstern,* Alfred Wegener Institute of Marine and Polar Research, this research would not have been possible. J. Thiede was the Chief Scientist during the first two Arctic sediment trap deployment cruises. Also, I thank K.-P. Koltermann, J. Meinke, and J. Olafsson for providing fine opportunities for deployment and recovery of many mooring arrays on board R/V *Meteor,* R/V *Valdivia,* and R/V *B. Saemundsson,* respectively. V. L. Asper, D. R. Ostermann, A. Karowe, and R. Krishfield assisted in all phases of the field programs among many colleagues from the University of Bremen, University of Kiel, and the Marine Research Institute, Reykjavik. C. Pudsey and P. Barker, British Antarctic Survey, kindly permitted me to cite their unpublished data. I owe my deep gratitude to S. J. Manganini, whose leadership made it possible to provide dependable analysis of the hundreds of flux samples used in this research. E. Evans ably assisted in editing this chapter.

Valuable suggestions from many colleagues, particularly from G. L. Johnson, T. Curtin, V. Ittekkot, P. Wassmann, P. Jones, J. D. Milliman, K. Takahashi, and T. Takahashi, were useful in writing this chapter. This research was supported by Contract N00014-C-001 from the U.S. Office of Naval Research and Grant DPP 83321472 from the National Science Foundation. This is contribution No. 6929 from the Woods Hole Oceanographic Institution.

References

Ackley, S. F., K. R. Buck & S. Taguchi. 1979. Standing crop of algae in the sea ice of the Weddell Sea region. *Deep-Sea Res.* **26**: 269–281.
Alexander, V. 1986. Arctic Ocean pollution. *Oceanus* **29**: 31–35.
Anderson, J. B., C. Blake & N. Myers. 1984. Sedimentation on the Ross Sea continental shelf. *Mar. Geol.* **57**: 295–333.
Anderson, L. G. & D. Dyrssen. 1981. Chemical constituents of the Arctic Ocean in the Svalbard area. *Oceanol. Acta* **4**: 304–311.
Andrews, J. T. (ed.). 1985. "Quaternary Environments: Eastern Canadian Arctic Baffin Bay and Western Greenland." Allen & Unwin, Boston, Massachusetts.
Arsen'ev, V. S. 1967. "Currents and Water Masses of the Bering Sea." Nauka, Moscow.
Bacon, M. P., C. A. Huh, A. P. Fleer & W. G. Deuser. 1985. Seasonality in the flux of natural radionuclides and plutonium in the deep Sargasso Sea. *Deep-Sea Res.* **32**: 273–286.
Berger, W. H., K. Fischer, C. Lai & G. Wu. 1987. "Ocean Productivity and Organic Carbon Flux. Part I. Overview and Maps of Primary Production and Export Production," SIO Tech. Rep. 87–30. Scripps Inst. Oceanogr., Univ. of California, San Diego.
Bialek, E. L. 1966. "Handbook of Oceanographic Tables" pp. 50–51. U.S. Nav. Oceanogr. Office, Washington, D.C.
Blindheim, J. 1987. Cascading of Barents Sea bottom water into the Norwegian Sea. *ICES Symp., Pap.* **61**: 1–17.
Brewer, P. G., K. W. Bruland, R. W. Eppley & J. J. McCarthy. 1986. The Global Ocean Flux Study (GOFS): Status of the U.S. GOFS program. *Eos* **67**: 827–832.
Butman, B. & J. A. Moody. 1983. Observations of bottom currents and sediment movement along the U.S. east coast continental shelf during winter. *U.S. Geol. Surv. Land Manage. Bur., Spec. Publ.* **83–824**: 1–60.
Calvert, S. E. 1983. Sedimentary geochemistry of silicon. *In* "Silicon Geochemistry and Biogeochemistry" (S. R. Aston, ed.), pp. 143–186. Academic Press, New York.
Chen, C.-T. A. 1986. Summer–winter comparison of Weddell Sea surface water and its productivity. *Antarct. J. U.S.* **21**: 128–129.
Cox, G. F. N. & W. F. Weeks. 1988. Numerical simulations of the profile properties of undeformed first-year sea-ice. *Eos* **68**: 1736 (abstr.).

Craig, H., W. S. Broecker & D. W. Spencer. 1981. "GEOSECS Pacific Expedition," Vol. 4. Natl. Sci. Found., Washington, D.C.
Danielsson, L. B. & S. Westerlund. 1983. Trace metals in the Arctic Ocean. *In* "Trace Metals in Sea Water" (C. S. Wong, E. Boyle, K. W. Bruland, J. D. Burton & E. D. Goldberg, eds.), pp. 85–96. Plenum, New York.
DeAngelis, M., N. I. Barkov & V. N. Petrov. 1987. Aerosol concentrations over the last climatic cycle (160 kyr) from an Antarctic ice core. *Nature (London)* **325**: 318–321.
Deuser, W. G., E. H. Ross & F. R. Anderson. 1981. Seasonality in supply of sediment to the deep Sargasso Sea and implications for the rapid transfer of matter to the deep ocean. *Deep-Sea Res.* **28**: 495–505.
Dietrich, G. 1969. "Atlas of the Hydrography of the Northern North Atlantic." International Council for the Exploration of the Sea, Copenhagen.
Duce, R. A. 1986. The impact of atmospheric nitrogen, phosphorus, and iron species on marine biological productivity. *In* "The Role of Air–Sea Exchange in Geochemical Cycling" (P. Buat-Menard, ed.), pp. 497–529. Reidel Publ. Co., Dordrecht, The Netherlands.
Dunbar, R. B. 1984. Sediment trap experiments on the Antarctic continental margin. *Antarct. J. U.S.* **19**: 70–71.
Dunbar, R. B. & A. R. Leventer. 1987. Sediment fluxes beneath fast ice: October, 1986 through February, 1987. *Antarct. J. U.S.* **22**: 112–114.
Dunbar, R. B., J. B. Anderson & W. E. Domack. 1985. Oceanographic influences on sedimentation along the Antarctic continental shelf. *In* "Oceanology of the Antarctic Continental Shelf Antarctic Research" (S. S. Jacobs, ed.), pp. 291–302. Am. Geophys. Union, Washington, D.C.
El-Sayed, S. Z. 1970. On the productivity of the Southern Ocean (Atlantic and Pacific Sectors). *In* "Antarctic Ecology" (M. W. Holdgate, ed.), pp. 119–135. Academic Press, London.
El-Sayed, S. Z. & J. T. Turner. 1977. Productivity of the Antarctic and tropical/subtropical regions: A comparative study. *In* "Polar Oceans" (M. J. Dunbar, ed.), pp. 463–503. Arctic Inst. North Am., Calgary, Alberta, Canada.
Elverhoi, A. & A. Solheim. 1983. The Barents Sea ice sheet—a sedimentological discussion. *Polar Res.* **1**: 23–42.
Eppley, R. W. & B. J. Peterson. 1979. Particulate organic matter flux and planktonic new production in the deep ocean. *Nature (London)* **282**: 677–680.
Fillon, R. H. & C. Duplessy. 1985. Northwest Labrador Sea stratigraphy, sand input, and paleoceanography during the last 160,000 years. *In* "Quaternary Environments: Eastern Canadian Arctic Baffin Bay and Western Greenland (J. T. Andrews, ed.), pp. 210–247. Allen & Unwin, Boston, Massachusetts.
Fillon, R. H., G. H. Miller & J. T. Andrews. 1981. Terrigenous sand in Labrador Sea hemipelagic sediments and paleo-glacial events on Baffin Island over the last 100,000 years. *Boreas* **10**: 107–124.
Fischer, G., D. Fuetterer, R. Gersonde, S. Honjo, D. R. Ostermann & G. Wefer. 1988. Seasonal variability of particle flux in the Weddell Sea and its relation to ice cover. *Nature (London)* **335**: 426–428.
Fogg, G. E. 1977. Aquatic primary production in the Antarctic. *Philos. Trans. R. Soc. London* **279**: 27–38.
Fowler, S. W. & L. F. Small. 1972. Sinking rates of euphausiid fecal pellets. *Limnol. Oceanogr.* **17**: 293–296.
Garrison, D. L., K. R. Buck & M. W. Silver. 1984. Microheterotrophs in the ice edge zone. *Antarct. J. U.S.* **19**: 109–111.
Gauld, D. T. 1957. A peritrophic membrane in calanoid copepods. *Nature (London)* **179**: 325–326.

Gow, A. J., S. F. Ackley, W. T. Weeks & J. W. Govoni. 1982. Physical and structural characteristics of Antarctic sea ice. *Ann. Glaciol.* **3**: 113–117.
Holm-Hansen, O., S. Z. El-Sayed, G. A. Francschini & R. L. Cuhel. 1977. Primary production and the factors controlling phytoplankton growth in the Southern Ocean. *In* "Adaptations Within Antarctic Ecosystems" (G. A. Llano, ed.), pp. 11–50. Gulf Publ. Corp., Houston, Texas.
Honjo, S. 1975. Dissolution of suspended coccoliths in the deep-sea water column and sedimentation of coccolith ooze. *Spec. Publ.—Cushman Found. Foraminiferal Res.* **13**: 114–128.
———. 1976. Coccoliths: Production, transportation and sedimentation. *Mar. Micropaleontol.* **1**: 65–79.
———. 1980. Material fluxes and modes of sedimentation in the mesopelagic and bathypelagic zones. *J. Mar. Res.* **38**: 53–97.
———. 1982. Seasonality and interaction of biogenic and lithogenic particulate flux at the Panama Basin. *Science* **218**: 883–884.
———. 1984. Study of ocean fluxes in time and space by bottom-tethered sediment trap arrays: A recommendation. *In* "Global Ocean Flux Study Workshop," pp. 305–324. National Research Council, Washington, D.C.
———. 1987. Oceanic particles and pelagic sediments in the western North Atlantic Ocean. *In* "Decade of North American Geology. Vol. M. The Western North Atlantic Region" (B. E. Tucholke, ed.), pp. 469–478. Geol. Soc. Am., Washington, D.C.
Honjo, S. & K. W. Doherty. 1987. Large aperture time-series sediment traps: Design objectives, construction, and application. *Deep-Sea Res.* **35**: 133–149.
Honjo, S. & M. R. Roman. 1978. Marine copepod fecal pellets: Production, transportation and sedimentation. *J. Mar. Res.* **36**: 45–57.
Honjo, S., S. J. Manganini & J. J. Cole. 1982. Sedimentation of biogenic matter in the deep ocean. *Deep-Sea Res.* **29**: 608–625.
Honjo, S., K. W. Doherty, Y. C. Agrawal & V. L. Asper. 1984. Direct optical assessment of large amorphous aggregates (marine snow) in the deep ocean. *Deep-Sea Res.* **31**: 67–76.
Honjo, S., S. J. Manganini, A. Karowe & B. J. Woodward. 1987a. "Particle Fluxes, Northeastern Nordic Seas: 1983–1986 (Nordic Seas Sedimentation Data File, Vol. 1)," WHOI Tech. Rep. 87–17. Mar. Biol. Lab., Woods Hole, Massachusetts.
Honjo, S., C. M. Wooding & G. Wefer. 1987b. "Current Measurements from the Northern Nordic Seas: 1983–1986 (Nordic Seas Sedimentation Data File, Vol. 2)," WHOI Tech. Rep. 87–18. Mar. Biol. Lab., Woods Hole, Massachusetts.
Honjo, S., B. J. Hay, S. J. Manganini, V. L. Asper, E. T. Degens, S. K. Kempe, V. Ittekkot, E. Izdar, Y. T. Konuk & H. Benli. 1987c. Seasonal cyclicity of lithogenic particle fluxes at a southern Black Sea sediment trap station. *In* "Ocean Particle Fluxes" (E. Degens, E. Izdar & S. Honjo, eds.), pp. 19–39. Univ. of Hamburg Press, Hamburg.
Honjo, S., S. J. Manganini & G. Wefer. 1988. Annual particle flux and a winter outburst of sedimentation in the northern Norwegian Sea. *Deep Sea Res.* **35**: 1223–1234.
Hood, D. W. & W. S. Reeburgh. 1974. Chemistry of the Bering Sea: An overview. *In* "Oceanography of the Bering Sea with Emphasis on Renewable Resources" (D. W. Hood & E. S. Kelly, eds.), pp. 191–204. Univ. of Alaska Press, Fairbanks.
Hopkins, T. L. & J. J. Torres. 1989. Midwater food web in the vicinity of a marginal ice zone in the western Weddell Sea. *Deep-Sea Res.* **36**: 543–560.
Hughes, F. W., L. K. Coachman & K. Aagaard. 1974. Circulation, transport and water exchange in the western Bering Sea. *In* "Oceanography of the Bering Sea with Emphasis on Renewable Resources" (D. W. Hood & E. J. Kelly, eds.), pp. 59–98. Univ. of Alaska Press, Fairbanks.

Hurd, D. C. 1983. Physical and chemical properties of siliceous skeletons. *In* "Silicon Geochemistry and Biochemistry" (S. R. Aston, ed.) pp. 187–244. Academic Press, London.
Hurdle, B. G. (ed.). 1986. "The Nordic Seas." Springer-Verlag, New York.
Jennings, J. C., L. I. Gordon & D. M. Nelson. 1984. Nutrient depletion indicates high primary productivity in the Weddell Sea. *Nature (London)* **308:** 51–54.
Johannessen, O. M. 1986. Brief overview of the physical oceanography. *In* "The Nordic Seas" (B. G. Hurdle, ed.), pp. 103–127. Springer-Verlag, New York.
Johnson, T. O. & W. O. Smith, Jr. 1986. Sinking rates of phytoplankton assemblages in the Weddell Sea marginal ice zone. *Mar. Ecol.: Prog. Ser.* **33:** 131–137.
Jousé, A. P., O. G. Kozlova & V. V. Muhina. 1971. Distribution of diatoms in the surface layer of sediment from the Pacific Ocean. *In* "The Micropalaeontology of Oceans" (B. M. Funnell & W. R. Riedel, eds.), pp. 263–269. Cambridge University Press, London.
Kempe, S., H. Nies, V. Ittekkot, E. T. Degens, K. O. Buesseler, H. D. Livingston, S. Honjo, B. J. Hay, S. J. Manganini, E. Izdar & T. Konuk. 1987. Comparison of Chernobyl nuclide deposition in the Black Sea and in the North Sea. *In* "Particle Flux in the Ocean" (E. T. Degens, E. Izdar & S. Honjo, eds.), SCOPE UNEP Publ. No. 62, pp. 165–178. Univ. of Hamburg, Hamburg.
Knauer, G. A., J. H. Martin & K. W. Bruland. 1979. Fluxes of particulate carbon, nitrogen, and phosphorus in the upper water column of the northeast pacific. *Deep-Sea Res.* **26A:** 97–108.
Kozlova, O. G. 1971. The main features of diatom and silicoflagellate distribution in the Indian Ocean. *In* "The Micropalaeontology of Oceans" (B. M. Funnell & W. R. Riedel, eds.), pp. 271–275. Cambridge University Press, London.
Ledford-Hoffman, P. A., D. J. DeMaster & C. A. Nittrouer. 1986. Biogenic silica accumulation in the Ross Sea and the importance of Antarctic continental shelf deposits in the marine silica budget. *Geochim. Cosmochim. Acta* **50:** 2099–2110.
Leventer, A. & R. B. Dunbar. 1987. Diatom flux in McMurdo Sound, Antarctica. *Mar. Micropaleont.* **12:** 49–64.
Lisitzin, A. L. 1972. "Sedimentation in the World Ocean." Special Publication No. 17. pp 1–218. Society of Economic Paleontologists and mineralogists, Tulsa, Oklahoma.
Marchant, H. J. & G. V. Nash. 1986. Electron microscopy of gut contents and faeces of *Euphausia superba* Dana. *Mem. Natl. Inst. Polar Res.* **40:** 167–177.
Martin, J. H. & R. M. Gordon. 1988. Northeast Pacific iron distributions in relation to phytoplankton productivity. *Deep-Sea Res.* **35:** 177–196.
Matsuda, O., S. Ishikawa & K. Kawaguchi. 1987. Seasonal variation of downward flux of particulate organic matter under the Antarctic fast ice. *Proc. NIPR Symp. Polar Biol.* **1:** 23–24.
Menzies, R. J. 1962. On the food and feeding habits of abyssal organisms as exemplified by the Isopoda. *Int. Rev. Gesamten Hydrobiol.* **47:** 339–358.
Miller, M. A., D. W. Krempin, D. T. Manahan & C. W. Sullivan. 1984. Growth rates, distribution, and abundance of bacteria in the ice edge zone of the Weddell and Scotia seas, Antarctica. *Antarct. J. U.S.* **19:** 103–105.
Mills, E. L. 1983. Problems of deep-sea biology: An historical perspective. *In* "The Sea" (G. T. Rowe, ed.), Vol. 8, pp. 1–79. Wiley, New York.
Morel, F. M. & R. J. M. Hudson. 1985. The geobiological cycle of trace elements in aquatic systems: Redfield revisited. *In* "The Chemical Processes in Lakes" (W. Stumm, ed.), pp. 251–281. Wiley, New York.
Motoda, S. & T. Minoda. 1974. Plankton of the Bering Sea. *In* "Oceanography of the Bering Sea with Emphasis on Renewable Resources" (D. W. Hood and E. J. Kelly, eds.), pp. 207–241. Univ. of Alaska Press, Fairbanks.

Murray, J. E. 1968. The drift deterioration and distribution of icebergs to the North Atlantic Ocean. In "Ice Seminar; A Conference Sponsored by the Petroleum Society of the Canadian Institute of Mining and Metallurgy," Vol. 10, pp. 3–18, Calgary.
Nair, R. R., V. Ittekkot, S. J. Manganini, V. Ramaswamy, B. Haake, E. T. Degens, B. N. Desai & S. Honjo. (1989) Monsoon related particle fluxes to the deep Arabian Sea. *Nature (London)* **338**: 749–751.
Noriki, S., K. Harada & S. Tsunogai. 1985. Sediment trap experiments in the Arctic Ocean. *In* "Marine and Estuary Geochemistry" (A. C. Sigleo & A. Hatton, eds.), pp. 161–170. Lewis Publ., Chelsea.
Okada, H. & S. Honjo. 1973. The distribution of oceanic coccolithophorids in the Pacific. *Deep-Sea Res.* **20**: 355–374.
Okada, H. & A. McIntyre. 1979. Seasonal distribution of modern coccolithophores in the western North Atlantic Ocean. *Mar. Biol. (Berlin)* **54**: 319–328.
Östlund, H. G. & G. Hut. 1984. Arctic Ocean water mass balance from isotope data. *J. Geophys. Res.* **89**: 6373–6381.
Paasche, E. 1968. Biology and physiology of coccolithophorids. *Annu. Rev. Microbiol.* **22**: 71–86.
Peinert, R., J. Bathmann, B. von Bodungen & T. Noji. 1987. The impact of grazing on spring phytoplankton growth and sedimentation in the Norwegian Current. *In* "Particle Flux in the Ocean" (E. T. Degens, E. Izdar & S. Honjo, eds.), pp. 149–164. SCOPE UNEP Publ. No. 62. Univ. of Hamburg, Hamburg.
Perry, R. K., H. S. Fleming, N. S. Cherkis, R. H. Felden & P. R. Vogi. 1980. Bathymetry of the Norwegian–Greenland and western Barents seas. *NRL Map Chart Ser.* **MC-21**.
Pfirman, S. L. 1985. Modern sedimentation in the northern Barents Sea: Input, dispersal and deposition of suspended sediments from glacial meltwater. Ph.D. Dissertation, MIT/WHOI, Cambridge, Massachusetts.
Pilskaln, C. H. & S. Honjo. 1987. The fecal pellet fraction of biogeochemical particle fluxes to the deep sea. *Global Biogeochem. Cycles* **1**: 31–48.
Quadfasel, D., B. Rudels & K. Kurz. 1988. Outflow of dense water from a Svalbard fjord into the Fram Strait. *Deep-Sea Res.* **35**: 1143–1150.
Rahn, K. W. 1981. Relative importance of North America and Eurasia as sources of Arctic aerosol. *Atmos. Environ.* **15**: 1447–1455.
Rey, F. 1981a. Primary production estimates in the Norwegian Coastal Current between 62°N and 72°N. *In* "The Norwegian Coastal Current" (R. Saetre & M. Mork, eds.), pp. 640–648. University of Bergen Press, Bergen.
———. 1981b. The development of the spring phytoplankton outburst at selected sites of the Norwegian Current. *In* "The Norwegian Coast Current" (R. Satre & M. Mork, eds.), pp. 649–680. University of Bergen, Bergen.
Reynolds, M. & C. H. Pease. 1984. Drift characteristics of northeastern Bering Sea ice during 1982. *NOAA Tech. Mer.* **ERL PMEL-55** p. 1–135.
Romankevich, E. A. 1984. "Geochemistry of Organic Matter in the Ocean." Springer-Verlag, Berlin.
Sakshaug, E. & O. Holm-Hansen. 1984. Factors governing pelagic production in polar oceans. *In* "Marine Phytoplankton and Productivity" (O. Holm-Hansen, ed.), pp. 1–18. Springer-Verlag, New York.
Shaw, D. M., N. D. Watkins & D. C. Huang. 1974. Atmospherically transported volcanic glass in deep sea sediments: Theoretical considerations. *J. Geophys. Res.* **79**: 3087–3094.
Shulenberger, E. 1983. Superswarms of antarctic krill (*Euphausia superba* Dana). *Antarct. J. U.S.* **18**: 194–197.
Silver, M. W. & A. L. Alldredge. 1981. Bathypelagic marine snow: Deep-sea algal and detrital community. *J. Mar. Res.* **39**: 501–530.

Small, L. F., S. W. Fowler & M. Y. Unlu. 1979. Sinking rates of natural copepod fecal pellets. *Mar. Biol. (Berlin)* **51**: 233–241.

Smith, W. O., Jr. & D. M. Nelson. 1986. The importance of ice edge phytoplankton blooms in the Southern Ocean. *BioScience* **36**: 251–257.

Subba Rao, D. V. & T. Platt. 1984. Primary production of Arctic waters. *Polar Biol.* **3**: 191–201.

Swift, J. H. 1986. The Arctic waters. *In* "The Nordic Seas" (B. G. Hurdle, ed.), pp. 129–153. Springer-Verlag, New York.

Tan, F. C., D.·Dyrssen & P. M. Strain. 1983. Sea ice meltwater and excess alkalinity in the East Greenland Current. *Oceanol. Acta.* **6**: 283–288.

Tolmachev, A. I. (ed.). 1970. "The Arctic Ocean and Its Coast in the Cenozoic Era." Amerind Publ. Co., New Delhi.

Tsunogai, S., M. Kusakabe, H. Iizumi, I. Koike & A. Hattori. 1979. Hydrographic features of the deep water of the Bering Sea—the sea of silica. *Deep-Sea Res.* **26**: 641–659.

Vinje, T. E. 1976. Sea ice conditions in the European sector of the marginal sea of the Arctic, 1966–1975. *Arb. Nor. Polarinst:* 163–174.

Vinogradov, M. E. 1962. Feeding of the deep-sea zooplankton. *Rapp. P.-V. Reun. Cons. Int. Explor. Mer* **153**: 114–120.

Wadhams, P. 1986. The ice cover. *In* The Nordic Seas" (B. G. Hurdle, ed.), pp. 21–84. Springer-Verlag, Berlin.

Wassmann, P. 1989. Sedimentation of organic matter and silicate from the euphotic zone of the Barents Sea. *Rapp. P.-V. Reun. Cons. Int. Explor. Mer.* **188**: 108–114.

Wefer, G., G. Fischer, D. Fuetterer & R. Gersonde. 1988. Seasonal particle flux in the Bransfield Strait, Antarctica. *Deep-Sea Res.* **35**: 891–898.

Wefer, G., E. Suess, W. Balzer, G. Liebezeit, P. J. Muller, C. A. Ungerer & W. Zenk. 1982. Fluxes of biogenic components from sediment trap deployment in circumpolar waters of the Drake Passage. *Nature (London)* **299**: 145–147.

Whiteker, T. M. 1977. Sea ice habitats of Signy Island (South Orkneys) and their primary production. *In* "Adaptation within Antarctic Ecosystem" (G. A. Llano, ed.), pp. 75–83. Gulf Publication Company, Houston.

Wong, C. S. & S. Honjo. 1984. Material flux at weather station PAPA: High frequency time-series observations through production cycles. *Eos* **65**: 225 (abstr.).

Young, R. A., & J. B. Southard. 1978. Erosion of fine-grained marine sediments: Sea floor and laboratory experiments. *Geol. Soc. Am. Bull.* **89**: 663–672.

Index

Absorption coefficients, for phytoplankton, 486
Acartia clausi, in marginal seas of the Arctic Ocean, 531
Acartia longiremis, in marginal seas of the Arctic Ocean, 531–533
Acartia spp., in the North Pacific, 535
Acodontaster conspicuus, Antarctic region, 660
Actinarians, Antarctic region, 659
Adamussium colbecki, 668
Adélie penguins, role in food web, 607–608
Aerosols, lithogenic, as source of dissolved Fe, 721
Aethia spp., *see* Auklets
Aetideopsis antarctica, in the Ross Sea, 565
Aetideopsis rostrata, in Canadian basin deep water, 531
Aetideopsis spp., in the Weddell Sea, 563
Agarum cribrosium, Arctic region, 652
AIWEX, *see* Arctic Internal Wave Experiment
Alaria esculenta, Arctic region, 652
Alcids, importance of oceanic fronts to, 616–617
Alcyonarians, Antarctic region, 665
Alcyonium, Antarctic region, 659
Algae, *see* Ice algae
Alkalinity, total, Arctic Ocean surface layer, 415–416
Aluminum, in Arctic Ocean waters, 427
Ammonium concentrations, island effects (Southern Ocean), 459
Amphipods
 Arctic land-fast ice zone, 645
 Beaufort Sea, 642–643
 below Antarctic ice shelves, 669
 in Bering–Chukchi seas, 646–647
 role in food web, northern hemisphere, 603
 in the Southern Ocean, 556
Anchor ice, disturbance of Arctic benthos, 644
Annelids, oligochaete, in Beaufort Sea, 642
Antarctic Bottom Waters
 chemical distributions, 443–448
 formation, 435, 445–447
 salinity, 443
 varieties of, 446
Antarctic Circumpolar Current
 biologically active chemical constituents, 451–453
 nutrients, sufficiency for phytoplankton, 451–453
Antarctic Divergence
 chemical distributions in, 453–454
 nutrients in, 433
Antarctic Intermediate Water
 chemical distributions in, 440–443
 formation, 442
Antarctic paradox, 454
Antarctic Peninsula
 sedimentation modes, 724
 zooplankton in adjacent waters, 557–559
Antarctic petrels, role in food web, 607–608
Antarctic region
 benthic habitats and species, 635–638
 benthic patterns and processes, 658–659
 below ice shelves, 668–669
 deep habitats and assemblages, 650, 663–665
 productivity, 665–668
 soft-bottom habitats, 660–663
 zonation and assemblages, 659–660
 benthic research in, 637
 biogeographic relationship to South America, 654

741

biogeography, 656-657
endemism, 654-655
exploration and collections, 634-635
oceanographic characteristics, 632-633
sedimentation, 657-658
Antarctic Surface Water
  nutrients in, 433
  silicate distribution, 434
Anthropods
  in the Beaufort Sea, 642
  in Bering-Chukchi seas, 647
Apherusa glacialis, role in food web, 603
Apparent oxygen utilization, 438
Aptenodytes forsteri, see Emperor penquins
Arctic cod
  in Arctic land-fast ice zone, 645
  in freshwater plumes, 619
  role in food web, 603, 605
Arctic haze, 721
Arctic Internal Wave Experiment, 408, 421, 429
Arctic Ocean
  accessiblity to dissolved/particulate matter, 719
  biological productivity, 427
  budgets for chemical constituents, 426-427
  chemical characteristics, 412-413
  artificial radionuclides, 429-431
  Atlantic layer, 420
  carbon dioxide assimilation, 426-427
  deep water, 420-421
  halocline, 412, 417-420
  sea ice effects, 431-432
  surface layer, 413-417
  temperature vs. salinity, 417
  trace metals, 427-429
  variability from basin to basin, 413
  waters masses below 2000 m, 416
  copepods in, 528-531
  data collection in, 408-410
  freshwater runoff, 641
  ice camps, 408-409
  major energy flows in upper trophic levels, 604
  mineral resources for, 718-720
  neritic sedimentation, 724-725
  photosynthesis/irradiance parameters, 500-501
  phytoplankton growth
    rates of, 482
    role of organic material, 491-492
  primary productivity, 496-499
  trophic pathways
    oceanographic complexity effects, 601-602
    physiographic setting effects, 601
    polynya effects, 601
    sea ice effects, 600-601
  trophic webs, 603-606
  water mass dating, 421-425
  zooplankton
    central area, 528-531
    marginal seas, 531-535
Arctic Ocean Deep Water, chemical characteristics, 420-421
Arctic region
  benthic habitats and species, 635-638
  benthic macroalgae, 651-653
  benthic patterns and processes, 641-643
  Bering Sea, 646-651
  Chukchi Sea, 646-651
  land-fast ice zone, 643-645
  benthic research in, 637
  biogeographic schemes for, 636
  Cenozoic events affecting, 636
  eurybathic species, 637
  exploration and collections, 634-635
  oceanographic characteristics, 632-633
  Quaternary period events affecting, 636
  stenobathic species, 637
Arctic shelf fronts, carbon sink in, 725, 730-731
Arctic tern, ice edge association, 610-611
Arctocephalus gazella, see Fur seals, Antarctic
Artemidactis, Antarctic region, 659
Arthur Harbour, benthic faunal abundance, 662
Ascidians, Antarctic region, 660, 665
Asterias amurensis, in Bering-Chukchi seas, 647
Asteroids
  Antarctic region, 659, 665
  in Bering-Chukchi seas, 647-648
Astrammina rara, Antarctic region, 668
Atlantic layer, Arctic Ocean
  chemical characteristics, 420
  residence times, 423
Atolla wyvillei
  triacylglycerols in, 575

Index

in the Weddell Sea, 562
ATP, ratio to other elemental constituents as index for phytoplankton activity, 509–511
Auklets
  along oceanic fronts, 617
  importance of insular fronts to, 616
  role in food web, 605

Baffin Bay
  benthic faunal abundance, 662
  benthic macroalgae, 652
  benthic research, 641
  nutrient uptake rates, 505–506
  photosynthesis/irradiance parameters, 500–501
Balaena mysticetus, see Bowhead whales
Balaenoptera acutorostrata, see Minke whales
Balaenoptera physalus, see Fin whales
Baleen whales
  ice edge association of, 610–611
  overexploitation of, 622
  role in food web
    northern hemisphere, 605
    southern hemisphere, 607–608
Ballia callitricha, Antarctic region, 671
Barents Sea
  benthic research, 639
  nutrient uptake rates, 505–506
  pelagic particle fluxes, 697–699
  primary production, 644
  winter (neritic) sedimentation in, 725–730
  zooplankton, 533–534
Bathylasma corolliforme, Antarctic region, 665
Beaked whales, role in food web, 607–608
Bearded seals
  dietary changes of, 622
  feeding patterns in Bering–Chukchi seas, 648
  role in food web, 603
Beaufort Sea
  benthic research, 640
  benthic species in, 642–643
  carbon sources, 644–645
  depth zonation, 642
  primary production, 644
  shelf, chemical and physical processes on, 410–412

zooplankton, 531–532
Benthic systems
  Antarctic
    below ice shelves, 668–669
    deep habitats, 650, 663–664
    disturbance, 665–667
    evolutionary situation, 674
    faunal abundance, 661–662
    patterns of decomposition, 673–674
    primary productivity and, 673
    productivity, 665–668
    soft-bottom habitats, 660–663
    zonation in, 659–660
  Arctic
    Beaufort Sea, 642–643
    Bering Sea, 646–650
    Chukchi Sea, 646–650
    disturbance, 665
    evolutionary situation, 674
    faunal abundance, 661–662
    land-fast ice zone, 643–645
    patterns of decomposition, 673–674
    primary productivity and, 673
Bering Sea
  benthic communities, 646–651
  benthic faunal abundance, 662
  benthic patterns and processes, 646–651
  dissolved organic carbon, 722
  nutrient uptake rates, 505–506
  zooplankton, 535–538
Biogenic particles, 692
  fluxes
    in Greenland Current area, 704–706
    in north-central Weddell Sea, 708–710
    in Norwegian Current area, 701–702
Biogenic silica
  Barents Sea, 726
  Southern Ocean, 714
Biogeography
  Antarctic region, 656–657
  McMurdo Sound, 656–657
Biological productivity, *see also* Primary productivity
  Arctic Ocean, 427
  over continental shelf waters, 411
Bivalves
  Antarctic region, 665
  Beaufort Sea, 642–643
  below Antarctic ice shelves, 669
  in Bering–Chukchi seas, 646–647, 649

Black guillemot, role in food web, 603
Blue whales, ice edge associations, 611
Boreogadus saida, see Arctic cod
Bowhead whales
  along freshwater plumes, 619
  role in food web, 604
Bransfield Strait
  annual particle flux, 722
  photosynthesis/irradiance parameters, 500–501
  zooplankton, 557–559
Bryozoans, Antarctic region, 665
Budgets, chemical constituents of Arctic Ocean, 426–427

Cadmium
  in Arctic Ocean waters, 427–429
  in the Southern Ocean, 449
Calanoides acutus
  in Antarctic Peninsula waters, 558
  in the Bransfield Strait, 558
  dietary trends with respect to ontogeny, 578
  lipids, 576
  in the Ross Sea, 565
  in the Scotia Sea, 558
  seasonal cycles and adaptations, 573–574
  in the Southern Ocean, 556
  in the Weddell Sea, 562–564
Calanus finmarchicus
  in the Arctic Ocean, 528
  digestive enzymes, 547
  distribution, 546
  ingestion rates, 546–547
  life cycle, 532, 534, 548
  lipid content, 543, 547–548
  overwintering strategy, 548–550
  respiration rates, 547
  seasonal cycles, 546–550
  specialized adaptations, 546–550
Calanus glacialis
  abundance throughout year, 529
  in archipelagos and fjords, 534
  in the Arctic Ocean, 528
  in Barents Sea, 534
  feeding, 532
  life cycle, 532, 539–540
  lipid content, 543
  in marginal seas of the Arctic Ocean, 531–532
  respiration rate, 541
  seasonal biomass variations in Canadian basin, 530
  seasonal cycles, 538–541
  specialized adaptations, 538–541
  in upper 200 m of water column, 530
Calanus hyperboreus
  in archipelagos and fjords, 534
  in the Arctic Ocean, 528
  development rate, 542
  digestive enzymes, 544
  ingestion rates, 544
  life cycle, 530, 542
  lipid content, 543, 546
  in marginal seas of the Arctic Ocean, 531–532
  nitrogen excretion, 544
  respiration rate, 542–543
  seasonal biomass variations in Canadian basin, 530
  seasonal cycles, 541–546
  specialized adaptations, 541–546
  in upper 200 m of water column, 530
  in upper 500 m of water column, 529
Calanus marshallae, in the Bering Sea and North Pacific, 535–538
Calanus propinquus
  in Antarctic Peninsula waters, 558
  in the Bransfield Strait, 558
  lipids, 576
  in the Scotia Sea, 558
  seasonal cycles and adaptations, 573–574
  in the Southern Ocean, 556
  triacylglycerols in, 577
  in the Weddell Sea, 562–564
Calanus simillimis
  in Antarctic Peninsula waters, 558
  in the Bransfield Strait, 558
  in the Scotia Sea, 558
Calanus spp.
  in Eurasian and Canadian basins, 530
  fecal pellets, 724
  role in Artic food web, 603
Callorhinus ursinus, see Fur seals
Calycopsis borchgrewinki
  lipids, 574
  in the Weddell Sea, 562
Canadian Archipelago, zooplankton, 534
Canadian Arctic
  benthic macroalgae, 652
  benthic research, 640–641

Index

Canadian Experiment to Study the Alpha
  Ridge, 408, 418, 421, 429
Capelin
  biomass, effects of commercial exploitation, 620
  role in food web, northern hemisphere,
    605, 613–615
Carbon
  inorganic, distribution in the Southern
    Ocean, 439
  organic
    fluxes
      in Antarctic Divergence, 454
      in the Barents Sea, 726–727
      in Greenland Current area, 704–706,
        716
      in Norwegian Current area,
        699–700, 716
      in the Southern Ocean, 713–716
      in Weddell Sea, 710–711
    Ross Sea
      cycling, 463–469
      neritic, content, 723
    sink in the Arctic shelf fronts, 730–731
    ratio to other elemental constituents as
      index for phytoplankton activity,
      509–511
Carbon-14
  in Antarctic surface waters, 435
  measurements, for Arctic Ocean water
    residence times, 425
Carbon/chlorophyll ratios, in phytoplankton
  populations, 503–504
Carbon dioxide
  assimilation in Arctic Ocean, 426–427
  excess
    estimation, 440
    in Southern Ocean surface waters,
      438–440
  in world oceans, 441
Carbon fluxes
  McMurdo Sound, 657
  in north-central Weddell Sea, 716
  Southern Ocean, 714
Carbon sink, in the Arctic shelf fronts, 725,
  730–731
Carbon sources
  Beaufort Sea, 644–645
  Bering–Chukchi seas, 647
Carbonate fluxes
  Arctic Ocean surface layer, 415–416

  in the Barents Sea, 726
  Greenland Current, 704–705, 716
  north-central Weddell Sea, 708, 716
  Norwegian Current, 701–702, 716
  in pelagic Weddell Sea, 713
Catch per unit effort, 621–622
Catharacta maccormicki, see Skua, south
  polar
Cepphus grylle, see Black guillemot
CESAR, *see* Canadian Experiment to Study
  the Alpha Ridge
Cesium-137, in Arctic Ocean waters,
  429–431
Chaetognaths
  in marginal seas of the Arctic Ocean,
    531–532
  in the Southern Ocean, 556
Chinstrap penguin, role in food web,
  607–608
Chionoecetes bairdi, in Bering–Chukchi
  seas, 647
Chionoecetes opilio, in Bering–Chukchi
  seas, 647
Chironomid larvae, Beaufort Sea, 642
Chlorofluoromethanes
  in central Arctic Ocean, 423–424
  in Southern Ocean surface waters, 438
Chorda filium, Arctic region, 652
Chorda tomentosa, Arctic region, 652
Chukchi Sea
  benthic communities, 646–651
  benthic faunal abundance, 662
  benthic macroalgae, 651
  benthic patterns and processes, 646–651
  benthic research, 640
  zooplankton, 531–532
Cibicides refulgens, 668
Cirripeds, Antarctic region, 665
Clangula hyemalis, see Oldsquaw ducks
Clausocalanus spp.
  in Antarctic Peninsula waters, 558
  in the Bransfield Strait, 558
  in the Scotia Sea, 558
Clavularia, Antarctic region, 659
Clio pyramidata, in the Weddell Sea,
  562
Clione spp., in the Ross Sea, 566
Clouds, light attenuation by, 484–485
Cnemidocarpa, Antarctic region, 659
Coccolithophorids, distribution in polar
  oceans, 716–718

Coccolithus pelagicus, in polar oceans, 717–718
Coelenterates, Antarctic region, 659
Continental shelves
  biological processes over, 411
  carbon sink in Arctic shelf fronts, 730–731
  chemical characteristics, 410–412
  effect of source waters, 410
  ice formation over, 412
  nutrient regeneration along, 412
  river runoff effects, 410–411
  sediment–water interface, 411
  trophic interactions along
    break fronts, 613–615
    insular fronts, 615–616
    midshelf fronts, 615
Copepods
  in Antarctic Peninsula waters, 558–559
  Arctic Ocean
    central, 528–531
    marginal seas, 531–535
  Beaufort Sea, 642
  in the Bransfield Strait, 558–559
  near insular fronts, 616
  role in food web
    northern hemisphere, 603–606
    southern hemisphere, 577–579, 609
  in the Ross Sea, 565–566
  in the Scotia Sea, 558–559
  in the Southern Ocean, 556
  in the Weddell Sea, 561–564
Copper
  in Arctic Ocean waters, 427, 429
  in Southern Ocean waters, 449
Corals, Antarctic region, 665
Crabeater seals
  importance of shelf-break fronts to, 613
  role in food web, southern hemisphere, 607–608
Crinoids
  Antarctic region, 665
  Beaufort Sea, 643
Crustacea
  Beaufort Sea, 642
  in Bering–Chukchi seas, 648
  role in food web, southern hemisphere, 607–608
Ctenocalanus spp.
  in Antarctic Peninsula waters, 558
  in the Bransfield Strait, 558

  in McMurdo Sound, 565
  in the Ross Sea, 565
  in the Scotia Sea, 558
  in the Weddell Sea, 562, 565
Cychlorhynchus psittaculus, see Auklets

Data collection, in Arctic Ocean, 408–410
Deep Shelf benthic assemblages, Antarctic, 664–665
Delphinapterus leucas, see White whale
Demersal fishes, role in food web, 605
Derjuginia tolli, in marginal seas of the Arctic Ocean, 532
Desmarestia menziesii, Antarctic region, 659
Desmarestia spp., Antarctic region, 670
Deuterium, distributions vs. salinity for Antarctic surface, deep, and bottom waters, 436
Diatom flux, McMurdo Sound, 657
Diphyes antarctica, lipids, 574
Dissolved organic carbon, Bering Sea, 722
Dissolved organic nitrogen, in the Southern Ocean, 449
Dissolved organic phosphorus, Southern Ocean, 449
  island effects, 459
Disturbances
  Antarctic benthos, 658–659
  Arctic benthos, land-fast ice zone, 644
Dovekies
  importance of shelf-break fronts to, 613
  role in food web, northern hemisphere, 605
Drepanopus bungei, in marginal seas of the Arctic Ocean, 531–535

E. huxleyi, in polar oceans, 717–718
East Siberian Sea, zooplankton, 532
Echinoderms
  Antarctic region, 668
  Beaufort Sea, 643
  in Bering–Chukchi seas, 647–648
Echinoids
  Antarctic region, 659–660
  Beaufort Sea, 643
Echiurus antarcticus, in Antarctic region, 662
Edswardsi meridionalis, McMurdo Sound, 662

Eider ducks, role in food web, 605
Eleginus gracilis, see Saffron cod
Elephant seals
  exploitation of, 622
  role in food web, 607-608
Emperor penguins, role in food web, 607-608
Endemism, Antarctic region, 654-655
Epibenthic species, in Bering-Chukchi seas, 647
Epimeriella, in the Ross Sea, 566
Epimeriella macronyx, in the Weddell Sea, 563
Erignathus barbatus, see Bearded seal
Esrichtius robustus, see Gray whales
Eucalanus bungii, in the Bering Sea and North Pacific, 535-538
Euchaeta antarctica
  in Antarctic Peninsula waters, 558
  in the Bransfield Strait, 558
  dietary trends with respect to ontogeny, 578
  lipids, 574-576
  in the Scotia Sea, 558
Euchirella rostromagna, in the Weddell Sea, 562
Eukrohnia hamata
  in the Southern Ocean, 556
  in the Weddell Sea, 562
Euphausia crystallorophias
  in Antarctic Peninsula waters, 559
  in the Bransfield Strait, 559
  lipids, 574-576
  role in food web, 607-608
  in the Ross Sea, 565-566
  in the Scotia Sea, 559
  in the Southern Ocean, 557
  in the Weddell Sea, 562-564
Euphausia frigida
  in Antarctic Peninsula waters, 559
  in the Bransfield Strait, 559
  in the Scotia Sea, 559
Euphausia superba
  in Antarctic Peninsula waters, 558
  in the Bransfield Strait, 558
  distribution, 566-567
  eggs, 569
  experimental harvesting in the Antarctic, 622
  fecal pellets, particle transport by, 724

  feeding patterns, 570-571
  growth rate, 571-572
  larvae, 569-570
  life cycles, 569
  life span, 568
  lipids, 576
  molt frequency and growth, 568
  reproduction, near ice edges, 610
  role in food web, 607-609
  in the Ross Sea, 566
  in the Scotia Sea, 558
  in the Southern Ocean, 556
  spawning, 568-569
  swarms of, 567-568
  triacylglycerols in, 577
  vertical migration, 567-568
  in the Weddell Sea, 562-564
Euphausia triacantha
  in Antarctic Peninsula waters, 559
  in the Bransfield Strait, 559
  in the Scotia Sea, 559
Euphausiids
  adaptations, 566-572
  along oceanic fronts
    in the Antarctic, 618
    in the North Atlantic, 617
  in Antarctic Peninsula waters, 558-559
  Beaufort Sea, 642
  below Antarctic ice shelves, 669
  in the Bransfield Strait, 558-559
  in marginal seas of the Arctic Ocean, 531
  role in food web
    northern hemisphere, 603, 605-606
    southern hemisphere, 577-579, 607-609
  in the Ross Sea, 565-566
  in the Scotia Sea, 558-559
  seasonal cycles, 566-572
  in the Southern Ocean, 556
  in the Weddell Sea, 562-564
Euphysa flammea, in marginal seas of the Arctic Ocean, 531
Eurytemora herdmani, in marginal seas of the Arctic Ocean, 531-532
Eusirus propeperdentatris, in the Weddell Sea, 563
Evadne nordmani, in marginal seas of the Arctic Ocean, 532
Evasterias echinosoma, in Bering-Chukchi seas, 647

Fecal pellets
  E. superba, particle transport by, 724
  gravitational settling, effect on silica cycling, 463
  particle incorporation into, 692-693
  peritrophic membrane, 724
  role in particle sedimentation in Weddell Sea, 711-713
  Southern Ocean, 714
Fin whales, role in food web, 605
Fish, mesopelagic, role in southern hemispheric food web, 607-608
Fisheries
  effect on
    Bering-Chukchi crab population, 647
    marine ecosystems, 620-622
    Greenland, 638
Fog, light attenuation by, 484-485
Food webs
  Antarctic, 577-579
  northern hemisphere, 603-606
  southern hemisphere, 606-609
Foraminifera
  below Antarctic ice shelves, 669
  McMurdo Sound, 668
  planktonic, in Weddell Sea carbonate fluxes, 713
Fram Strait, estuarine hypothesis, 719
Fram Strait, nutrient uptake rates, 506
Franz Josef Land, benthic research, 640
Freon 11
  in central Arctic Ocean, 423-424
  in Southern Ocean surface waters, 438
Freon 12
  in central Arctic Ocean, 423-424
  in Southern Ocean surface waters, 438
Freshwater plumes, trophic interactions, 618-619
Fronts
  Arctic shelf, carbon sink in, 730-731
  trophic interactions
    insular, 615-616
    midshelf, 615
    oceanic, 616-618
    shelf-break, 613-615
Frustration Bay, benthic faunal abundance, 662
Fulmarus glacialis, role in food web, 605

Fur seals
  exploitation of, 622
  importance of insular fronts to, 615-616
  role in food web
    northern hemisphere, 605
    southern hemisphere, 607-608

Gaidius previspinus, in Canadian basin deep water, 531
Gaidius spp., in the Weddell Sea, 562
Galiteuthis glacialis, see Squid
Gammarus setosus, in the Beaufort Sea, 642
Gas transfer, at Southern Ocean sea surface, 434-438
Gastropods
  Antarctic region, 655-656
  below Antarctic ice shelves, 669
  in Bering-Chukchi seas, 648
Gersemia rubiformis, Beaufort Sea, 645
Gersemia spp., Beaufort Sea, 643
Glaucous gulls, role in food web, 605
Gonatus antarcticus, see Squid
Gorgonians, Antarctic region, 665
Gray whales
  importance of midshelf fronts to, 613
  role in food web, 605
Grazing, effects on phytoplankton distribution, 511-512
Greenland
  benthic macroalgae, 652
  benthic research, 637-638
Greenland Current
  carbonate fluxes, 716
  coccolithophorid distribution, 717-718
  lithogenic particle fluxes, 703-704
  organic carbon flux, 716
  pelagic particle fluxes, 699-706
Greenland Sea
  coccolithophorid distribution, 717-718
  pelagic particle fluxes, 697-699
  photosynthesis/irradiance parameters, 500-501

Habitats, benthic
  Antarctic region, 653-658
  soft-bottom, 660-663
  Arctic region, 635-641

Halecium, Antarctic region, 660
Halitholus cirratus, in marginal seas of the Arctic Ocean, 531
Halocline
 central Arctic Ocean, 412
 chemical characteristics, 412, 417–420
 maintenance of, 418–419
 nutrient concentrations, 417, 420
 types of water in, 419–420
 effects of shelf processes, 412
Haloptilus ocellatus
 in the Ross Sea, 565
 in the Weddell Sea, 562
Haloptilus oxycephalus, in the Weddell Sea, 562
Haplospora globusu, Arctic region, 652
Harp seals, role in food web, 605
Harvests, commercial, effect on marine resources, 620–623
Helium-3
 distributions in Weddell Gyre, 437
 in Southern Ocean surface waters, 438
Helium-4, supersaturation in Southern Ocean ice shelf water, 438
Herring, role in food web, 605
Heterophoxus videns, McMurdo Sound, 663
Heterorhabdus austrinus, in the Ross Sea, 565
Heterorhabdus farrani, in the Ross Sea, 565
Heterorhabdus spp., in the Weddell Sea, 562
Hiatella arctica, 639
Himantothallus grandifolium, Antarctic region, 659, 670
Histriophoca fasciata, see Ribbon seal
Holothurians
 Antarctic region, 665
 Beaufort Sea, 642–643
Homaxinella balfourensis, Antarctic region, 660
Hydroids, Antarctic region, 659–660, 665
Hydrurga leptonyx, see Leopard seals
Hyperiella dilatata, in the Ross Sea, 566

Ice algae
 in Arctic land-fast ice zone, 645
 communities
  characterization, 512–513
  phytoplankton in water column and, 515–516
 growth
  light effects, 513–514
  nutrient effects, 514
  temperature effects, 513
 production in polar regions, 479
 productivity, irradiance–nutrient relationships, 514–515
 temporal development relative to phytoplankton, 480
Ice camps, Arctic Ocean, 408–409
Ice edges
 development, maximum particle flux and, 710–711
 trophic interactions, 610–612
Ice formation, over continental shelves, 412
Ice keels, disturbance of Arctic benthos, 644
Ice shelves, life below (Antarctic), 668–669
Icebergs
 glacial sediments in, 706–707
 production, 706
Iceland, benthic research, 639
Iceland gulls, role in food web, 604
Insular fronts, trophic interactions, 615–616
Invertebrates, epifaunal, in Bering–Chukchi seas, 647
Iridaea obovata, Antarctic region, 671
Iron, role in ocean primary production, 720–722
Irradiance, solar
 annual cycle, 479
 attenuation
  by clouds and fog, 484–485
  by sea ice, 485
  variability, 485–486
 effect on
  ice algal growth, 513–515
  phytoplankton growth, 482–489
 measurement, 502
 parameters in polar waters, 500–501
Islands
 insular fronts, trophic interactions along, 615–616
 Southern Ocean, effects on biogeochemical cycles, 457–459
 wake phenomenon, 616
Isopods
 Antarctic region, 654
 Arctic land-fast ice zone, 645
 Beaufort Sea, 642
 below Antarctic ice shelves, 669

Isotealia, Antarctic region, 659
Isotopic tracers, for nutrient uptake, 505–506
Ivory gulls, role in food web, 604
  northern hemisphere, 603

Jan Mayen
  benthic research, 638–639
  macroalgal flora, 652

Kara Sea
  benthic macroalgae, 652
  zooplankton, 532
Kelps
  Arctic region, 652
  carbon from, Beaufort Sea, 645
Kerguelen Island, photosynthesis/irradiance parameters, 500–501
Killer whales, role in food web, 607–608
King Edward Cove, benthic faunal abundance, 662
Kittiwakes
  nesting success, effect of pollock fisheries, 622
  role in food web, 605
Krill, Antarctic, *see Euphausia superba*

Laminaria saccharina, Arctic region, 652
Laminaria solidungula, Arctic region, 652
Laminaria spp.
  Arctic region, 652
  at Hut Point, 671
Lampra, Antarctic region, 659–660
Lancaster Sound, photosynthesis/irradiance parameters, 500–501
Laptev Sea, zooplankton, 532
Larus glaucescens, *see* Glaucous gulls
Larus leucopterus, *see* Iceland gulls
Larus schistisagus, *see* Slaty-backed gulls
Lead, in Arctic Ocean waters, 427, 429
Lead-210, in central Arctic Ocean halocline, 419
Leopard seals, role in food web, southern hemisphere, 607–608
Leptasterias polaris acervata, in Bering–Chukchi seas, 647

Leptonychotes weddelli, *see* Weddell seals
Lethasterias nanimensis, in Bering–Chukchi seas, 647
Limacina helicina
  in the Ross Sea, 566
  in the Weddell Sea, 563
Limnocalanus grimaldi, in marginal seas of the Arctic Ocean, 531–532
Lipids, in polar zooplankton, 574–576
Lithogenic aerosols, as source of dissolved Fe, 721
Lithogenic particles, 692
  Barents Sea, 726–727, 729
  Greenland Current area, 703–704
  Norwegian Current area, 701–702
Lobodon carcinophagus, *see* Crabeater seals
Lomonosov Ridge Experiment, 408, 418, 421, 427, 429
Longipedia, in the Ross Sea, 566
LOREX, *see* Lomonosov Ridge Experiment
Lucicutia spp., in Canadian basin deep water, 531

Macroalgae, benthic
  Antarctic region, 669–672
  Arctic region, 651–653
Macrocystis pyrifera, Antarctic region, 670
Maldane sarsi antarctic, 663
Mammals, marine
  in Bering–Chukchi seas, 648
  migration routes along ice edges, 611–612
  polynya-related distribution, 612
Manganese
  in Arctic Ocean waters, 427–429
  in the Southern Ocean, 449
Marginal ice zones
  phytoplankton primary production, 492–496
  Southern Ocean, biologically active chemical constituents, 456–457
  trophic interactions
    large-scale ice edges, 610–612
    polynyas, 612–613
Marine ecosystems
  closed systems, 620
  structured by oceanic processes, 620
Marine snow, 693
Matochkin Sea, benthic macroalgae, 652

Index

McMurdo Sound
  benthic faunal abundance, 661
  biogeography, 656-657
  carbon flux, 658
  diatom flux data, 657
  infauna differences in, 663
  photosynthesis/irradiance parameters, 500-501
  primary production, 666
Medusae, in marginal seas of the Arctic Ocean, 531-532
Metridia curticauda, in the Ross Sea, 565
Metridia gerlachei
  in Antarctic Peninsula waters, 558
  in the Bransfield Strait, 558
  in the Ross Sea, 565
  in the Scotia Sea, 558
  in the Southern Ocean, 556
  in the Weddell Sea, 562-564
Metridia longa
  in archipelagos and fjords, 534
  characterization, 550
  distribution, 551
  feeding, 532
  growth, 550-551
  life cycle, 532
  lipid content, 552-553
  in marginal seas of the Arctic Ocean, 531-532
  nutrition of, 552
  respiration rate, 543, 551-552
  seasonal cycles, 550-553
  specialized adaptations, 550-553
  in upper 200 m of water colum, 530
Metridia lucens, in the Arctic Ocean, 528
Metridia pacifica, in the Bering Sea and North Pacific, 535-538
Microcalanus pygmaeus
  in Antarctic Peninsula waters, 558
  in the Bransfield Strait, 558
  in Canadian basin, 530
  in marginal seas of the Arctic Ocean, 531-532
  in the Scotia Sea, 558
  in upper 500 m of water column, 529
Migrations, along ice edges, 610-612
Minke whales
  pack ice-edge associations, 611
  role in food web, 607-608
Mirounga leonian, *see* Elephant seals

Mollusks
  Beaufort Sea, 642
  Bering-Chukchi seas, 647-648
  role in food web, 605
Monodon monoceros, *see* Narwhale
Monostroma hariotii, Antarctic region, 671
Murman Sea, benthic macroalgae, 651
Murres
  along oceanic fronts, 617
  population decline, catch per unit effort of pollock fishery and, 621-622
  role in food web, 605
  thick-billed, role in food web, 605
Myctophids, role in food web
  northern hemisphere, 606
  southern hemisphere, 607-609
Myoxoephalus quadricornis, in Arctic land-fast ice zone, 645
Mysids
  Arctic land-fast ice zone, 645
  Beaufort Sea, 642
  below Antarctic ice shelves, 669
Mysis littoralis, Beaufort Sea, 645

Nain Bay, benthic faunal abundance, 662
Narwhale, role in food web, northern hemisphere, 604
Neocalanus cristatus, in the Bering Sea and North Pacific, 535-538
Neocalanus plumchrus, in the Bering Sea and North Pacific, 535-538
Neogloboquadrina pachyderma, in Weddell Sea carbonate fluxes, 713
Neritic sedimentation, *see* Sedimentation, neritic
Neritic taxa, in marginal seas of the Arctic Ocean, 531-532
Nickel, in Arctic Ocean waters, 427-429
Nitrate
  annual cycle, 479
  removal from Southern Ocean, 506-507
  surface distribution in the Southern Ocean, 452
Nitrogen
  dissolved organic, in the Southern Ocean, 449
  inorganic, uptake in marginal ice zone, 505-506

particulate organic, in Antarctic Divergence, 454
ratio to other elemental constituents as index for phytoplankton activity, 509–511
total inorganic, in Antarctic Divergence, 453
Nitrogen monoxide (NO) tracers
Arctic Ocean waters, 420
Warm Deep Waters, 443–444
Nitzschia curta
in fecal pellets in Weddell Sea, 712–713
in ice edge phytoplankton blooms in Ross Sea, 468
Nordic Seas
coccoliths, 716–718
pelagic particle fluxes, 697–699
physical and biogeochemical differences with other polar oceans, 718–722
North Pacific Ocean, zooplankton, 535–538
Norwegian Current
carbonate fluxes, 716
coccolithophorid distribution, 717–718
organic carbon flux, 716
pelagic particle fluxes, 699–706
total mass fluxes, 701
Norwegian Sea, pelagic particle fluxes, 697–699
Nototanais dimorphus, McMurdo Sound, 663
Novaya Zemlya, benthic macroalgae, 652
Novosibirskye Islands, benthic research, 640
Nutrients
effect on ice algae growth and productivity, 514–515
island effects on, 459
regeneration along continental shelves, 412
in Southern Ocean surface waters, 433–434
sufficiency for phytoplankton
in Antarctic Circumpolar Current, 451–453
in Arctic Ocean, 489–491
in the Southern Ocean, 489–491
uptake relationships, 505–508

Oceanographic complexity, effects on trophic pathways

Antarctic, 602
Arctic, 601–602
Odobenus rosmarus, see Walrus
Odontaster validus
Antarctic region, 659–660, 668
McMurdo Sound, 656
Oikopleura vanhoeffeni, in marginal seas of the Arctic Ocean, 532
Oithona similis
abundance throughout year, 529
in archipelagos and fjords, 534
in Canadian basin, 530
in marginal seas of the Arctic Ocean, 531–532
in the Ross Sea, 565
in upper 500 m of water column, 529
in the Weddell Sea, 565
Oithona spp.
in the North Pacific, 535
in the Ross Sea, 566
in the Weddell Sea, 562
Oldsquaw ducks, role in food web, northern hemisphere, 605
Oligochaete annelids, Beaufort Sea, 642
Ommatophoca rossi, see Ross seals
Oncaea borealis
in the Canadian basin, 530
in marginal seas of the Arctic Ocean, 531–532
Oncaea curvata
in Antarctic Peninsula waters, 559
in the Bransfield Strait, 559
in the Ross Sea, 565
in the Scotia Sea, 559
in the Weddell Sea, 565
Oncaea spp., in the Weddell Sea, 562
Onchocalanus wolfendeni, in the Ross Sea, 565
Onisimus littoralis, in the Beaufort Sea, 642
Opal, particle dissolution, 723
Ophionotus, Antarctic region, 668
Ophiuroids
Antarctic region, 665
Beaufort Sea, 642–643
Orchomene, below Antarctic ice shelves, 669
Orchomenella, in the Ross Sea, 566
Orcinus orca, see Killer whales
Organic flux, in the ocean interior, expression for, 694

Index

Organic matter, regional and global cycles of, 461–463
Ostracod skeletons, below Antarctic ice shelves, 669
Oxygen
 apparent oxygen utilization, 438
 minimum, Warm Deep Water, 443, 445
 in Southern Ocean surface layers, 435–438
 Weddel Sea, distribution, 437
Oxygen-18
 distributions vs. salinity for Antarctic surface, deep, and bottom waters, 436
 measurements, for Arctic Ocean water residence times, 425

Pachyptila spp., *see* Prions
Pacific Ocean, North, zooplankton, 535–538
Pagodroma nivea, *see* Snow petrels
Pagophila eburnea, *see* Ivory gulls
Pagophilus groenlandicus, *see* Harp seals
Pandalus borealis, *see* Pink shrimp
Paralabidocera antarctica
 in Antarctic Peninsula waters, 559
 in the Bransfield Strait, 559
 in the Ross Sea, 565
 in the Scotia Sea, 559
Paralithodes spp., as predators in Bering–Chukchi seas, 647–648
Parathemisto libellula, role in food web, 605
Parathemisto spp., role in food web, 603
Pareuchaeta glacialis, in upper 200 m of water colum, 530
Particle fluxes
 methods of study
  bottom-tethered moorings, 695–696
  time fractionation, 694–695
 neritic
  Antarctic Ocean, 722–723
  Arctic Ocean, 724–725
  Barents Sea, 725–730
  carbon sink in Arctic shelf fronts, 730–731
  Ross Sea, 724
 pelagic
  Arctic vs. Antarctic, 715–722
  Greenland Current area, 699–706
  ice edge development and, 710–711
  ice-rafted sediment, 706–707

Nordic seas, 697–699
Norwegian Current area, 699–706
organic carbon flux in the Southern Ocean, 713–715
Weddell Sea
 carbonate fluxes in, 713
 modes of sedimentation, 711–713
 north-central area, 708–710
Particles, oceanic
 biogenic sources, 692
 derived from phytoplankton, 716
 fast descent of, 693–694
 incorporation into fecal pellets, 692–693
 lithogenic sources, 692
 scavenging of, 693
 settling mechanisms, 692–693
 snow aggregates, 693
Particulate organic carbon, in Antarctic Divergence, 454
Particulate organic nitrogen, in Antarctic Divergence, 454
Patinigera polaris, Antarctic region, 659
Pelagobia longicirrata
 in Antarctic Peninsula waters, 559
 in the Bransfield Strait, 559
 in the Scotia Sea, 559
Pelecanoides spp., role in food web, 609
Penguins, importance of insular fronts to, 616
Peritrophic membrane, of krill fecal pellets, 724
Petrels, *see also* specific petrel
 role in food web, 607–609
pH, Southern Ocean water masses, 439
Phoca hispida, *see* Ringed seals
Phosphorus
 dissolved organic
  in the Southern Ocean, 449
  Southern Ocean, island effects, 459
 flux, in the Barents Sea, 726
 ratio to other elemental constituents as index for phytoplankton activity, 509–511
 total inorganic, in Antarctic Divergence, 453
Photoadaptation, phytoplankton, 488–489, 498
 index of, 499
Photoinhibition, phytoplankton, 499

Photosynthesis
  by epotic algae in Arctic land-fast ice
    zone, 645
  parameters in polar waters, 500–501
  phytoplankton, 499–505
Photosynthetic response
  conversion to carbon-specific response,
    503
  phytoplankton, 499
Phyllophora antarctica, 671
Phytoplankton
  carbon/chlorophyll ratios, 503–505
  distribution, 511–512
    grazing effect, 512
    sinking effects, 511–512
  effect of nutrient levels, 489–491
  growth
    elemental ratios and, 509–511
    irradiance effects, 482–489
    regulation, role of organic material,
      491–492
    settling of biogenic particles and, 694
    temperature effects, 481–482
  growth rates
    calculation, 503–505
    in polar regions, 482
  ice algae development relative to, 480
  iron-limited habitat in the Southern
    Ocean, 457
  nutrient sufficiency for
    in Antarctic Circumpolar Current,
      451–453
    in Arctic Ocean, 489–491
    in the Southern Ocean, 489–491
    in Southern Ocean marginal ice zones,
      456–457
    in Southern Ocean Polar Front,
      454–456
  nutrient uptake, 505–508
  oceanic particles derived from, 716
  photoadaptation, 488–489, 498
  photoinhibition, 499
  photosynthesis, 499–505
    Antarctic, 502–503
    Arctic, 502–503
  photosynthetic response, 499
  in the Polar Front, 455
  primary production, 496–498
    marginal ice zones, 492–496
    open water, 492–496
    space–time variations in, 498–499
  primary productivity in polar regions,
    478–479, 508, 509
  specific absorption coefficient for, 486
  vertical mixing of, 487
  in water column, ice algae communities
    and, 515–516
Pink shrimp, in Bering–Chukchi seas, 647
Pinnipeds, *see also* Seals; Walruses
  near ice edges, 610
Plautus alle, see Dovekies
Pleuragramma antarcticum, see Silverfish
Plumes, freshwater, trophic interactions,
  618–619
Polar bears, role in food web, 603
Polar Front
  biologically active chemical constituents,
    454–456
  nutrients in, 433–434
Polar oceans
  coccolithophorid distribution, 717–718
  physical and biogeochemical settings,
    718–722
Polar systems, primary productivity in,
  478–479
Pollock
  commercial fishery effects
    on Kittiwake nesting success, 622
    on Murre population, 621–622
  role in food web, 606
  walleye, role in food web, 605
Polonium-210, in central Arctic Ocean
  halocline, 419
Polychaetes
  Antarctic Peninsula waters, 559
  Antarctic region, 662, 665
  Beaufort Sea, 642–643
  below Antarctic ice shelves, 669
  Bransfield Strait, 559
  Scotia Sea, 559
Polynyas
  effects on trophic pathways, 601
  enhanced productivity at ice edges of,
    612–613
  importance to seabirds, 612–613
Poralithodes camtschatica, in Bering–
  Chukchi seas, 647
Poralithodes platypus, in Bering–Chukchi
  seas, 647
Predators, on benthic organisms, in the

Index

Bering–Chukchi seas, 647–649
Primary productivity, *see also* Biological productivity
  Antarctic benthic communities and, 665–668, 673
  Arctic Ocean, 641, 673
  Barents Sea, 644
  Beaufort Sea, 644
  carbon production, Norwegian Current area, 699
  Norwegian Current area, 701–702
  phytoplankton, *see* Phytoplankton
  polar system patterns, 478–479
  Ross Sea, 463–469
  Southern Ocean, 714
  Weddell Sea, 710
    ice edge development and maximum flux period, 710–711
    north-central, 709–710
Primno macropa, in the Weddell Sea, 562
Prions, role in food web, 609
Pseudocalanus elongatus
  life cycle, 533
  in marginal seas of the Arctic Ocean, 532–533
Pseudocalanus minutus, in marginal seas of the Arctic Ocean, 531
Pseudocalanus spp., 553
  in the Bering Sea and North Pacific, 535–538
  in marginal seas of the Arctic Ocean, 532
Pseudolibrotus littoralis, in Arctic land-fast ice zone, 645
Psychroteuthis glacialis, *see* Squid
Puffinus spp., role in food web, 605
Pycnogonids, Antarctic region, 665
Pygoscelis adelia, *see* Adélie penguins
Pygoscelis antarctica, *see* Chinstrap penguin
Pyrostephos vanhoeffeni
  lipids, 574
  in the Ross Sea, 566
Pyseter macrocephalus, *see* Sperm whales

Racovitzanus antarcticus
  in the Ross Sea, 565
  in the Weddell Sea, 562
Radiolarians
  in Antarctic Peninsula waters, 559
  in the Bransfield Strait, 559
  in the Scotia Sea, 559
Radionuclides
  artificial
    in Arctic Ocean waters, 429–431
    in continental shelf waters, 410–411
  natural, in continental shelf waters, 410–411
Redfield ratios, 411
Residence times
  Arctic Ocean water masses, 421–425
  Circumpolar Deep Water/Warm Deep Water, 443
Resolute Bay, photosynthesis/irradiance parameters, 500–501
Rhincalanus gigas
  in Antarctic Peninsula waters, 558–559
  in the Bransfield Strait, 558–559
  in the Ross Sea, 565
  in the Scotia Sea, 558–559
  seasonal cycles and adaptations, 573
  in the Southern Ocean, 556
  in the Weddell Sea, 562
Rhodostethia rosea, *see* Ross's gulls
Ribbon seals, role in food web, 604–605
Right whales
  migration, 610
  role in food web, southern hemisphere, 607–609
Ringed seals
  in Arctic land-fast ice zone, 645
  role in food web, 603
Rissa brevirostris, *see* Kittiwakes
Rissa tridactyla, *see* Kittiwakes
River runoff
  to the Arctic Ocean, 641, 719–720
  effect on chemical characteristics of Arctic Ocean surface layer, 413–415
  continental shelf waters, 410–411
Ross Sea
  benthic faunal abundance, 661
  Bottom Water formation, 447
  carbon cycling, 463–469
  fecal pellet-mediated sedimentation, 724
  nutrient uptake rates, 506
  phytoplankton growth rates in, 482
  silica accumulation by settling of biogenic opal, 723
  silica cycling, 463–469

silica flux, 724
zooplankton, 564–566
Ross seals, role in food web, 607–608
Ross's gulls, role in food web, 604

Saccorhiza dermatodea, Arctic region, 652
Saduria entomon, Arctic land-fast ice zone, 645
Saffron cod, murre diets and, 621
Sagitta gazellae
  lipids, 574
  seasonal cycles and adaptations, 572
  in the Southern Ocean, 556
  in the Weddell Sea, 562
Sagitta marri
  lipids, 574
  in the Weddell Sea, 562
Salinity
  Antarctic Bottom Waters, 443
  Arctic land-fast ice zone, 644
  Arctic Ocean surface layer, 413
  maximum, Warm Deep Water, 443–444
Salpa thompsoni
  in Antarctic Peninsula waters, 558–559
  in the Bransfield Strait, 558–559
  in the Scotia Sea, 558–559
  seasonal cycles and adaptations, 573
  in the Southern Ocean, 556
  in the Weddell Sea, 562
Sandlance
  murre diets and, 621
  role in food web, northern hemisphere, 605
Sarsia princeps, in marginal seas of the Arctic Ocean, 531
Scallops, Beaufort Sea, 643
Scaphocalanus vervoorti, in the Weddell Sea, 562
Scaphopods, Antarctic region, 665
Scavenging, of oceanic particles, 693
Scotia Arc, gastropods, 654–655
Scotia Sea
  nutrient uptake rates, 506
  photosynthesis/irradiance parameters, 500–501
  zooplankton, 557–559
Scott Inlet, photosynthesis/irradiance parameters, 500–501
Sea ice
  Arctic Ocean, chemical effects, 431–432
  communities, types of, 512–513
  effects on trophic pathways, 600–601
  light attenuation by, 485
Seabirds
  along freshwater plumes, 619
  diving, importance of insular fronts to, 616
  importance of oceanic fronts to
    in the Antarctic, 618
    in the North Atlantic, 616–617
  importance of polynyas to, 612–613
  importance of shelf-break fronts to, 613–615
  marginal ice zone
    Antarctic, 610
    Arctic, 611–612
  role in food web
    northern hemisphere, 605
    southern hemisphere, 607–608
Seals, *see also* Pinnipeds; specific seal
  along freshwater plumes, 619
  along ice edges, 610
  feeding patterns in Bering–Chukchi seas, 648
Sediment, ice-rafted
  in Arctic Oceans and marginal seas, 706–707
  in Barents Sea, 729
Sediment traps
  locations of, 689
  particle flux experiments with, 689–690
Sediment–water interface, effect on chemical characteristics of continental shelf waters, 411
Sedimentation, *see also* Particle fluxes
  in the Antarctic Peninsula area, 724
  Antarctic region, 657–658
  in Arctic land-fast ice zone, 645
  neritic
    in the Antarctic Ocean, 722–724
    in the Arctic Ocean, 724–725
    in the Barents Sea, 725–730
  in pelagic Weddell Sea, 711–713
Senecella calanoides, in marginal seas of the Arctic Ocean, 532
Serolis trilobitoides, below Antarctic ice shelves, 669
Shetland Islands, benthic faunal abundance, 661
Shrimps
  Beaufort Sea, 643

pasiphaeid, role in food web, 607–608
Silica
  accumulation in Ross Sea, 723
  biogenic
    accumulation in marine sediments, 461–463
    cycling
      Ross Sea, 463–469
      Southern Ocean vs oceanic, 463
    effects of gravitational settling of fecal pellets, 463
    grazing effects, 463
    in the Polar Front, 455–456
    production rates in Southern Ocean surface waters, 462
    temperature dependency, 462–463
    uptake of silicic acid into, 508
  flux in Ross Sea, 724
  ratio to other elemental constituents as index for phytoplankton activity, 509–511
Silicate
  in Antarctic Divergence, 453–454
  distribution in Antarctic Surface Water, 434
  surface distribution in the Southern Ocean, 452
Siliceous matter, regional and global cycles of, 461–463
Silicic acid, uptake into biogenic silica, 508
Silicon, biogeochemical cycles in the Southern Ocean, 459–469
Silverfish
  dietary composition, 577–578
  role in food web, 607–608
  in the Ross Sea, 566
  in the Southern Ocean, 557
  in the Weddell Sea, 563
Sinking, effects on phytoplankton distribution, 511–512
Sipunculids, Antarctic region, 665
Skua, south polar
  attraction to polynyas, 612
  role in food web, 607–608
Slaty-backed gulls, role in food web, 604
Snow crabs, in Bering–Chukchi seas, 647
Snow petrels, role in food web, 607–608
Solar irradiance, *see* Irradiance, solar
Somateria spp., *see* Eider ducks

Source waters, effect on chemical characteristics of continental shelf waters, 410
South America, biogeographic relationship of Antarctic with, 654
Southern Ocean
  accessiblity to dissolved/particulate matter, 719
  chemical characteristics
    Antarctic Bottom Waters, 443–448
    Antarctic Intermediate Water, 440–443
    carbon/chlorophyll ratio, 504
    Ca/Si ratio in biogenic fluxes, 717–718
    dissolved organic matter, 449
    inorganic carbon distributions, 439
    organic carbon flux, 716
    pH distributions, 439
    surface water
      biogenic silica production rates, 462
      carbon dioxide in, 438–440
      gas transfer, 434–438
      nutrients, 433–434
      trace metals, 449
    Warm Deep Water, 443
  elemental cycles, 449–450
    in the Antarctic Circumpolar Current, 451–453
    in the Antarctic Divergence, 453–454
    island effects on, 457–459
    in marginal ice zones, 456–457
    in the Polar Front, 454–456
    silicon cycling, 459–469
  mineral resources for, 721
  nitrate removal from, 506–507
  photosynthesis/irradiance parameters, 500–501
  physical and biogeochemical differences with other polar oceans, 718–722
  phytoplankton growth
    rates of, 482
    role of organic material, 481–492
  primary productivity, 496–499
  trophic pathways
    oceanographic complexity effects, 602
    physiographic setting effects, 600–601
    polynya effects, 601
    sea ice effects, 600–601
    trophic webs, 606–609
    zooplankton, 554–557
      intermediate zone, 556

northern zone, 556
southern zone, 556-557
Soviet seas, northern, benthic research in, 640
Sperm whales, role in food web, 607-608
Spinocalanus spp., in Canadian basin deep water, 531
Spitsbergen, benthic macroalgae, 652-653
Sponges, Antarctic region, 660, 665
Squid
  along oceanic fronts, 617
  role in food web, southern hemisphere, 607-609
Stephus longipes
  in Antarctic Peninsula waters, 558
  in the Bransfield Strait, 558
  in the Scotia Sea, 558
Sterechinus neumayeri, Antarctic region, 659-660, 671
Sterna paradisaea, see Arctic tern
Strongylocentrotus droebachiensis, 639
Strontium-90, in Arctic Ocean waters, 429-431
Subtropical Convergence, nutrients in, 433
Summer Antarctic Surface Water, formation, 435
Surface layer
  Arctic Ocean
    chemical characteristics, 413-417
    residence times, 422-423
  Southern Ocean
    carbon dioxide in, 438-440
    gas transfer at sea surface, 434-438
    nutrients, 433-434
    nutrient sufficiency for phytoplankton, 451-453
    production rates of biogenic silica, 462
Svalbard
  benthic macroalgae, 652-653
  benthic research, 639

Tanner crabs
  Beaufort Sea, 643
  in Bering-Chukchi seas, 647
  as predators in Bering-Chukchi seas, 647-648
Tealia felina, 639
Temora longicornis, in marginal seas of the Arctic Ocean, 532-533

Temperature effects
  dissolution rate of biogenic silica, 462-463
  on ice algae growth, 513
  on phytoplankton growth, 481-482
Thalassiosira gracilis, in fecal pellets in Weddell Sea, 712-713
Thalassoica antarctica, see Antarctic petrels
Tharybis, in the Ross Sea, 566
Themisto gaudichaudii
  in Antarctic Peninsula waters, 559
  in the Bransfield Strait, 559
  in the Scotia Sea, 559
  seasonal cycles and adaptations, 574
  in the Southern Ocean, 556
Thysanoessa inermis
  in Barents Sea, 534
  growth, 553
  life cycle, 534
  lipid content, 543
  overwintering, 553-554
  seasonal cycles, 553-554
  specialized adaptations, 553-554
Thysanoessa macrura
  in Antarctic Peninsula waters, 559
  in the Bransfield Strait, 559
  in the Ross Sea, 566
  in the Scotia Sea, 559
  in the Weddell Sea, 562-564
Thysanoessa raschi
  in Barents Sea, 534
  in the Bering Sea and North Pacific, 535-538
  growth, 553
  life cycle, 534
  lipid content, 543
  overwintering, 553-554
  seasonal cycles, 553-554
  specialized adaptations, 553-554
Thysanoessa spp., in the Southern Ocean, 556
Tisbe, in the Ross Sea, 566
Tomopteris carpenteri
  in Antarctic Peninsula waters, 559
  in the Bransfield Strait, 559
  lipids, 574
  in the Scotia Sea, 559
Trace metals
  Antarctic Divergence, 454
  Arctic Ocean, 427-429

Bering Sea, 722
  in continental shelf waters, 410–411
  role in ocean primary production, 720–722
  Southern Ocean, 449
  Southern Ocean, island effects, 459
Transpolar Drift
  fluxes of ice-rafted sediment, 706
  total mass fluxes, 701
Triacylglycerols, in zooplankton, 575–576
Tritium
  in Antarctic surface waters, 435
  measurements, for Arctic Ocean water residence times, 424–425
Trophic interactions
  competitive, 620–623
  freshwater plumes, 618–619
  insular fronts, 615–616
  large-scale ice edges, 610–612
  midshelf fronts, 615
  oceanic fronts, 616–618
  polynyas, 612–613
  shelf-break fronts, 613–615
Trophic pathways
  oceanographic complexity effects, 601–602
  physiographic setting effects, 601
  polynya effects, 601
  sea ice effects, 600–601
Tubularia, Antarctic region, 660

Urea
  concentrations in the Southern Ocean, island effects, 459
  uptake in Arctic waters, 505
Uria aalge, see Murres, thick-billed
Uria lomvia, see Murres, thick-billed
Ursus maritimus, see Polar bear
Urticinopsis, Antarctic region, 659

Vanadis antarctica, lipids, 574
Viblila antarctica
  in Antarctic Peninsula waters, 559
  in the Bransfield Strait, 559
  in the Scotia Sea, 559
Vogtia serrata, in the Weddell Sea, 562

Walleye pollock, role in food web, 605
Walruses, see also Pinnipeds
  feeding patterns in Bering–Chukchi seas, 648–649
  importance of midshelf fronts to, 615
  population size changes, 622
  role in food web, 604–605
Warm Deep Water
  chemical distributions, 443
  input to Antarctic surface layers, 435
  oxygen minimum, 443, 445
  residence time, 443
  salinity maximum, 443–444
Weddell Gyre
  helium-3 distributions, 437
  surface waters, origin, 435
Weddell Sea
  Bottom Water formation, 447
  north-central
    biogenic particle fluxes, 708–710
    carbonate fluxes, 708, 716
    organic carbon flux, 716
    primary production, 709–710
  oxygen distribution in surface mixed layer, 437
  phytoplankton growth rates in, 482
  zooplankton communities, 559–564
    northeastern shelf, 562–563
    oceanic, 562
    southern shelf, 563–564
Weddell seals, role in food web, 607–608
White Sea
  benthic macroalgae, 652
  benthic research, 639–640
  zooplankton, 532–533
White whales, role in food web, 604
Winter Water, formation, 435

Xanthocalanus, in the Ross Sea, 566

Zinc
  in Arctic Ocean waters, 427–429
  in the Southern Ocean, 449
Zonation
  Antarctic benthic assemblages, 659–660
  Antarctic macroalgae, 670

Zooplankton, *see also* specific species
  Arctic Ocean
    central, 528–531
    marginal seas, 531–535
  in the Bering Sea, 535–538
  growth, settling of biogenic particles and, 694
  North Pacific Ocean, subarctic, 535–538
  Ross Sea, 564–566
  Southern Ocean, 554–557
    intermediate zone, 556
    northern zone, 556
    southern zone, 556–557
  Weddell Sea, 559–564
    northeastern shelf community, 562–563
    oceanic community, 562
    southern shelf community, 563–564